Lecture Notes in Bioinformatics 10847

Subseries of Lecture Notes in Computer Science

More information about this series at http://www.springer.com/series/5381

Fa Zhang · Zhipeng Cai
Pavel Skums · Shihua Zhang (Eds.)

Bioinformatics Research and Applications

14th International Symposium, ISBRA 2018
Beijing, China, June 8–11, 2018
Proceedings

 Springer

Editors
Fa Zhang
Chinese Academy of Sciences
Beijing
China

Pavel Skums
Georgia State University
Atlanta, GA
USA

Zhipeng Cai
Georgia State University
Atlanta, GA
USA

Shihua Zhang
Chinese Academy of Sciences
Beijing
China

ISSN 0302-9743 ISSN 1611-3349 (electronic)
Lecture Notes in Bioinformatics
ISBN 978-3-319-94967-3 ISBN 978-3-319-94968-0 (eBook)
https://doi.org/10.1007/978-3-319-94968-0

Library of Congress Control Number: 2018947451

LNCS Sublibrary: SL8 – Bioinformatics

Printed on acid-free paper

This Springer imprint is published by the registered company Springer Nature Switzerland AG
The registered company address is: Gewerbestrasse 11, 6330 Cham, Switzerland

Preface

On behalf of the Program Committee, we would like to welcome you to the proceedings of the 14th edition of the International Symposium on Bioinformatics Research and Applications (ISBRA 2018), held in Beijing, China, June 8–11, 2018. The symposium provides a forum for the exchange of ideas and results among researchers, developers, and practitioners working on all aspects of bioinformatics and computational biology and their applications.

This year we received 138 submissions in response to the call for extended abstracts. The Program Committee decided to accept 24 of them for full publication in the proceedings and oral presentation at the symposium. We also accepted 30 for oral presentation; a list of these contributions can be found in this front matter. Furthermore, we received ten submissions in response to the call for short abstracts. The technical program also featured two keynote and two invited talks by four distinguished speakers: Prof. Ying Xu from the University of Georgia presented on mining omic data of large numbers of cancer tissue samples; Prof. Xuegong Zhang from Tsinghua University gave a primary view on single-cell bioinformatics; Prof. Xin Gao from King Abdullah University of Science and Technology introduced a graph-based biclustering method for mining phenotype data; Prof. Min Li from Central South University spoke on de novo genome assembly by using statistical characteristics of paired-end reads.

We would like to thank the Program Committee members and the additional reviewers for volunteering their time to review and discuss symposium papers. We would like to extend special thanks to the steering and general chairs of the symposium for their leadership, and to the finance, publicity, workshops, local organization, and publications chairs for their hard work in making ISBRA 2018 a successful event. Last but not least, we would like to thank all authors for presenting their work at the symposium.

April 2018

Fa Zhang
Min Li
Xiaohua Wan
Zhipeng Cai

Organization

Steering Committee

Dan Gusfield   University of California Davis, USA
Ion Mandoiu   University of Connecticut, USA
Yi Pan (Chair)   Georgia State University, USA
Marie-France Sagot   Inria, France
Ying Xu   University of Georgia, USA
Aidong Zhang   State University of New York, USA

General Chairs

Xilin Chen   Institute of Computing Technology, Chinese Academy
                                   of Sciences, China
Alexander Zelikovsky   Georgia State University, USA
Shuigeng Zhou   Fudan University, China

Program Chairs

Fa Zhang   Institute of Computing Technology, Chinese Academy
                                   of Sciences, China
Zhipeng Cai   Georgia State University, USA
Pavel Skums   Georgia State University, USA
Shihua Zhang   Academy of Mathematics and Systems Science,
                                   Chinese Academy of Sciences, China

Publications Chairs

Min Li   Central South University, China
Xiaohua Wan   Institute of Computing Technology, Chinese Academy
                                   of Sciences, China
Le Zhang   Southwest University, China
Quan Zhou   Tianjin University, China

Publicity Chairs

Xuan Guo   University of North Texas, USA
Shaoliang Peng   National University of Defense Technology, China

Workshop Chairs

Yudong Liu   Institute of Computing Technology, Chinese Academy
                           of Sciences, China
Fei Ren   Institute of Computing Technology, Chinese Academy
                           of Sciences, China

Webmasters

Sergey Knyazev   Georgia State University, USA
Vyacheslav Tsivina   Georgia State University, USA

Program Committee

Max Alekseyev   George Washington University, USA
Mukul S. Bansal   University of Connecticut, USA
Paola Bonizzoni   Università di Milano-Bicocca, Italy
Zhipeng Cai   Georgia State University, USA
Hongmin Cai   South China University of Technology, China
Doina Caragea   Kansas State University, USA
Xing Chen   National Center for Mathematics and Interdisciplinary
                           Sciences, Chinese Academy of Sciences, China
Xuefeng Cui   Tsinghua University, China
Ovidiu Daescu   University of Texas at Dallas, China
Daming Zhu   Shandong University, China
Fei Deng   University of California, Davis, USA
Lei Deng   Central South University, China
Pufeng Du   Tianjin University, China
Oliver Eulenstein   Iowa State University, USA
Lin Gao   Xidian University, China
Xin Gao   King Abdullah University of Science and Technology,
                           Saudi Arabia
Olga Glebova   Georgia State University, USA
Xuan Guo   University of North Texas, USA
Jiong Guo   Shandong University, China
Zengyou He   Dalian University of Technology, China
Steffen Heber   North Carolina State University, USA
Jinling Huang   East Carolina University, USA
Shihao Ji   Georgia State University, USA
Mingon Kang   Kennesaw State University, USA
Wooyoung Kim   University of Washington Bothell, USA
Danny Krizanc   Wesleyan University, USA

Abstracts of Invited Talks

A Primary View on Single-Cell Bioinformatics

Xuegong Zhang

School of Life Sciences and School of Medicine, Tsinghua University, Beijing
100084, China
zhangxg@tsinghua.edu.cn

Abstract. Cells are not created equal. The Human Cell Atlas (HCA) project aims to build the atlas of all human cell types and cell states with their molecular signatures. Single-cell sequencing especially single-cell RNA-sequencing (scRNA-seq) is the key technology for obtaining the molecular signatures of a large amount of single cells at the whole transcriptome scale. It is a fundamental step toward the complete understanding of the human body, a super complex system composed of tens of trillions of cells that are all developed from a single cell. This opens the new broad field of single-cell biology. Single-cell biology converts each cell to a mathematical vector in the high-dimensional spaces of the expression of all genes and other molecular features. Therefore, single-cell bioinformatics, or the computational analyses of single-cell data, become the key component of all single-cell biology studies. This talk will give an overview of some key bioinformatics tasks in single-cell bioinformatics, and present examples of our on-going work on new methods for differential expression analysis and dimension reduction.

Searching for Roots of Cancer Development through Mining Large Scale Cancer Tissue Data and Modeling the Chemistry of Cellular Base-Acid Homeostasis

Ying Xu

Department of Biochemistry and Molecular Biology, University of Georgia,
Athens, GA, USA
xyn@uga.edu

Abstract. Over one million research articles have been published about cancer, but yet our understanding about cancer is undeniably little. We are yet to understand some of the most basic questions such as: (1) why some cancers such as pancreatic or liver cancers are so deadly while other cancers such as basal cell carcinoma are rarely life-threatening? or (2) why some cancers are highly drug resistant while other cancers are not? In this talk I will present some of our recent discoveries made through mining omic data of large numbers of cancer tissue samples. Our analyses strongly suggest that all cancer tissue cells have high levels of Fenton reactions, due to increased iron accumulation and H2O2 concentration at the disease sites, both being the result of persistent immune responses. A key consequence of the reaction is: it continuously produces OH-, to which the affected cells respond fiercely to maintain the pH homeostasis as changes in the intracellular pH would have profound impacts to the viability of the cells. We will demonstrate that cancer cells immobilize a wide range of metabolic activities through metabolic reprogramming, to keep the intracellular pH stable, including inhibition of the urea cycle, nucleotide synthesis, glycolytic ATP generation (Warburg effect) and even selection of mutations in specific amino acids. Some of the long-standing open questions can be answered naturally using our new model.

Gracob: A Graph-Based Constant-Column Biclustering Method for Mining Growth Phenotype Data

Xin Gao

Computational Bioscience Research Center, King Abdullah University
of Science and Technology, Thuwal, Saudi Arabia
xin.gao@kaust.edu.sa

Abstract. Growth phenotype profiling of genome-wide gene-deletion strains over stress conditions can offer a clear picture that the essentiality of genes depends on environmental conditions. Systematically identifying groups of genes from such high-throughput data that share similar patterns of conditional essentiality and dispensability under various environmental conditions can elucidate how genetic interactions of the growth phenotype are regulated in response to the environment. In this talk, I will first demonstrate that detecting such "co-fit" gene groups can be cast as a less well-studied problem in biclustering, i.e., constant-column biclustering. Despite significant advances in biclustering techniques, very few were designed for mining in growth phenotype data. I will then propose Gracob, a novel, efficient graph-based method that casts and solves the constant-column biclustering problem as a maximal clique finding problem in a multipartite graph. We compared Gracob with a large collection of widely used biclustering methods that cover different types of algorithms designed to detect different types of biclusters. Gracob showed superior performance on finding co-fit genes over all the existing methods on both a variety of synthetic data sets with a wide range of settings, and three real growth phenotype data sets for E. coli, proteobacteria, and yeast.

De novo Genome Assembly by Using Statistical Characteristics of Paired-end Reads

Min Li

School of Information Science and Engineering, Central South University,
Changsha, China
limin@mail.csu.edu.cn

Abstract. DNA sequence is the carrier of genetic information, which guides the development of biological and functions of life. De novo genome assembly is aimed at acquiring a complete and accurate genome sequence, so it has become one of the fundamental issues in genome research for understanding the organization and process of life activities. However, de novo genome assembly still faces the challenges of repetitive regions in genome, sequencing errors, and uneven sequencing depth. In this talk, I will present our recent work as follows: (1) a sequence assembler based on the distributions of insert size and read, called EPGA. Through assessing the variation of the distribution of insert size, EPGA can solve problems introduced by some complex repetitive regions. And an improved assembler EPGA2 adopts error corrections and memory-efficient DSK to count k-mers; (2) a scaffolding method based on iterative strategy and linear programming to detect spurious edges, called BOSS. And scaffolding algorithm SCOP, which is the first method to classify the contigs and utilize the vertices and edges to optimize the scaffold graph; (3) a gap filling method called GapReduce, which aligns the paired-end reads to the scaffolds. For each gap, GapReduce determines two read sets, and then constructs De Bruijn graphs. GapReduce extracts paths from De Bruijn graphs to cover the gaps by using the characteristics of insert size and k-mer frequencies based on the partitioned read sets. Finally, the future development and challenges of de novo genome assembly will also be discussed.

List of Oral Presentations not Included in this Volume

Drug Repositioning based on Individual Bi-random Walks on a Heterogenous Network
Yuehui Wang, Maozu Guo, Yazhou Ren, Lianyin Jia and Guoxian Yu

Detecting Differential Consistency Network Modules
Jianwei Lu, Yao Lu, Yusheng Ding, Qingyang Xiao, Linqing Liu, Qingpo Cai, Yunchuan Kong, Yun Bai and Tianwei Yu

Joint SNMF Method for Finding Modules of Multiple Brain Networks
Lingkai Tang, Yulian Ding, Jie Zhang and Fang-Xiang Wu

Identifying Driver Genes Involving Gene Dyregulated Expression, Tissue-Specic Expression and Gene-Gene Network
Junrong Song, Feng Wang, Wei Peng and Jianxin Wang

Region-Based Interaction Detection in Genome-Wide Case-Control Studies
Sen Zhang, Wei Jiang, Ronald Cw Ma and Weichuan Yu

HiSSI: High-order SNP-SNP Interactions Detection based on Efficient Significant Pattern and Differential Evolution
Xia Cao, Jie Liu, Maozu Guo and Jun Wang

Detecting Horizontal Gene Transfer: A Probabilistic Approach
Gur Sevillya, Orit Adato and Sagi Snir

Predicting Comorbid Diseases with Geometric Embeddings of Human Interactome
Pakeeza Akram and Li Liao

A Novel Network Based Approach for Predicting Survivability of Breast Cancer Patients
Sheikh Jubair, Luis Rueda and Alioune Ngom

Directional Association Test Reveals High-Quality Putative Cancer Driver Biomarkers Including Noncoding RNAs
Hua Zhong and Mingzhou Song

Mining Information of Co-expression Network based on TGCA Data
Mi-Xiao Hou, Ying-Lian Gao, Jin-Xing Liu, Jun-Liang Shang, Rong Zhu and Sha-Sha Yuan

Identifying MicroRNA-Gene Networks Specific to Pathologic Stages in Colon Cancer
Benika Hall and Xinghua Shi

Cancer Driver Genes Discovery by Integrating Expression and Mutation Data
Ying Hui, Pi-Jing Wei, Junfeng Xia and Chun-Hou Zheng

Contents

Structure and Interaction

HPC and CryoEM

Machine and Deep Learning

Data Analysis and Methodology

Analysis and Visualization Tools

RNA-Seq Data Analysis

Network Analysis and Modelling

Prediction of Drug Response with a Topology Based Dual-Layer Network Model

Suyun Huang[1] and Xing-Ming Zhao[2(✉)]

[1] Department of Computer Science, Tongji University, Shanghai 201804, China
zero_hsy@hotmail.com
[2] Institute of Science and Technology for Brain-Inspired Intelligence,
Fudan University, Shanghai 200433, China
xmzhao@fudan.edu.cn

Abstract. Identifying the response of a cancer patient to a particular therapeutic agent is critical in drug discovery and will significantly facilitate the development of personalized medicine. The publicly available drug response profiles across cell lines provide an alternative way for predicting the response of cancer drugs. In this work, we propose a topology based dual-layer network (TDLN) model to predict drug response based on large-scale cell line experiments. With the Cancer Cell Line Encyclopedia (CCLE), Genomics of Drug Sensitivity in Cancer (GDSC) and Cancer Therapeutic Response Portal (CTRP) datasets as benchmark datasets, our proposed topology based dual-layer network model outperforms other existing popular approaches and identify some novel indications of known drugs for cancer.

Keywords: Drug responses · Large-scale data · Dual-layer network model
Novel indications

1 Introduction

With the accumulation of omics data thanks to the advance of high-throughput technologies, it is becoming possible to implement precision medicine in clinic [1–4] and design personalized therapy for patients [5, 6]. Therefore, it is much necessary to predict the response of patients to drugs in advance based on their molecular profiles. However, it is expensive to generate molecular profiles for patients and it is not feasible to test drugs in patients, which makes it difficult to define the molecular signatures for drug response and hinders the progress of precision medicine.

Recently, the emerging drug response profiles across cell line makes it possible to predict drug response in vitro experiments, where gene expression profiles will be generated with or without drug treatment. Specifically, several large-scale datasets have been generated by monitoring the sensitivity of cancer cell lines to various drugs along with genomics and transcriptomics profiles of cancer cell lines, such as the Cancer Cell Line Encyclopedia (CCLE) [7], the Genomics of Drug Sensitivity in Cancer (GDSC) [8] and the Cancer Therapeutic Response Portal (CTRP) [9]. With these valuable data, some computational approaches, e.g. elastic net and random forest, have been developed to drug responses [1, 6, 10–14] with the assumption that genetically similar cell

© Springer International Publishing AG, part of Springer Nature 2018
F. Zhang et al. (Eds.): ISBRA 2018, LNBI 10847, pp. 3–12, 2018.
https://doi.org/10.1007/978-3-319-94968-0_1

lines will respond to the same drug in a similar way or structurally similar drugs will affect the same cell line with similar response. In particular, with the gene expression profiles of cells treated without drugs, a dual-layer drug-cell line network (DLDCN) model have been proposed to predict drug response [15], where two networks were respectively constructed based on drug-drug and cell line-cell line similarities. However, only the local neighborhood was explored in the dual-layer drug-cell line network model, while the global network topological information was not considered.

In this work, we present a topology based dual-layer network (TDLN) model constructed with gene expression profiles of cells treated without drugs to predict the response of cells to novel drugs. Compared with dual-layer drug-cell line network model (DLDCN), our approach is novel in the following parts. First, we define a new weight accompanying the edge in the dual-layer network by taking into account the topological structures of networks; Second, our proposed approach predicts drug response by integrating both the local neighborhood of the node of interest and the global network topological structure. Benchmarking on CCLE, GDSC and CTRP datasets, our approach significantly outperforms other popular approach, where the global topology information contributes a lot to the improvement of our approach. In addition, we predict some responses of cell line to novel drugs, some of which can be validated with evidence from literature. For example, we predict that EWS-FLI1-muted cell lines are sensitive to PARP inhibitors.

2   Materials and Methods

2.1   Data Sources

The cancer genomic and drug response profiles in this study are available from the Cancer Cell Line Encyclopedia (CCLE), Genomics of Drug Sensitivity in Cancer (GDSC) and Cancer Therapeutic Response Portal (CTRP).

For the CCLE data set, cancer cell line gene expression, copy number, and mutation profiles along with drug sensitivity data were downloaded from the Cancer Cell Line Encyclopedia (CCLE) website (http://www.broadinstitute.org/ccle/data/browseData). Pharmacological profiles of 24 small molecules and gene expression data for common 491 cancer cell lines were used in our analysis. For the GDSC data set, gene expression profiles and drug sensitivity measurements for the common 985 cancer cell lines and 265 drugs were downloaded from GDSC (http://www.cancerrxgene.org/downloads) and used to generate the performance. We also obtained drug sensitivity data and gene expression profiles from the CTRP website (www.broadinstitute.org/ctrp), which included common data for 481 drugs and 823 cell lines.

The chemical structural information for each drug was obtained from PubChem [16], a data base of chemical molecules and their activities in different biological assays. It contained validated chemical depiction information for 19 million unique compounds contributed from over 70 depositing organizations. We downloaded raw chemical property profiles (SDF files) for 23 drugs in the CCLE study, 179 drugs in the GDSC study and 431 drugs in the CTRP study from the PubChem website. The 1-D and 2-D chemical structural features of each drug were retrieved using the PaDEL [17] software program (v2.11, downloaded from the project website http://padel.nus.edu.sg/software/padeldescriptor/) with default settings.

2.2 Identification of Local-Based Weight, Global-Based Weight and Combined-Based Weight

In order to construct topology based dual layer network model (TDLN), it is essential to define the combined-based weight consisting of the local-based weight and global-based weight.

Local-Based Weight. For the purpose to examine the similarity between different cell lines or between drugs, the local-based weight parameter should be an increase function with their correlation using gene expression profiles or chemical structural data. The formula:

$$w_{local}^C(C, C_i) = e^{-\frac{[1-\rho(C,C_i)]^2}{2\sigma^2}}$$
$$w_{local}^D(D, D_j) = e^{-\frac{[1-\rho(D,D_j)]^2}{2\tau^2}}$$

$$(1)$$

It could be seen that $\rho(C, C_i)$ is the gene expression correlation between cell line C and C_i and $\rho(D, D_j)$ is the correlation between the chemical structural feature of drug D and D_j. σ and τ are determined as the bandwidth parameters. It means that if $\rho(C, C_i)$ is close to 1, $W_{local}^C(c, c_i)$ will also be close to 1, implying that these two cell lines share high concordance. On the contrary, if $\rho(C, C_i)$ is small (e.g. close to 0), $W_{local}^C(c, c_i)$ will be relatively small too. It is the same with the drugs.

Global-Based Weight. The DLDCN model only focuses on the local-based weight ignoring the sensitive information between the drug and cell lines. Based on the assumption that if drug A and drug B share many same sensitive cell lines, they would be highly similar, implying that $W_{global}^D(D, D_j)$ is equally high. For the cell line C that is being investigated, neighboring cell lines which get high similarity with C should respond to a large number of same sensitive drugs [18]. In the literature [18], it was found that the global topological structure can help improve prediction accuracy. According to this hypothesis, we propose the calculation formula of global-based weight as follows:

$$w_{global}^C(C, C_i) = \frac{1}{\max(k_c, k_{c_i})} \sum_{D_l} \frac{A_{D_l C} A_{D_l C_i}}{k_{D_l}}$$
$$K_c = \sum_{D_l} A_{D_l C}$$
$$K_{D_l} = \sum_{C_u} A_{D_l C_u}$$

$$(2)$$

where A_{DC} is bipartite network and can be presented as an adjacent matrix where $a_{ji} = 1$ if C_i is sensitive to D_j, otherwise $a_{ji} = 0$.

$$w^D_{global}(D, D_j) = \frac{1}{\max(k_D, k_{D_j})} \sum_{C_l} \frac{A_{DC_l} A_{D_j C_l}}{k_{C_l}}$$

$$K_D = \sum_{C_l} A_{DC_l} \tag{3}$$

$$K_{C_l} = \sum_{D_u} A_{D_u C_l}$$

Combined-Based Weight. After acquiring the formula of the local-based weight and global-based weight, it is vital to combine them together using the following formula:

$$w^D_{combined}(D, D_j) = \beta_{drug} w^D_{global}(D, D_j) + (1 - \beta_{drug}) w^D_{local}(D, D_j)$$
$$w^C_{combined}(C, C_i) = \beta_{cell} w^C_{global}(C, C_i) + (1 - \beta_{cell}) w^C_{local}(C, C_i) \tag{4}$$

By combining the local-based weight and global-based weight together, we could guarantee that in most conditions, we could get the similarity of two drugs or two cell lines since if there is a lack of one weight (e.g. global-based weight), exists the other one, making DLDCN model more robust.

2.3   Prediction of Drug Response to a New Drug or New Cell Line Based on the Similarity Network

In order to predict the response of a new drug D to a known cell line C, we take full use of the combined-based weight defined in the previous section in the drug similarity network (DSN). Based on the assumption that drugs with the high combined-based weight will respond similarly to the cell line, we develop a linear weighted model to calculate the sensitivity of the drug D to cell line C as follows:

$$\hat{Sens}_{DSN}(D, C) = \gamma_{drug} \sum_{D_j \neq D} \frac{w^D_{combined}(D, D_j)}{k_{D_j}} \cdot Sens(D_j, C) \tag{5}$$

where $Sens(D_j, C)$ is the sensitivity data of the drug D_j against cell line C and $\hat{Sens}_{DSN}(D, C)$ is the predicted value using the drug similarity network. It could be observed that if $w^D_{combined}(D, D_j)$ is relatively high and the ratio the weight occupied of out degrees called as k_{D_j} is large, the corresponding drug will have a strong contribution to the predicted value of the unknown drug.

In the cell line similarity network (CSN), we could develop a similar model to predict the response of a known drug to a new cell line as follows:

$$\hat{Sens}_{CSN}(D, C) = \gamma_{cell} \sum_{C_i \neq C} \frac{w_{combined}^C(C, C_i)}{k_{C_i}} \cdot Sens(D, C_i) \tag{6}$$

2.4 Drug Response Prediction Based on the Network Model

In the above section, we propose the two separate models for predicting the response of a cell line to a drug based on the cell line similarity network (CSN) and the drug similarity network (DSN). To take the advantage of the two models, the integrated network is produced using the following formula:

$$\hat{Sens}(D, C) = \lambda \hat{Sens}_{DSN}(D, C) + (1 - \lambda)\hat{Sens}_{CSN}(D, C) \tag{7}$$

where λ is the combination weight to determine the weight of the DSN and CSN and it could be optimized by the leave-one-out cross validation. $\hat{Sens}_{DSN}(D, C)$ is the predicted value using the drug similarity network and $\hat{Sens}_{CSN}(D, C)$ is the predicted value using the cell line similarity network.

2.5 Leave-One-Out Cross-Validation

To determine the parameters and test the performance, we consider the leave-one-out cross validation method. It means that each possible drug-cell line pair would be left as the test data set while the remaining data are trained. Firstly, we obtain local-based weight parameters σ and τ using the CSN and DSN separately. Then the combined-based weight parameters β_{drug} and β_{cell} are also fixed for all drugs and cell lines in CSN and DSN separately. After integrating the CSN and DSN, the parameter λ should be defined by individual drug, considering that different drugs have different relative proportion of two layers.

In general, all the parameters are optimized by minimizing the sum of squared errors. Take the optimized parameter σ for example, the overall error function is defined as:

$$J(\sigma) = \sum_{D,C} \left(Sens(D, C) - \hat{Sens}(D, C)\right)^2 \tag{8}$$

where $Sens(D, C)$ is the observed sensitivity value of cell line C to drug D, and $\hat{Sens}(D, C)$ is the predicted value using all other drug-cell lines as training set. Thus the best parameter could be obtained by minimizing the sum of the squared errors.

After achieving the parameters, the prediction performance of the models is evaluated by the Pearson correlation coefficient between predicted and observed responses in every drug. A higher correlation coefficient shows a better performance of a model.

3 Results

3.1 Prediction of Drug Response on Benchmark Datasets

In order to evaluate the performance of TDLN model, it was applied to three benchmark datasets, i.e. CCLE, GDSC and CTRP, and all these three datasets contain gene expression profiles generated from cells treated without drugs. We also compared TDLN model with existing popular approaches, including random forest, elastic net regression, where the leave-one-out cross-validations were performed for each approach. For each computational approach, it will predict drug response values. To see the performance of those computational approaches, the Pearson correlation coefficients will be calculated between their predicted values and observed drug responses, where the higher the correlation coefficient is the better the performance will be. Figure 1 shows the results obtained by TDLN model, random forest and elastic net on the three benchmark datasets.

For both random forest and elastic net model, the gene expression profiles were used to predict drug response. From the prediction results, we can see that our TDLN model significantly outperforms the other two approaches on all three datasets, where the elastic net model performs the second best. The good performance of TDLN model on all three benchmark datasets demonstrates the predictive power of our proposed approach.

In particular, we compared TDLN model with previously proposed DLDCN model, which is also a dual-layer network model [15]. Figure 2 shows the results on three benchmark datasets, where the Pearson correlation coefficients between predicted and observed responses were used to evaluate the performance of the two approaches. Since the activity area (abbreviated as AArea hereafter) was used for quantifying drug response in some datasets while the IC50 was used in other datasets, we compared TDLN against DLDCN with respect to either AArea or IC50 depending on the dataset of interest. From Fig. 2, we can see that our proposed TDLN significantly outperforms DLDCN according to AArea on all three benchmark datasets. We take the top left picture for example, the activity area data is used to predict the drug responses in CCLE datasets using leave-one-out cross validation. For each drug, we could get the Pearson correlation coefficient between predicted and observed responses. Since there are 23 drugs, it means there are 23 values and we use a box plot to describe the result. We could find the TDLN model achieves a correlation coefficient of 0.75 on average and improves about 10% over the DLDCN model. The good performance of TDLN is attributed to the weights we defined for the links in the network.

For both CCLE and GDSC datasets, there are also IC50 values for drug responses. From Fig. 2, our TDLN outperforms DLDCN model on both CCLE and GDSC datasets. The comparison of TDLN and DLDCN with either AArea or IC50 demonstrates that our proposed TDLN model is indeed effective to predict drug responses, especially where more cell lines were tested for a certain drug.

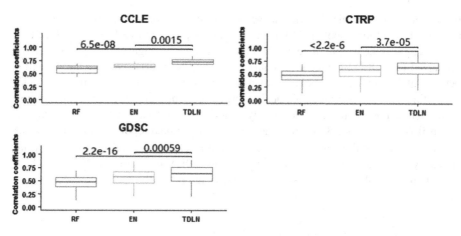

Fig. 1. Comparison of prediction performance of TDLN against RF, EN using activity area data, where the p-values were determined with t-test based on the distribution of correlation coefficients. RF denotes random forest and EN denotes elastic net while TDLN denotes topology based dual-layer network.

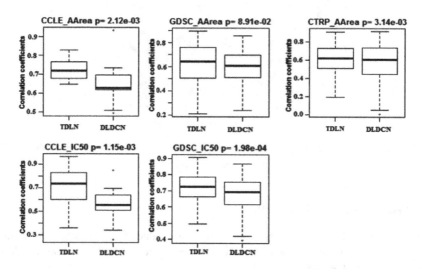

Fig. 2. Comparison of prediction performance between TDLN and DLDCN on the three datasets based on the correlation coefficients between predicted and observed drug responses on three datasets, where p-values were calculated with t-test. Here, "AArea" denotes "activity area".

3.2   Global Topological Information Boosts the Prediction Performance

Compared with DLDCN, our proposed TDLN utilized both the local and global topological information of the heterogeneous drug-cell line network, whereas the DLDCN model considers only the local neighborhood information. In TDLN, the global and local information were balanced with two weights, i.e. β_{drug} and c. The two parameters were optimized during leave-one-out cross-validations. To see the contributions of the global topological information to the prediction performance of TDLN, we investigated the weight of beta across the three datasets. Figure 3 shows the weights optimized for TDLN over the three datasets. It could be clearly seen that the weight of beta for global information is larger than 0.5 on most of the datasets considered here, suggesting that the global topological information contribute more to the prediction performance. For example, in the activity area data in the CCLE dataset, we could see β_{cell} is 0.7 and β_{drug} is 0.8 meaning the weight of the global topological information is greater than the local neighborhood whether in the cell line similarity network or drug similarity network. That is also the reason that TDLN outperforms DLDCN which exploits only local neighborhood information of the network.

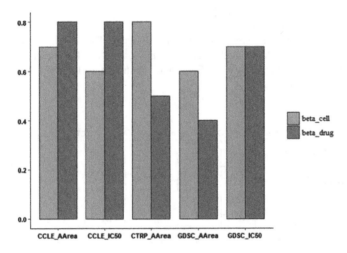

Fig. 3. The optimized weight parameters of β_{drug} and β_{cell} that are determined in the CCLE, GDSC and CTRP datasets. Here, "AArea" denotes "activity area".

3.3   Prediction of Novel Drug Responses

In the datasets, there are some missing drug response values between certain drugs and some cell lines. Here, we applied our TDLN model to predict drug responses for those missing values in the GDSC dataset.

For the GDSC dataset, we focused on the PRAP inhibitors, AG-014699 and Olaparib, for which the response values are missing in a number of cell lines. Especially, we predicted the responses of the two inhibitors in the EWS-FLI1-muted cell lines, where the IC50 values were predicted. As a result, we predicted five cell lines that are sensitive to both AG-014699 and Olaparib. Figure 4 shows the distribution of the IC50 values predicted by TDLN in both EWS-FLI1-muted cell lines and other cell lines. From the distribution profiles, we can see that the EWS-FLI1-muted cell lines show an increased sensitivity to PARP inhibitors, which is in agreement with previous findings [8], implying the effectiveness of TDLN in predicting drug responses.

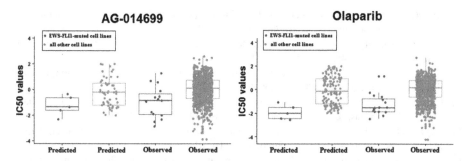

Fig. 4. The distribution of IC50 profiles for PRAP inhibitors, AG-014699 and Olaparib, across EWS-FLI1-muted cell lines and other cell lines in the GDSC dataset.

4   Conclusions

In this work, we have proposed a topology based dual-layer network model (TDLN) to predict drug response based on a drug-cell line heterogeneous network. In particular, we defined the weights for the links within and between networks based on both local neighborhood and global topological information of the networks. Benchmarking on three datasets, our TDLN significantly outperforms other popular computational approaches and we find that global topological information boosts the prediction performance. In particular, we predicted that the EWS-FLI1-muted cell lines are sensitive to PARP inhibitors. The finding has been validated by evidences from literature, indicating the predictive power of our proposed TDLN model in predicting drug response.

Acknowledgments. This work was partly supported by National Natural Science Foundation of China (61772368, 61572363, 91530321, 61602347) and Natural Science Foundation of Shanghai (17ZR1445600).

References

1. Costello, J.C., et al.: A community effort to assess and improve drug sensitivity prediction algorithms. Nat. Biotechnol. **32**, 1202–1212 (2014)
2. Mirnezami, R., et al.: Preparing for precision medicine. N. Engl. J. Med. **366**, 489–491 (2012)
3. Simon, R., Roychowdhury, S.: Implementing personalized cancer genomics in clinical trials. Nat. Rev. Drug Discov. **12**, 358–369 (2013)
4. Wistuba, I.I., et al.: Methodological and practical challenges for personalized cancer therapies. Nat. Rev. Clin. Oncol. **8**, 135–141 (2011)
5. Qin, G., Zhao, X.M.: A survey on computational approaches to identifying disease biomarkers based on molecular networks. J. Theoret. Biol. **362**, 9–16 (2014)
6. Zeng, T., et al.: Big-data-based edge biomarkers: study on dynamical drug sensitivity and resistance in individuals. Brief. Bioinform. **17**, 576–592 (2016)

7. Barretina, J., et al.: The Cancer Cell Line Encyclopedia enables predictive modelling of anticancer drug sensitivity. Nature **483**, 603–607 (2012)
8. Garnett, M.J., et al.: Systematic identification of genomic markers of drug sensitivity in cancer cells. Nature **483**, 570–575 (2012)
9. Rees, M.G., et al.: Correlating chemical sensitivity and basal gene expression reveals mechanism of action. Nat. Chem. Biol. **12**, 109–116 (2016)
10. Cui, J., et al.: An integrated transcriptomic and computational analysis for biomarker identification in gastric cancer. Nucleic Acids Res. **39**, 1197–1207 (2011)
11. Li, Z., et al.: Two-stage flux balance analysis of metabolic networks for drug target identification. BMC Syst. Biol. **5**(Suppl 1), S11 (2011)
12. Liu, X., et al.: Identifying disease genes and module biomarkers by differential interactions. J. Am. Med. Inf. Assoc. **19**, 241–248 (2012)
13. Menden, M.P., et al.: Machine learning prediction of cancer cell sensitivity to drugs based on genomic and chemical properties. PLoS ONE **8**, e61318 (2013)
14. Zeng, T., et al.: Prediction of dynamical drug sensitivity and resistance by module network rewiring-analysis based on transcriptional profiling. Drug Resist. Updat. **17**, 64–76 (2014)
15. Zhang, N., et al.: Predicting anticancer drug responses using a dual-layer integrated cell line-drug network model. PLoS Comput. Biol. **11**, e1004498 (2015)
16. Wang, Y., et al.: PubChem: a public information system for analyzing bioactivities of small molecules. Nucleic Acids Res. **37**, W623–W633 (2009)
17. Yap, C.W.: PaDEL-descriptor: an open source software to calculate molecular descriptors and fingerprints. J. Comput. Chem. **32**, 1466–1474 (2011)
18. Cheng, F., et al.: Prediction of drug-target interactions and drug repositioning via network-based inference. PLoS Comput. Biol. **8**, e1002503 (2012)

GRTR: Drug-Disease Association Prediction Based on Graph Regularized Transductive Regression on Heterogeneous Network

Qiao Zhu, Jiawei Luo[✉], Pingjian Ding, and Qiu Xiao

College of Computer Science and Electronic Engineering,
Collaboration and Innovation Center for Digital Chinese Medicine
in Hunan Province, Hunan University, Changsha 410082, China
luojiawei@hnu.edu.cn

Abstract. Computational drug repositioning helps to decipher the complex relations among drugs, targets, and diseases at a system level. However, most existing computational methods are biased towards known drugs-disease associations already verified by biological experiments. It is difficult to achieve excellent performance with sparse known drug-disease associations. In this article, we present a graph regularized transductive regression method (GRTR) to predict novel drug-disease associations. The proposed method first constructs a heterogeneous graph consisting of three interlinked sub-graphs including drugs, diseases and targets from multiple sources and adopts preliminary estimation of drug-related disease to initial unknown drug-disease associations for unlabeled drugs. Since the known drug-disease associations are sparse, graph regularized transductive regression is used to score and rank drug-disease associations iteratively. In the computational experiments, the proposed method achieves better performance than others in terms of AUC and AUPR. Moreover, the varying of parameters is shown to verify the importance of preliminary estimation in GRTR. Case studies on several selected drugs further confirm the practicality of our method in discovering potential indications for drugs.

Keywords: Transductive regression · Drug repositioning
Drug-disease association · Graph regularization · Heterogeneous network

1 Introduction

Traditional drug development faces difficulties relating to the expensive, time consuming and high risk of failure. Studies have demonstrated that drug repositioning, which aims to discovery new indications for existing drugs, offers a promising alternative to drug development. Some successful repositioned drugs (e.g. Sildenafil, thalidomide, raloxifene) have historically generated high revenues for their patent holders or companies [1]. Compared to in vivo experimental methods for drug repositioning, in silico approaches are efficient at identifying potential drug-disease association, and thus significantly reduce research costs. Therefore, it is necessary to develop a computational method for identifying drug-disease associations.

© Springer International Publishing AG, part of Springer Nature 2018
F. Zhang et al. (Eds.): ISBRA 2018, LNBI 10847, pp. 13–25, 2018.
https://doi.org/10.1007/978-3-319-94968-0_2

To date, much effort has been allocated to developing computational approaches for predicting drug-disease associations. Conventional computational methods mainly depend on two strategies, the network-based method and feature-based method. A key idea behind network-based algorithms is the construction of complex biological networks with large-scale biological data. Wang et al. [2] proposed a drug-disease heterogeneous network model termed Heterogeneous Graph Based Inference (HGBI) and extended the algorithm to a three-layer network (HL_HGBI), adding a new layer of the target information [3]. However, the assumption was that drugs should have diverse indications and diseases should have diverse treatments. Martínez et al. [4] constructed a complex network which included drugs, diseases and proteins. Protein interactions were used as a bridge to perform DrugNet, a general network-based prioritization based on a propagation flow algorithm. Luo et al. [5] exploited known drug-disease associations to devise the drug-drug and disease-disease similarity measures, then building a drug-disease heterogeneous network, on which a bi-random walk algorithm was adopted to predict novel potential associations between drugs and diseases.

Much attention has also been devoted to introducing feature-based methods. Bleakley et al. provided a supervised learning approach [i.e. support vector machine (SVM)] on a bipartite local model (BLM) from chemical and genomic data [6]. Mei et al. [7] proposed BLM-NII, combining BLM with a neighbor-based interaction-profile inferring(NII) procedure. Gottlieb et al. [8] conducted multiple drug-drug and disease-disease similarity measures as classification features, implementing a classification algorithm named PREDICT to infer potential drug indications. Yang et al. [9] calculated relevance scores between drugs and diseases from a drug-target-pathway-gene-disease network and learnt a probabilistic matrix factorization model (PMF) based on known drug-disease associations to classify drug-disease associations. However, most of these approaches rely on the known association information and directly set the weight of unknown disease-drug associations to zero. This is perhaps the major reason that existing methods can't obtain a satisfactory performance based on sparse known associations validated by biological experiments.

In this article, we propose a graph regularized transductive regression (GRTR) method to deal with the problem of the sparse known associations for drug-disease association prediction. A three-layer heterogeneous network composed of drugs, diseases and targets is constructed from multiple datasets. Then we approximately calculate drug related diseases from local neighborhood information and adjust the weight of links with diseases based on it. Through a transductive regression model with graph regularization, the relevance score for potential drug-disease associations will be iteratively updated and all drugs ranked by their scores and judged whether they are related to a disease. Compared to the previous nine advanced prediction methods, GRTR performs better in terms of AUC and AUPR. Furthermore, the effect of varying weighted parameters and the effect of preliminarily estimating drug-related disease are analyzed. Case studies on the selected drugs and targets further exhibit the predictive ability of drug-disease association.

2   Methods

The overall process of predicting new drug-disease associations by GRTR is displayed in Fig. 1. GRTR first constructs a three-layer heterogeneous network composed of drugs, diseases and targets. Next, local information is obtained based on preliminary estimates for drug-related disease from the distribution of diseases associated with neighbor nodes in the heterogeneous network. Finally, using the heterogeneous network, the known relationships with diseases and preliminary estimation results as inputs, GRTR adopt graph regularization transductive regression to score and rank drug-disease associations iteratively.

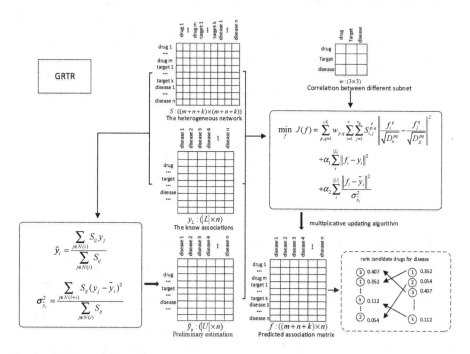

Fig. 1. GRTR workflow. Given the inputs of the heterogeneous network matrix $\mathbf{S}$ and the matrix of known association $\mathbf{y}_L$, we first obtain preliminary estimates for drug related diseases $\mathbf{y}_u$ using neighbor distribution information. We then score and rank drug-disease associations iteratively based on graph regularization transductive regression. The top rank drugs for each disease in the predicted association matrix $\mathbf{f}$ are treated as the candidate drugs for those diseases for further experimental investigation.

2.1   Heterogeneous Network Construction

The three-layer heterogeneous network consists of three nodes types: drug nodes, disease nodes and target nodes. Suppose that m, n and k are the number of drugs, diseases and targets, respectively. $S^{11} = \left\{ S_{i,j}^{11} \right\}_{i=1,j=1}^{m,m}$ is an adjacency matrix of the drugs similarity network, $S^{22} = \left\{ S_{i,j}^{22} \right\}_{i=1,j=1}^{k,k}$ is an adjacency matrix of the protein interaction network and $S^{33} = \left\{ S_{i,j}^{33} \right\}_{i=1,j=1}^{n,n}$ is an adjacency matrix of the disease similarity network. Drug similarities can be calculated based on their chemical structures. Disease similarities and protein-protein interactions can be obtained from online datasets. We connect the above three subnetworks using experimentally verified drug-disease associations ($S^{13} = \left\{ S_{i,j}^{13} \right\}_{i=1,j=1}^{m,n}$), target-disease associations ($S^{23} = \left\{ S_{i,j}^{23} \right\}_{i=1,j=1}^{k,n}$) and drug-target associations ($S^{12} = \left\{ S_{i,j}^{12} \right\}_{i=1,j=1}^{m,k}$) to form a heterogeneous network. The adjacency matrix of the heterogeneous network can be represented as follows:

$$
S = \begin{pmatrix} S^{11} & S^{12} & S^{13} \\ \left(S^{12} \right)^T & S^{22} & S^{23} \\ \left(S^{13} \right)^T & \left(S^{23} \right)^T & S^{33} \end{pmatrix}
$$

where $(\cdot)^T$ represents the transpose of a matrix.

2.2   Preliminary Estimation of Drug Related Disease

In our research, the node with no known associations with a disease is unlabeled while other nodes are labeled. Preliminary estimation for the related diseases for an unlabeled drug is a local estimation. According to the assumption that drugs which are 'close together' will have associations with the same disease [10], we will consider neighborhood information based on the equal combination of diseases which have association with neighbor nodes in the heterogeneous network. Firstly, the neighbors of a drug i can be defined by the nearest labeled nodes N in the heterogeneous network.

$$
N(i) = \left\{ j \, | \, S_{ij} > \sigma, 1 \leq i \leq m+n+k, 1 \leq j \leq m+n+k \right\} \tag{1}
$$

where σ is a threshold and in this paper $\sigma = 0.5$. Then we use the mean distribution of the neighbor's disease to describe the biological network's local information and obtain preliminary estimations for the related diseases ($\tilde{y}$).

$$
\tilde{y}_i = \frac{\sum\limits_{j \in N(i)} S_{i,j} y_j}{\sum\limits_{j \in N(i)} S_{i,j}} \tag{2}
$$

where y_j denotes the known associations between nodes j and diseases. Here, diseases can be understood as discrete variables. Hence, the variance of a neighbor's disease distribution ($\sigma_{\tilde{y}}^2$) can be obtained as follows:

$$\sigma_{\tilde{y}_i}^2 = \frac{\sum\limits_{j \in N(l+i)} S_{i,j}(y_j - \tilde{y}_i)^2}{\sum\limits_{j \in N(i)} S_{i,j}} \tag{3}$$

2.3   Graph Regularized Transductive Regression

The main idea of our prediction method is based on transductive regression which is one of the most popular methods for imbalanced (sparse) data analysis, because prediction though transductive regression can lead to good knowledge extraction of the hidden network structure [11]. Wan et al. [12] presented a graph regularization-based transductive regression (Grempt) method using a symmetry meta-path to deal with label prediction on heterogeneous information networks, which have performed satisfactorily. In order to address the limitations of the symmetry meta-path, we revise the objective function's first term to directly consider different links classes in the heterogeneous network. The revised objective function is defined as follows:

$$\begin{aligned}
J(f) = &\sum_{p,q=1}^{|A|} w_{p,q} \sum_{i=1}^{v_p} \sum_{j=1}^{v_q} S_{i,j}^{p,q} \left\| \frac{f_i^p}{\sqrt{D_{ii}^{pq}}} - \frac{f_j^q}{\sqrt{D_{jj}^{pq}}} \right\|^2 \\
&+ \alpha_1 \sum_{i}^{|L|} \| f_i - y_i \|^2 \\
&+ \alpha_2 \sum_{i}^{|U|} \frac{\| f_i - \tilde{y}_i \|^2}{\sigma_{\tilde{y}_i}^2}
\end{aligned} \tag{4}$$

where α_1 and α_2 are two regularization coefficients which balance the different components of the model. A = {drug, disease, target} is the network nodes category, $w_{p,q}$ is the correlation between categories A_p and A_q (p, q $\in \{1, 2, \ldots, |A|\}$), v_p is the number of nodes which belong to category A_p, $S_{i,j}^{pq}$ is the relevance between object i $\in A_p$ and object j $\in A_q$ in the network, f_i and f_i^p are the prediction results of node i where p denotes i $\in A_p$, D_{ii}^{pq} is the sum of the i-th row in S^{pq}, L is the labeled nodes set which has an association with disease and U is the unlabeled node set. The model consists of 3 functions and each one corresponds to different meaning:

- The first part of the objective function is the global smoothness item, which formulates that similar nodes are likely to be associated with similar diseases.
- The second term of the objective function minimizes the difference between the predicted results and the known association.
- The last term formulates a regularization item to minimize the difference between the predicted results and the preliminary estimation from local characteristics.

The global minimum is calculated by differentiating (4) with respect to f_L^p and f_U^p respectively, which gives:

$$
\frac{\partial J(f)}{\partial f_L^p} = \sum_{p,q,p \neq q}^{|A|} 2w_{p,q}(f_L^p - R_{LL}^{pq}f_L^q - R_{LU}^{pq}f_U^q)
$$
$$
+ 4w_{p,p}(f_L^p - R_{LL}^{pp}f_L^p - R_{LU}^{pp}f_U^p) + 2\alpha_1(f_L^p - y_L^p)
$$
(5)

$$
\frac{\partial J(f)}{\partial f_U^p} = \sum_{p,q,p \neq q}^{|A|} 2w_{p,q}(f_U^p - R_{UL}^{pq}f_L^q - R_{UU}^{pq}f_U^q)
$$
$$
+ 4w_{p,p}(f_U^p - R_{UL}^{pp}f_L^p - R_{UU}^{pp}f_U^p) + 2\frac{\alpha_1}{\sigma_{\tilde{y}^p}^2}(f_L^p - \tilde{y}^p)
$$
(6)

where f_L^p denotes the prediction result of labeled nodes belonging to A_p and f_U^p denotes the prediction result of unlabeled nodes belonging to A_p. $R^{pq} = (D^{pq})^{-\frac{1}{2}}S^{pq}(D^{qp})^{-\frac{1}{2}}$ is the integration of the whole heterogeneous network, which can be rearranged based on labeled and unlabeled objectives.

$$
R^{pq} = \begin{pmatrix} R_{LL}^{pq} & R_{LU}^{pq} \\ R_{UL}^{pq} & R_{UU}^{pq} \end{pmatrix} \cdot p, q \in \{1, 2, \ldots, |A|\}
$$

Suggested that $\frac{\partial J(f)}{\partial f_L^p} = 0$ and $\frac{\partial J(f)}{\partial f_U^p} = 0$, the closed-form solution is obtained. However, the iterative solution is sometimes preferable [13]. The detail steps of GRTR to predict potential associations are described in Algorithm 1.

Algorithm 1. GRTR

Input: $R, D, T, G_{RD}, G_{TD}, G_{RT}, w, y, m, n, k, \sigma$;

Output: f

1. Use R, D, T, G_{RD}, G_{TD} and G_{RT} to build the heterogeneous network S;
2. Get $\tilde{y}$ and $\sigma_{\tilde{y}}^2$ from the preliminary estimation of drug-related disease based on (2) and (3);
3. Initialize $f(0) = (y_L^T, \tilde{y}_L^T)^T$ and t = 0;
4. Repeat
5. $p, q \in \{1, 2, \ldots, |A|\}$;

6.
$$f_L^p(t+1) = \frac{\alpha_1 y_L^p + \sum\limits_{p,q,p\neq q}^{|A|} w_{p,q}\left(R_{LL}^{pq} f_L^q(t) + R_{LU}^{pq} f_U^q(t)\right) + 2w_{p,p}\left(R_{LL}^{pp} f_L^p(t) + R_{LU}^{pp} f_U^p(t)\right)}{\sum\limits_{p,q,p\neq q}^{|A|} w_{p,q} + 4w_{p,p} + \alpha_1};$$

7.
$$f_U^p(t+1) = \frac{\dfrac{\alpha_2 \tilde{y}^p}{\sigma_{\tilde{y}^p}^2} + \sum\limits_{p,q,p\neq q}^{|A|} w_{p,q}\left(R_{UL}^{pq} f_L^q(t) + R_{UU}^{pq} f_U^q(t)\right) + 2w_{p,p}\left(R_{UL}^{pp} f_L^p(t) + R_{UU}^{pp} f_U^p(t)\right)}{\sum\limits_{p,q,p\neq q}^{|A|} w_{p,q} + 4w_{p,p} + \dfrac{\alpha_2}{\sigma_{\tilde{y}^p}^2}};$$

8. $t = t + 1$;
9. Repeat until convergence;
10. $f = \left\{f^{1^T}, f^{2^T}, \ldots, f^{|A|^T}\right\}^T$. Every unlabeled drug is assigned to the disease label which is on top r scores in each row of f.

---

3 Experiment and Results

3.1 Dataset

Experimentally confirmed drug-disease associations and drug–target associations are both downloaded from the supplementary material of [8]. Gottlieb et al. have collected 1933 known drug-disease associations involving 593 drugs registered in DrugBank [14] and 313 diseases listed in the Online Mendelian Inheritance in Man (OMIM) database [15]. At last, we get 2814 known drug–target associations between 593 drugs and 777 proteins.

The interactions between diseases and proteins are obtained from DisGeNET [16], for a total of 10010 relationships between 3221 proteins and 313 diseases.

The disease–disease similarity network is downloaded from Online Mendelian Inheritance in Man Mining Tool (MimMiner) [17]. According to the MimMiner database, disease–disease similarities have already been normalized to the range [0, 1].

The protein–protein interaction network is built using 37039 binary interactions among 9465 genes in the Human Protein Reference Database (HPRD) [18].

The drug–drug similarities are calculated based on their chemical structures. First, the chemical structures of all drug compounds in the Canonical Simplified Molecular Input Line-Entry System (SMILES) format [19] are downloaded from DrugBank. Then, the Chemical Development Kit [20] is used to calculate a binary fingerprint for each drug. Finally, Tanimoto score [21] of two drugs was calculated based on their fingerprints, which was in the range of [0, 1].

3.2   Parameters Selection and the Effect of Preliminary Estimation for Drug-Related Disease

There are three parameters w, α_1 and α_2 in our prediction. w controls the importance of different network. α_1 and α_2 control the contribution of known labeled objects and preliminary estimation, respectively. We set w $= 1$ for easy. To determine the optimal configuration of α_1 and α_2, we firstly let both increase from 0 to 1 in increments of 0.05 and record the change in AUC. The results can be seen in Fig. 2(a), in which AUC value increases rapidly as both α_1 and α_2 increase, and then became steadily reaching the maximum AUC value. However, in order to determine the general future trend as α_1 and α_2 become larger, we also vary them from 1 to 200, as demonstrated in Fig. 2(b). AUC value rapidly decreases in the range $0 \leq \alpha_1 \leq 10$ and then remains almost constant in the range $10 < \alpha_1 \leq 200$ which shows the result is not improved for these regions. But, there is an opposite trend for α_2, which first rises rapidly in the range of $0 \leq \alpha_2 \leq 10$ and after keeping a short constant, it decreases in the range $30 \leq \alpha_2 \leq 200$. Finally, we select $\alpha_1 = 1$ and $\alpha_2 = 20$ for getting a better prediction result. Although α_2 is much larger than α_1, it fits with the reality that preliminary estimation for drug-related disease information is more important than it is for predicting new relations.

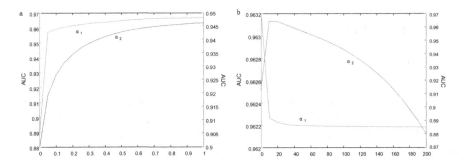

Fig. 2. The influence of different α_1 and α_2 values on AUC. (a): 0–1 in 0.05 increments. (b): 0–200 in 10 increments.

If we don't use the preliminary estimation for drug-related disease ($\alpha_2 = 0$, $\alpha_1 \neq 0$), the largest AUC is 0.9139. As α_2 gets larger, AUC turns to be larger until reaches the best value. To a certain extent, preliminary estimation for drug-related disease is significant, and can improve predictive ability.

3.3   Compared with Existing Methods

Systematic experiments are performed to evaluate the performance of the presented approach with nine other methods: Weighted Nearest Neighbor-Gaussian Interaction Profile(WNN-GIP) [22], Collaborative Matrix Factorization(CMF) [23], Kernelized Bayesian matrix factorization(KBMF) [24], Neighborhood Regularized Logistic Matrix Factorization(NRLMF) [25], a bipartite local model (BLM) [12], BLM with neighbor-based interaction profile inferring (BLM-NII) [7], comprehensive similarity measures and Bi-Random Walk algorithm (MBiRW) [5], standard LapRLS improved by incorporating a new kernel (NetLapRLS) [26]. We use 10-fold validation to compare GRTR performance. The area under the receiver operating characteristic (ROC) curve (AUC) [27] and the area under the precision recall (PR) curve (AUPR) are used to measure the quality of the predicted drugs for diseases. Figure 3 shows the ROC and PR curves of the 10-fold validation experiments. Table 1 gives the AUC and AUPR values. As expected, the GRTR's AUC value is 0.9668, which outperforms all other competitive methods significantly. GRTR is 2.10% better than the second-best method, NRLMF, which also achieved an impressive result of 0.9465. For AUPR, we observe that the values are lower than those in the original papers. The main reason for this is that the data we used is larger and comparatively sparser. But GRTR also performs well, obtaining the second best in the dataset with the AUPR value of 0.5925. Though GRTR is slightly lower than NRLMF, it is still very competitive among the methods.

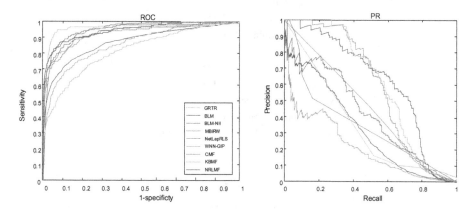

Fig. 3. The ROC and PR curves of GRTR and nine existing methods.

Table 1. AUC and AUPR values of GRTR and nine existing method.

Metric	GRTR	BLM	BLM-NII	MBiRW	NetLapRLS	WNN-GIP	CMF	KBMF	NRLMF
AUC	**0. 9668**	0. 8719	0.9442	0. 9179	0.9444	0. 8584	0.9309	0.8713	0. 9465
AUPR	0.5925	0.3256	0.4075	0.0469	0.5750	0.205	0.3455	0.3463	**0.6790**

3.4   Case Study

Here, the capability of our method in predicting novel drug-disease associations is further examined here. One well-known biological database CTD [30] and some references are used to verify the predicted novel drug-disease associations. For each disease, the candidate drugs are ranked based on the prediction scores and the top-10 predicted drugs as prediction results are collected. For instance, 8 of the top 10 potentially related drugs have been directly shown to be linked with Diabetes Mellitus type II (see Table 2), a endocrine system disease and metabolic disease. Lovastatin (DrugBank: DB00227) is predicted to treat it and has been recorded in CTD. Figure 4 presents lovastatin's neighbor drugs and the diseases they can treat. Vitamin d-dependent rickets, osteoporosis and hyperlipoproteinemia are metabolic disease. And barakat syndrome is an endocrine system disease. In addition, we also find many associated genes between those that lovastatin can interact with to treat diabetes mellitus and the neighbors can act on to treat corresponding disease, e.g. there are 1307 genes shared with the pravastatin treating hyperlipoproteinemia, 1460 genes shared with the calcitriol treating vitamin d-dependent Rickets and 762 genes shared with the Ergocalciferol treating barakat syndrome, etc.

Table 2. The top 10 predicted results for diabetes mellitus associated drugs.

Rank	Drug	Evidence
1	Guanfacine	Literature [28]
2	Nalbuphine	
3	Lovastatin	CTD
4	Tamoxifen	CTD
5	Bicalutamide	
6	Promethazine	CTD
7	Risperidone	CTD
8	Dinoprostone	CTD
9	Spironolactone	CTD
10	Carvedilol	Literature [29]

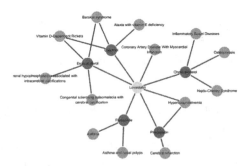

Fig. 4. Lovastatin (DB00227)'s neighbors and diseases can be treated. The yellow circle is the predicted drug, the red circles are the neighbor drugs of the predicted drug and the green circles are the diseases its neighbor can treated. (Color figure online)

4 Conclusions

Identifying drug-disease associations is helpful in reducing the difficulty of drug development and contributing to improved understanding of the underlying complex relations among drugs, targets and diseases. In this work, we systematically studied the problem of predicting drug-disease associations. Conventional methods for drug-disease association prediction mainly achieved unsatisfactory performance for the sparse known associations. However, the number of drug-disease associations verified by biological experiments is far less than that of the potential drug-disease associations. Therefore, GRTR based on graph regularized transductive regression was developed to predict potential drug-disease associations. At first a three-layer heterogeneous network consisting of drugs, diseases and targets was constructed. Afterwards, preliminary estimation for drug-related diseases was conceived from neighbor information. Ultimately, transductive regression strategy was adopted a to predict drug-disease associations on the heterogeneous network. The superior performance of GRTR was validated by cross validation and the top-ranked predictions. Experiment results indicate that our method can predict better than nine other approaches. Furthermore, case studies on several drugs indicated that potential drug-disease association predicted by GRTR could assist in the biomedical research.

Despite the efficiency of GRTR, there are still some limitations which require further optimization. Firstly, our method involved multiple parameters and the establishment of the optimal parameter values is still a challenging problem. Secondly, more biological information can be used to improve predictions. Finally, although higher reliability has been achieved, the current capability of GRTR remains unsatisfactory and necessitates further improvement.

Acknowledgment. This work has been supported by the National Natural Science Foundation of China (Grant No. 61572180).

References

1. Ashburn, T.T., Thor, K.B.: Drug repositioning: identifying and developing new uses for existing drugs. Nat. Rev. Drug Discov. **3**, 673–683 (2004)
2. Wang, W., Yang, S., Li, J.: Drug target predictions based on heterogeneous graph inference. Biocomputing 2013, pp. 53–64. World scientific, Kohala Coast, Hawaii, USA (2012)
3. Wang, W., Yang, S., Zhang, X., Li, J.: Drug repositioning by integrating target information through a heterogeneous network model. Bioinformatics **30**(20), 2923–2930 (2014)
4. Martínez, V., Navarro, C., Cano, C., Fajardo, W., Blanco, A.: DrugNet: network-based drug–disease prioritization by integrating heterogeneous data. Artif. Intell. Med. **63**(1), 41–49 (2015)
5. Luo, H., Wang, J., Li, M., Luo, J., Peng, X., Wu, F.-X., Pan, Y.: Drug repositioning based on comprehensive similarity measures and bi-random walk algorithm. Bioinformatics **32**(17), 2664–2671 (2016)
6. Bleakley, K., Yamanishi, Y.: Supervised prediction of drug–target interactions using bipartite local models. Bioinformatics **25**(18), 2397–2403 (2009)

7. Mei, J.P., Kwoh, C.K., Yang, P., Li, X.L., Zheng, J.: Drug–target interaction prediction by learning from local information and neighbors. Bioinformatics **29**(2), 238–245 (2013)
8. Gottlieb, A., Stein, G.Y., Ruppin, E., Sharan, R.: PREDICT: a method for inferring novel drug indications with application to personalized medicine. Mol. Syst. Biol. **7**(1), 496 (2011)
9. Yang, J., Li, Z., Fan, X., Cheng, Y.: Drug-disease association and drug-repositioning predictions in complex diseases using causal inference-probabilistic matrix factorization. J. Chem. Inf. Model. **54**(9), 2562–2569 (2014)
10. Dudani, S.A.: The distance-weighted K-nearest-neighbor rule. IEEE Trans. Syst. Man. Cybern. SMC **6**(4), 325–327 (1976)
11. Luo, J., Ding, P., Liang, C., Cao, B., Chen, X.: Collective prediction of disease-associated miRNAs based on transduction learning. IEEE/ACM Trans. Comput. Biol. Bioinform. **14** (6), 1468–1475 (2017)
12. Wan, M., Ouyang, Y., Kaplan, L., Han, J.: Graph regularized meta-path based transductive regression in heterogeneous information network. In: Proceedings of the 2015 SIAM International Conference on Data Mining 2015, pp. 918–926 (2015)
13. Xiao, Q., Luo, J., Liang, C., Cai, J., Ding, P.: A graph regularized non-negative matrix factorization method for identifying microRNA-disease associations. Bioinformatics **34**(2), 239–248 (2018)
14. Knox, C., Law, V., Jewison, T., Liu, P., Ly, S., Frolkis, A., Pon, A., Banco, K., Mak, C., Neveu, V., Djoumbou, Y., Eisner, R., Guo, A.C., Wishart, D.S.: DrugBank 3.0: a comprehensive resource for Omics research on drugs. Nucleic Acids Res. **39**(suppl_1), D1035–D1041 (2011)
15. Hamosh, A., Scott, A.F., Amberger, J.S., Bocchini, C.A., McKusick, V.A.: Online mendelian inheritance in man (OMIM) a knowledgebase of human genes and genetic disorders. Nucleic Acids Res. **33**, D514–D517 (2005)
16. Piñero, J., Bravo, À., Queralt-Rosinach, N., Gutiérrez-Sacristán, A., Deu-Pons, J., Centeno, E., García-García, J., Sanz, F., Furlong, L.I.: DisGeNET: a comprehensive platform integrating information on human disease-associated genes and variants. Nucleic Acids Res. **45**(D1), D833–D839 (2017)
17. Van Driel, M.A., Bruggeman, J., Vriend, G., Brunner, H.G., Leunissen, J.A.M.: A text-mining analysis of the human phenome. Eur. J. Hum. Genet. **14**, 535 (2006)
18. Keshava Prasad, T.S., Goel, R., Kandasamy, K., Keerthikumar, S., Kumar, S., Mathivanan, S., Telikicherla, D., Raju, R., Shafreen, B., Venugopal, A., Balakrishnan, L., Marimuthu, A., Banerjee, S., Somanathan, D.S., Sebastian, A., Rani, S., Ray, S., Harrys Kishore, C.J., Kanth, S., Ahmed, M., Kashyap, M.K., Mohmood, R., Ramachandra, Y.L., Krishna, V., Rahiman, B.A., Mohan, S., Ranganathan, P., Ramabadran, S., Chaerkady, R., Pandey, A.: Human protein reference database—2009 update. Nucleic Acids Res. **37**(suppl_1), D767–D772 (2009)
19. Weininger, D.: SMILES, a chemical language and information system. 1. introduction to methodology and encoding rules. J. Chem. Inf. Comput. Sci. **28**(1), 31–36 (1988)
20. Steinbeck, C., Hoppe, C., Kuhn, S., Floris, M., Guha, R., Willighagen, E.L.: Recent developments of the chemistry development kit (CDK) - an open-source java library for chemo- and bioinformatics. Curr. Pharm. Des. **12**(17), 2111–2120 (2006)
21. Tanimoto, T.T.: Elementary mathematical theory of classification and prediction. IBM Internal report, pp. 1–10 (1958)
22. Van Laarhoven, T., Marchiori, E.: Predicting drug-target interactions for new drug compounds using a weighted nearest neighbor profile. PLoS One **8**(6), e66952 (2013)

23. Zheng, X., Ding, H., Mamitsuka, H., Zhu, S.: Collaborative matrix factorization with multiple similarities for predicting drug-target interactions. In: Proceedings of the 19th ACM SIGKDD International Conference on Knowledge Discovery and Data Mining, pp. 1025–1033. ACM, Chicago, Illinois, USA (2013)

24. Gönen, M.: Predicting drug–target interactions from chemical and genomic kernels using Bayesian matrix factorization. Bioinformatics 28(18), 2304–2310 (2012)

25. Liu, Y., Wu, M., Miao, C., Zhao, P., Li, X.-L.: Neighborhood regularized logistic matrix factorization for drug-target interaction prediction. PLoS Comput. Biol. 12(2), e1004760 (2016)

26. Xia, Z., Wu, L.-Y., Zhou, X., Wong, S.T.: Semi-supervised drug-protein interaction prediction from heterogeneous biological spaces. BMC Syst. Biol. 4(2), S6 (2010)

27. Li, G., Luo, J., Xiao, Q., Liang, C., Ding, P., Cao, B.: Predicting MicroRNA-disease associations using network topological similarity based on deepwalk. IEEE Access 5, 24032–24039 (2017)

28. Coves, M.J., Gomis, R., Goday, A., Casamitjana, R., Rivera, F., Vilardell, E.: Antihypertensive treatment with guanfacine in patients with type II diabetes mellitus. Med Clin (Barc) 88(8), 315–317 (1987)

29. Ahmad, A.: Carvedilol can replace insulin in the treatment of type 2 diabetes mellitus. J. Diab. Metab. 8(2), (2017)

30. Davis, A.P., Grondin, C.J., Johnson, R.J., Sciaky, D., King, B.L., McMorran, R., Wiegers, J., Wiegers, T.C., Mattingly, C.J.: The comparative toxicogenomics database: update 2017. Nucleic Acids Res. 45(D1), D972–D978 (2017)

An Improved Particle Swarm Optimization with Dynamic Scale-Free Network for Detecting Multi-omics Features

Huiyu Li[1], Sheng-Jun Li[1(✉)], Junliang Shang[1,2(✉)], Jin-Xing Liu[1], and Chun-Hou Zheng[3]

[1] School of Information Science and Engineering, Qufu Normal University, Rizhao 276826, China
lihuiyu20@163.com, qfnulsj@163.com,
shangjunliang110@163.com, sdcavell@126.com
[2] School of Statistics, Qufu Normal University, Qufu 273165, China
[3] School of Computer Science and Technology, Anhui University, Hefei 230601, China
zhengch99@126.com

Abstract. Along with the rapid development of high-throughput sequencing technology, a large amount of multi-omics data sets are generated, which provide more opportunities to understand the mechanism of complex diseases. In this study, an improved particle swarm optimization with dynamic scale-free network, named DSFPSO, is proposed for detecting multi-omics features. The highlights of DSFPSO are the introduced scale-free network and velocity updating strategies. The scale-free network is employed to DSFPSO as its population structure, which can dynamically adjust the iteration processes. Three types of velocity updating strategies are used in DSFPSO for fully considering the heterogeneity of particles and their neighbors. Both gene function analysis and pathway analysis on colorectal cancer (CRC) data show that DSFPSO can detect CRC-associated features effectively.

Keywords: Particle swarm optimization · Dynamic scale-free network
Colorectal cancer · Multi-omics · Mutual information

1 Introduction

With the development of high-throughput sequencing technology, a vast amount of biological data of different categories have been generated by The Cancer Genome Atlas (TCGA). They provide us more opportunities to learn the biological mechanism of complex diseases [1].

Detecting features from biological data is an effective way to illuminate the underlying mechanism of diseases. A variety of feature extraction methods have been widely used to analyze the gene expression data. For instance, least absolute shrinkage and selection operator (LASSO), penalized matrix decomposition (PMD) and sparse principal component analysis (SPCA) are commonly used methods of feature extraction. Roth V. used the generalized LASSO method to feature selection problems for microarray data [2]. Liu carried differential expression analysis on RNA-seq count data based on PMD [3]. Lass *et al.* applied SPCA to clustering and feature selection problems [4]. Although LASSO, PMD and SPCA have achieved satisfactory performance on explaining the gene expression, they still have some defects in multi-omics feature extraction. These conventional feature extraction methods which can only identify genomic feature from single type of genomic feature cannot handle the integrated TCGA datasets.

Recently, many particle swarm optimization (PSO) based methods have been proposed for determining SNP-SNP interactions [5], gene features selection [6], and cancer classifications [7]. PSO is a population-based search algorithm of adaptive evolution, which proposed by Kennedy and Eberhart in 1995 [8]. Owing to its simple structure and fast convergence, PSO has become an important evolutionary algorithm. In recent years, numerous studies have been carried out to improve the performance of PSO. Kennedy and Mendes have conducted a deep research on population structure and particle behavior, founding that topology has a profound impact on particle behavior [9]. Liu *et al.* proposed SFPSO (Scale-Free PSO) [10]. Gao proposed SIPSO (Selectively-informed Particle Swarm Optimization), which employed scale-free network to simulate the population structure and greatly improved the optimization process [11]. The DMSPSO proposed by Zhao, used random dynamic changed population structure which greatly improved the ability of local search [12].

However, conventional improvement on PSO algorithm suffers from the limited particle population structure. For example, SFPSO and SIPSO generate the population structure before experiments which cannot embody the dynamic changes in the process of iteration. DMSPSO achieves the dynamic changes in population structure to a certain extent, but the population structure building becomes a completely random process which is unable to fit in with the actual optimization problems.

In this paper, we propose an improved PSO-based algorithm with dynamic scale-free network, named DSFPSO, to detect multi-omics features. The innovations of DSFPSO are the introduction of scale-free network and velocity updating strategies. We employ scale-free network as its population structure which can be dynamically adjusted in the process of iteration. Three types of velocity updating strategies are used in DSFPSO for fully considering the heterogeneity of particles and the connecting between neighbors. Specifically, to utilize the difference of gene expression based on different levels of multi-omics data, we employ the ranking function to extract the most effective gene features. To evaluate the validity of DSFPSO, experiments applied on CRC are handled by DSFPSO and other compared methods. The identified genes are appraised by gene function analysis and pathway analysis. Results show that the novel method can identify CRC-associated features effectively.

2   Methods

2.1   Standard PSO Algorithm

PSO is similar to other evolutionary algorithms which use the concepts of "groups" and "evolution" [13]. The speed of each particle can be dynamically adjusted according to the particle itself and its peers' experience based on the fitness value. Based on the fitness of the position, each particle will move to a better place and obtain the optimal solution of optimization problems.

Standard PSO algorithm can be illustrated as follows.

Step1: Initialize the particle velocity and position;
Step2: Evaluate the fitness of each particle;
Step3: Decide whether to update personal and group best positions by comparing the fitness;
Step4: Update the position and speed of the particles;
Step5: If not meet the ending condition, then return to Step2.

2.2   DSFPSO on Multi-omics Data

The flowchart of the proposed method is shown in Fig. 1. We will describe DSFPSO in details on six aspects.

2.2.1   Initializing Particles with Multi-omics Data

According to the characteristics of the omics data, we integrate the data as genomics and clinical information matrices. The whole genome matrix is the search space of particles while the clinical information matrix is used for the test of particle fitness.

Based on the above mapping of multi-omics data, the position of particle i at iteration t can be illustrated as

$$
\begin{aligned}
Position_t(i) &= (x_{i1}^t, \cdots, x_{ik}^t, \cdots, x_{iK}^t) \\
i &\in \{1, 2, \cdots, I\} \\
k &\in \{1, 2, \cdots, K\} \\
t &\in \{1, 2, \cdots, T\} \\
x_{ik}^t &\in \{1, 2, \cdots, M\}
\end{aligned}
\tag{1}
$$

where I, K, T, M represents the number of particles, combination dimension of genomic features, iteration, and gene features in the genome datasets, respectively. x_{ik}^t is the selected genomic feature of particle i at iteration t in k dimensional space.

The speed of particle i at iteration t can be defined as

$$
\begin{aligned}
Velocity_t(i) &= (v_{i1}^t, \cdots, v_{ik}^t, \cdots, v_{iK}^t) \\
v_{ik}^t &\in [1 - M, M - 1]
\end{aligned}
\tag{2}
$$

where v_{ik}^t is the speed of x_{ik}^t.

Similarly, before the first iteration, $Position_t(i)$, $Velocity_t(i)$, $Pbest_t(i)$, $Neibest_t(i)$, $Gbest_t(i)$ are assigned a random value in their domain respectively.

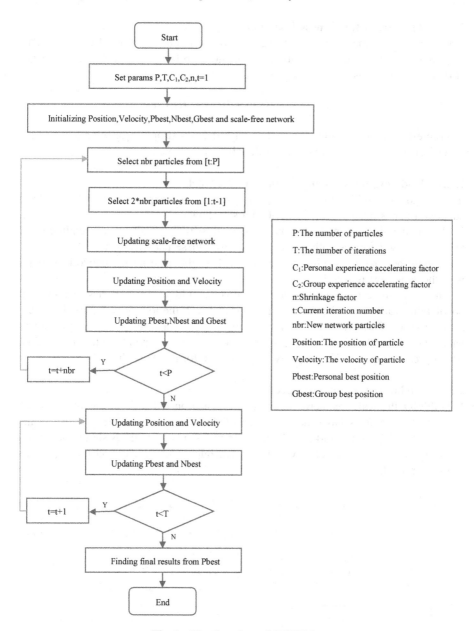

Fig. 1. The flowchart of DSFPSO.

2.2.2 Analysis of the Fitness Function

Since mutual information does not need to assume the distribution of genomics data and can effectively measure the nonlinear relationship between genetic characteristics [14], we employ it as fitness function, which can be formulated as

$$MI(X;Y) = H(X) + H(Y) - H(X,Y) \tag{3}$$

Therefore, higher mutual information value denotes strong association between the genetic characteristic combination and the clinical information.

2.2.3 Updating the Dynamic Scale-Free Network

In order to fully utilize the properties of particles and experimental data, we have adopted a new strategy of link growth and selecting.

In one iteration, we make the out of network particles in fitness descending order and select new particles with higher fitness from these particles to join the network. Then these new particles will choose excellent neighbors from the network particles with the same sort processing.

In the dynamic process of scale-free network building, the particles position and population structure will be dynamically updated with the join of new particles in the solution space. Furthermore, we select the excellent new particles according to fitness value instead of the basic scale-free network adding new points without selection, which greatly improve the reliability of particles information exchange.

2.2.4 Updating the Particle Speed

In DSFPSO, the scale-free network building is synchronized with the solving iteration. Accordingly, particles have the difference of "in" and "out" of the network in the process of scale-free network building, so the two kinds of particles should be treated differently using different velocity updating strategies. The velocity updating equations can be formulated as

$$
\begin{aligned}
v_{ik}^{t+1} &= \begin{cases} \eta \cdot (v_{ik}^t + \frac{1}{k_i} \sum\limits_{j \in N(i)} rand(0,\phi) \cdot (pbx_{jk}^t - x_{pk}^t)), & "in" \\ w_{ik}^t \cdot v_{ik}^t + rand(0,c1) \cdot (pbx_{ik}^t - x_{ik}^t) + rand(0,c2) \cdot (gbx_{ik}^t - x_{ik}^t), & "out" \end{cases} \\
v_{ik}^{t+1} &= \begin{cases} v_{ik}^{t+1} & v_{ik}^{t+1} \in [1-M, M-1] \\ rand(1-M, M-1) & v_{ik}^{t+1} \notin [1-M, M-1] \end{cases} \\
\eta &= \frac{2}{\left|2 - \phi - \sqrt{\phi^2 - 4\phi}\right|} \\
\phi &= c_1 + c_2 > 4 \\
w_{ik}^t &= b - iter \cdot (b-a)/n
\end{aligned}
\tag{4}
$$

where η is learning rate, c_1 and c_2 are acceleration coefficients. w_{ik}^t is dynamic inertia weight balancing the capability between global and local search, $rand(a,b)$ is random number between a and b, $N(i)$ denotes the neighbors of the particle i, K_i is the number of neighbors for particle i.

Based on the speed updating of particles, the position updating equation can be formulated as

$$x_{ik}^{t+1} = x_{ik}^t + v_{ik}^{t+1}$$

$$x_{ik}^{t+1} = \begin{cases} x_{ik}^{t+1} & x_{ik}^{t+1} \in [1, M] \\ int(rand(1, M)) & x_{ik}^{t+1} \notin [1, M] \end{cases} \tag{5}$$

2.2.5 Updating Personal Best Position, Neighbor Best Position and Group Best Position

In DSFPSO, particle's personal best position will be updated by the position with the maximum mutual information. The specific equations can be formulated as

$$Pbest_{t+1} = \begin{cases} Position_t(i) & MI(Position_t(i); Y) = Val \\ Pbest_t(i) & MI(Pbest_t(i); Y) = Val \end{cases}$$
$$Val = max(MI(Position_t(i); Y), MI(Pbest_t(i); Y)) \tag{6}$$

Similarly, the group best position updating equations can be written as

$$Gbest_{t+1} = \begin{cases} Pbest_{t+1}(i) & MI(Pbest_{t+1}(i); Y) = Val \\ Gbest_t(i) & MI(Gbest_t(i); Y) = Val \end{cases}$$
$$Val = max(Pbest_{t+1}(i); Y), MI(Gbest_t(i); Y)) \tag{7}$$

And the neighbor best position updating equations can be written as

$$Neibest_{t+1} = \begin{cases} Position_t(j) & MI(Position_t(j); Y) = Val \\ Neibest_t(i) & MI(Neibest_t; Y) = Val \end{cases}$$
$$Val = max(MI(Position_t(j); Y), MI(Neibest_t(i); Y))$$
$$j \in N(i) \tag{8}$$

2.2.6 Finding Final Results

In genomics data, each gene may have several genetic characteristics due to the differences of gene expression. In the results of DSFPSO, a gene may have a variety of genomic characteristics or may not. In this paper, we resort scoring strategies to extract gene features based on the score of gene expression [15]. The scoring function can be described as

$$Score1(i) = rank(i) \cdot (n - i + 1)$$
$$Score2(j) = \sum_{i \in G} Score1(i) \tag{9}$$

where *rank*(*i*) represents the rank value of genomic features *i*, *n* is the total order value of all the gene characteristics, *G* is the expression set of each gene.

3   Results

3.1   TCGA CRC Data

TCGA CRC data can be obtained from its web portal (https://tcga-data.nci.nih.gov/docs/publications/tcga/). Data used in this paper is the integrated data which has been preprocessed by Lee [16] (http://genomeportal.stanford.edu/tcga-crc/pages/datainformation). Considering the experiment needs, we carry discretization on somatic mutations and methylation data,which greatly improved the stability of the experiment.

The CRC data of TCGA used in this paper from 197 samples contains 5,188 genomic features of 1325 genes, including copy number variation, somatic mutations, methylation data and gene expression data (Fig. 2).

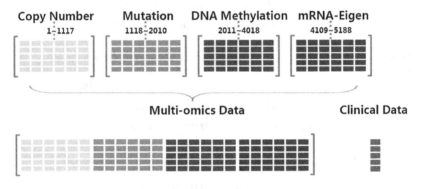

Fig. 2. The CRC data of TCGA.

3.2   Gene Enrichment Analysis

ToppGene is a one-stop portal for gene list enrichment analysis and candidate gene prioritization based on functional annotations and protein interactions network.

To show the effectiveness of DSFPSO, we carry out GO enrichment analysis using ToppGene (https://toppgene.cchmc.org/enrichment.jsp) and compare the results on the same data set, including PSO,SIPSO, LASSO, PMD and SPCA. We input the top 500 genes identified by these methods into the ToppGene Suite, respectively, whose threshold value of the p-value is set to 0.001 and other parameters are set as default. Table 1 lists the top 10 closely related GO terms found by ToppGene. From this table, we can see that the term of "positive regulation of gene expression" has the lowest P-Value (9.38E-19), so it is considered as the most probable enrichment item. Furthermore, we notice that in the term of "regulation of multicellular organismal development" PSO outperforms DSFPSO and in the term of "regulation of transcription by

Table 1. The closely related GO terms found by toppgene.

GO terms	P-Value					
	DSFPSO	SIPSO	PSO	LASSO	PMD	SPCA
GO:0010628	**9.38E-19**	3.45E-16	1.13E-13	7.59E-8	8.64E-15	3.83E-15
GO:0045595	**4.43E-18**	8.10E-14	3.18E-17	/	1.26E-11	8.01E-11
GO:2000026	2.35E-17	1.14E-13	**2.04E-17**	/	/	7.88E-11
GO:0051254	**2.42E-16**	4.92E-13	2.36E-13	3.64E-8	6.51E-15	5.43E-14
GO:1902680	**2.75E-16**	2.24E-13	4.02E-14	2.52E-8	2.78E-15	2.41E-14
GO:1903508	**3.69E-16**	1.13E-13	5.39E-14	3.64E-8	1.33E-15	1.19E-14
GO:0045893	**3.69E-16**	1.13E-13	5.39E-14	3.64E-8	1.33E-15	1.19E-14
GO:0006357	4.30E-16	1.59E-14	1.09E-13	2.49E-7	**9.90E-19**	3.35E-18
GO:0045935	**9.39E-16**	5.37E-13	8.58E-12	9.38E-8	1.03E-15	8.47E-15
GO:0051172	**1.65E-15**	/	1.95E-13	/	1.48E-10	2.93E-11

GO:0010628: positive regulation of gene expression; GO:0045595: regulation of cell differentiation; GO:2000026: regulation of multicellular organismal development; GO:0051254: positive regulation of RNA metabolic process; GO:1902680: positive regulation of RNA biosynthetic process; GO:1903508: positive regulation of nucleic acid-templated transcription; GO:0045893: positive regulation of transcription, DNA-templated; GO:0006357: regulation of transcription by RNA polymerase II; GO:0045935: positive regulation of nucleobase-containing compound metabolic process; GO:0051172: negative regulation of nitrogen compound metabolic process.

RNA polymerase II" PMD outperforms DSFPSO. In general, DSFPSO shows better performance than SIPSO, PSO, LASSO, PMD and SPCA in majority results.

3.3   KEGG Pathway Analysis

KEGG (Kyoto Encyclopedia of Genes and Genomes) is a database which systematacially analyzes the function of gene to reveal the genetic and chemical blueprint of life [17].

In this study, we use DAVID (https://David-d.ncifcrf.gov/) on KEGG pathway to analyze the results. The top 10 CRC-associated pathways are shown in Table 2. Among them, Pathways in cancer and Colorectal cancer are obviously correlated with cancers. [18] indicates that PI3 K-Akt signaling pathway play an important role in inflammation-induced colorectal carcinogenesis. PI3 K-Akt signaling pathway links intimately with cellular metabolism and has great influence on cancer biological behavior [19]. The FoxO signaling pathway plays a central role in diverse physiological processes including cellular energy storage, growth and survival, among others [20]. [21] suggests that FOXO3a is a relevant mediator of the cytotoxic effects of cisplatin in colon cancer cells. Adherens junction pathway plays a critical role in cellular adhesion, glandular differentiation, and cellular proliferation. The function of this pathway correlated proteins is compromised in a number of intestinal diseases, including ulcerative colitis that has an increased incidence for colorectal cancer [22].

Table 2. The top 10 CRC-associated pathways.

Rank	Pathway	Count	P-Value
1	Pathways in cancer	46	1.1E-13
2	Colorectal cancer	12	5.2E-6
3	PI3 K-Akt signaling pathway	27	7.6E-5
4	Viral carcinogenesis	19	1.6E-4
5	MicroRNAs in cancer	22	5.2E-4
6	Cell cycle	13	8.5E-4
7	Focal adhesion	17	1.4E-3
8	Hepatitis B	13	3.3E-3
9	FoxO signaling pathway	12	5.1E-3
10	Adherens junction	7	3.1E-2

3.4   Analysis of Gene Function

In order to evaluate the algorithm's performance and explore the correlation between genes and the pathogenesis of colorectal cancer, we carry out detailed analysis on 10 CRC-related genes among top identified 50 genes. The gene function descriptions are shown in Table 3.

Table 3. The function of genes identified by DSFPSO.

Rank	Gene	Gene function
1	CSMD1	CSMD1 alterations can correlate with earlier clinical presentation in colorectal tumors
2	KBTBD11	KBTBD11 significantly associated with CRC susceptibility
3	WRN	WRN promoter methylation is common in colorectal cancer with the CpG island methylator phenotype (CIMP)
4	SUZ12	SUZ12 mRNA expression in the CRC tissues was significantly increased than in the non-cancerous tissue
5	CDX2	CDX2 is mutated in a colorectal cancer with normal APC/β-catenin signaling
6	NRIP2	NRIP2 in colorectal cancer initiating cells modulates the Wnt pathway
7	CUX1	CUX1 could represent an important regulator of colonic epithelium homeostasis
8	ASB4	ASB4 was higher expressed in CRC tissue than corresponding normal tissue
9	CDK6	CDK6 plays a key role in the cycle of colorectal cancer cells
10	PDK4	PDK4 are highly expressed in human CRC cells

CSMD1 alterations can correlate with earlier clinical presentation in colorectal tumors, thus further implicating CSMD1 as a tumor suppressor gene [23]. Loss of CSMD1 may contribute to the poor prognosis of colorectal cancer patients [24]. [25] indicates that KBTBD11 influences colorectal cancer risk, especially in interaction with an MYC-regulated SNP rs6983267. WRN promoter methylation connects mucinous differentiation, microsatellite instability and CpG island methylator phenotype in colorectal cancer [26]. SUZ12 mRNA expression in the CRC tissues is significantly increased than in the non-cancerous tissue. Increased SUZ12 mRNA expression is directly correlated with primary tumor size, regional lymph node metastases, distant metastasis and AJCC stage. Furthermore, CRC patients with higher level of SUZ12 showed a worse disease-free survival (DFS) [27]. CDX2 is mutated in a colorectal cancer with normal APC/β-catenin signaling [28, 29] shows that CDX2 specifies intestinal development and homeostasis and is considered a tumor suppressor in colorectal carcinogenesis.

4 Conclusions

Considering traditional PSO algorithms usually take equal treatment of all particles and ignore the disadvantages related to the heterogeneity of population structure, we propose an improved PSO algorithm named as DSFPSO to identify gene features of complex diseases. This algorithm dynamically adjusts population structure according to the particles status in the process of iteration.

With fitness of particles as a standard for preferred link selection, DSFPSO realizes the true meaning of PSO for dynamic scale-free network. Moreover,this is the first time for PSO algorithm introduced into multi-omics data analysis with CRC data provided by TCGA as the experiment data and filtering results through scoring strategies. Experimental results show that DSFPSO can be convergent to global optimization quickly and find CRC-associated genes, which will provide valid references for early diagnosis, effective treatment and prognostic guidance of colorectal cancer. To explore correlations among differentially expressed genes is left as our future work.

Acknowledgments. This work was supported by the National Natural Science Foundation of China (Grant No. 61502272, 61572284); Project of Shandong Province Higher Educational Science and Technology Program (J18KA373); the Scientific Research Foundation of Qufu Normal University (BSQD20130119); the Science and Technology Planning Project of Qufu Normal University (xkj201524).Conflict of InterestsThe authors declare that there is no conflict of interests regarding the publication of this paper.

References

1. Bersanelli, M., Mosca, E., Remondini, D., et al.: Methods for the integration of multi-omics data: mathematical aspects. BMC Bioinformatics 17(2), S15 (2016)
2. Roth, V.: The generalized LASSO: a wrapper approach to gene selection for microarray data. Sekretariat für Forschungsberichte, Inst. für Informatik III (2002)
3. Liu, J.X., Gao, Y.L., Xu, Y., et al.: Differential expression analysis on RNA-Seq count data based on penalized matrix decomposition. IEEE Trans. Nanobiosci. 13(1), 12–18 (2014)
4. Luss, R., d'Aspremont, A.: Clustering and feature selection using sparse principal component analysis. Optim. Eng. 11(1), 145–157 (2010)

5. Zhang, W., Shang, J., Li, H., Sun, Y., Liu, J.X.: SIPSO: Selectively Informed Particle Swarm Optimization Based on Mutual Information to Determine SNP-SNP Interactions. In: Huang, D.-S., Bevilacqua, V., Premaratne, P. (eds.) ICIC 2016. LNCS, vol. 9771, pp. 112–121. Springer, Cham (2016). https://doi.org/10.1007/978-3-319-42291-6_11

6. Chuang, L.Y., Chang, H.W., Tu, C.J., et al.: Improved binary PSO for feature selection using gene expression data. Comput. Biol. Chem. **32**(1), 29–38 (2008)

7. Kar, S., Sharma, K.D., Maitra, M.: Gene selection from microarray gene expression data for classification of cancer subgroups employing PSO and adaptive K-nearest neighborhood technique. Expert Syst. Appl. **42**(1), 612–627 (2015)

8. He, R.: An improved particle swarm optimization based on self-adaptive escape velocity. J. Softw. **16**(12), 2036–2044 (2005)

9. Kennedy, J., Mendes, R.: Population structure and particle swarm performance. In: Proceedings of the 2002 Congress on evolutionary computation 2002 CEC 2002, vol. 2, pp. 1671–1676. IEEE (2002)

10. Liu, C., Du, W.B., Wang, W.X.: Particle swarm optimization with scale-free interactions. PLoS ONE **9**(5), e97822 (2014)

11. Gao, Y., Du, W., Yan, G.: Selectively-informed particle swarm optimization. Sci. Rep. **5**, 9295 (2015)

12. Zhao, S.Z., Liang, J.J., Suganthan, P.N., et al.: Dynamic multi-swarm particle swarm optimizer with local search for large scale global optimization. In: IEEE Congress on 2008 Evolutionary Computation (IEEE World Congress on Computational Intelligence), pp. 3845–3852. IEEE (2008)

13. Shi, Y.: Particle swarm optimization: developments, applications and resources evolutionary computation. In: Proceedings of the 2001 Congress on 2001, vol. 1, pp. 81–86. IEEE (2001)

14. Shang, J., Sun, Y., Li, S., et al.: An improved opposition-based learning particle swarm optimization for the detection of SNP-SNP interactions. Biomed Res. Int. **2015**, 524821 (2015)

15. Liu, J.X., Gao, Y.L., Zheng, C.H., et al.: Block-constraint robust principal component analysis and its application to integrated analysis of TCGA data. IEEE Trans. Nanobiosci. **15**(6), 510–516 (2016)

16. Lee, H.J., Flaherty, P., Ji, H.P.: Systematic genomic identification of colorectal cancer genes delineating advanced from early clinical stage and metastasis. BMC Med. Genomics **6**(1), 1 (2013)

17. Kanehisa, M., Araki, M., Goto, S., et al.: KEGG for linking genomes to life and the environment. Nucleic Acids Res. **36**(suppl_1), D480–D484 (2007)

18. Li, N., Bu, X., Tian, X., et al.: Fatty acid synthase regulates proliferation and migration of colorectal cancer cells via HER2-PI3 K/Akt signaling pathway. Nutr. Cancer **64**(6), 864–870 (2012)

19. Josse, C., Bouznad, N., Geurts, P., et al.: Identification of a microRNA landscape targeting the PI3 K/Akt signaling pathway in inflammation-induced colorectal carcinogenesis. Am. J. Physiol. Gastrointest. Liver Physiol. **306**(3), G229–G243 (2014)

20. Zhang, Y., Gan, B., Liu, D., et al.: FoxO family members in cancer. Cancer Biol. Ther. **12**(4), 253–259 (2011)

21. De Mattos, S.F., Villalonga, P., Clardy, J., et al.: FOXO3a mediates the cytotoxic effects of cisplatin in colon cancer cells. Mol. Cancer Ther. **7**(10), 3237–3246 (2008)

22. Mees, S.T., Mennigen, R., Spieker, T., et al.: Expression of tight and adherens junction proteins in ulcerative colitis associated colorectal carcinoma: upregulation of claudin-1, claudin-3, claudin-4, and β-catenin. Int. J. Colorectal Dis. **24**(4), 361–368 (2009)

23. Webb, E.L., Rudd, M.F., Sellick, G.S., et al.: Search for low penetrance alleles for colorectal cancer through a scan of 1467 non-synonymous SNPs in 2575 cases and 2707 controls with validation by kin-cohort analysis of 14 704 first-degree relatives. Hum. Mol. Genet. **15**(21), 3263–3271 (2006)

24. Zhang, R., Song, C.: Loss of CSMD1 or 2 may contribute to the poor prognosis of colorectal cancer patients. Tumor Biol. **35**(5), 4419–4423 (2014)

25. Gong, J., Tian, J., Lou, J., et al.: A polymorphic MYC response element in KBTBD11 influences colorectal cancer risk, especially in interaction with a MYC regulated SNP rs6983267. Ann. Oncol. **29**(3), 632–639 (2018)

26. Kawasaki, T., Ohnishi, M., Suemoto, Y., et al.: WRN promoter methylation possibly connects mucinous differentiation, microsatellite instability and CpG island methylator phenotype in colorectal cancer. Mod. Pathol. **21**(2), 150 (2008)

27. Liu, Y.L., Gao, X., Jiang, Y., et al.: Expression and clinicopathological significance of EED, SUZ12 and EZH2 mRNA in colorectal cancer. J. Cancer Res. Clin. Oncol. **141**(4), 661–669 (2015)

28. Da Costa, L.T., He, T.C., Yu, J., et al.: CDX2 is mutated in a colorectal cancer with normal APC/β-catenin signaling. Oncogene **18**(35), 5010 (1999)

29. Brabletz, T., Spaderna, S., Kolb, J., et al.: Down-regulation of the homeodomain factor Cdx2 in colorectal cancer by collagen type I: an active role for the tumor environment in malignant tumor progression. Cancer Res. **64**(19), 6973–6977 (2004)

PBMarsNet: A Multivariate Adaptive Regression Splines Based Method to Reconstruct Gene Regulatory Networks

Siyu Zhao[1], Ruiqing Zheng[1], Xiang Chen[1], Yaohang Li[3], Fang-Xiang Wu[2], and Min Li[1(✉)]

[1] School of Information Science and Engineering, Central South University, Changsha 410083, China
limin@mail.csu.edu.cn
[2] Department of Mechanical Engineering and Division of Biomedical Engineering, University of Saskatchewan, Saskatoon, SK S7N 5A9, Canada
[3] Department of Computer Science, Old Dominion University, Norfolk, VA, USA

Abstract. Gene Regulatory Network (GRN) is a directed graph which describes the regulations between genes. The problem of reconstructing GRNs has been studied for decades. Most of existing methods infer the GRNs from gene expression data. Previous studies use random forest, partial least squares or other feature selection techniques to solve it. In this paper, we propose a Multivariate Adaptive Regression Splines (MARS) based method to estimate the feature importance and reconstruct the GRNs. MARS can catch the nonlinear relationships between genes. To avoid the overfitting and make the estimation robust, we apply an ensemble model of MARS based on bootstrap and weighted features by PMI (Part mutual information), called PBMarsNet. The results show that PBMarsNet performs better than the state-of-the-art methods.

Keywords: Gene Regulatory Network · Gene expression · MARS PCA-PMI

1 Introduction

Gene Regulatory Network (GRN) is one of the most essential biological networks to expose the mechanism of gene expression. The reconstruction of GRNs is significant to the research of cell differentiation, body development and pathogenesis of disease. However, the biological techniques such as ChIP [1], RIP [2] are time-consuming and expensive to discover gene regulations. The computational methods have been proposed and applied in the problem.

Mutual information (MI) is a typical correlation analysis method to catch nonlinear regulations between pairs of genes [3,4]. Although the pairwise mutual information is flexible, it can't distinguish the indirect regulations. The following MI based methods focus on filtering the indirect regulations by considering data processing inequality or information redundancy [5,6]. Recently, Zhang et al. [7]

© Springer International Publishing AG, part of Springer Nature 2018
F. Zhang et al. (Eds.): ISBRA 2018, LNBI 10847, pp. 38–48, 2018.
https://doi.org/10.1007/978-3-319-94968-0_4

propose PCA-CMI, which infers the GRNs by conditional mutual information (CMI) and path consistency algorithm. PCA-CMI applies conditional dependency test to detect the indirect regulations. Several modified methods [8,9] also have been proposed based on PCA-CMI. These methods overcome the problems of underestimation in PCA-CMI, which further improve the precision of inferring GRNs. Due to the symmetry of mutual information, the mutual information based methods are generally used to infer the indirected structure of GRNs.

Bayesian Networks (BNs) is another popular method to reconstruct GRNs. BNs is a directed acyclic graph, which quantifies the regulations by probabilistic conditional dependency [10,11]. With the development of time-series gene expression data, a series of adaptive models for time-series data are proposed, called Dynamic Bayesian Networks (DBNs) [12,13]. One of the major problems in BNs and DBNs is to learn the optimal structure of networks. The biological information can reduce the learning time efficiently [14]. Besides, Liu et al. [15] present Local Bayesian Network (LBN) to handle large-scale networks. LBN divides the whole gene set into several subsets by clustering which based on mutual information, then applies BNs in each subset of genes.

In addition to mutual information and BNs, regression methods with feature importance estimation are emerging recently. Regression models divide the problem of reconstructing GRNs into multiple independent regressions. In each regression, the expression of a specific gene is modeled by the effect of other regulatory genes expression. Then, feature selection methods are applied to estimate the importance of the candidate regulatory genes. The regression methods include two main types: the model-based and the model-free. Model-based methods generally present a specific form of the regression, mainly generalizing the linear model. There are various effective feature selection techniques, such as lasso [16], least angle regression [17,18] and partial least squares [19]. These methods also modify the traditional models to be applicable in real biological data [20,21]. Model-free methods avoid the predefinition of the model. The most typical model-free method is GENIE3 [22], which is based on random forest. Although the model-free methods are flexible and effective, it is difficult to interpret the details of model and the mechanism of regulations. Furthermore, NIMEFI [23] combines both model-based and model-free feature selection methods. Huynh-Thu and Sanguinetti [24] propose Jump3 to bridge the gap between model-based and model-free methods by combining tree-based and dynamical systems.

In this paper, we introduce a novel ensemble method, called PBMarsNet. The method is based on multivariate adaptive regression splines (MARS), which is more effectively to model continuous data and estimate the feature importance compared with random forest. We apply the bootstrap with MARS to obtain the ensemble result, which is similar with GENIE3. To avoid overfitting, in each bootstrap run, PBMarsNet only selects one subset of candidate regulatory genes. The probability of genes to be selected as candidate regulators are pre-calculated by PCA-PMI. Comparing with other state-of-the-art methods on DREAM (Dialogue for Reverse Engineering Assessments and Methods) [25,26] datasets, the results show that PBMarsNet could get better performance.

2   Materials and Methods

2.1   Data

DREAM4 [25] (Dialogue for Reverse Engineering Assessments and Methods) is one of the common datasets used to evaluate the performance of GRNs reconstruction. DREAM4 provides simulated mRNA expression data and gold standard networks generated by GeneNetWeaver [27]. GeneNetWeaver generates the networks based on GRNs pattern of E.coli and S.cerevisiae. The dataset contains different type of biological data, including wild type, time series, multifactorial, etc. In this paper, we focus on inferring the GRNs from multifactorial gene expression data. DREAM4 provides five multifactorial gene expression datasets (Table 1) with different gold standard networks. Each dataset contains 100 genes and 100 samples. The samples simulate gene expressions under different perturbation experiments. In the result section, we compare PBMarsNet with other state-of-the-art methods on the five datasets.

To further verify the effectiveness of PBMarsNet on large scale of GRNs, we select another in silico benchmark from DREAM5 [26]. The dataset contains 1643 genes, 805 samples under perturbation, and the gold standard network has 4012 edges. Moreover, the dataset also provides 195 genes as candidate regulatory genes, which is more similar to the real biological data. The details of all the datasets are shown in Table 1.

Table 1. The details of datasets

Networks	Genes	Regulatory genes	Samples	Edges
DREAM4 multifactorial network 1	100	100	100	176
DREAM4 multifactorial network 2	100	100	100	249
DREAM4 multifactorial network 3	100	100	100	195
DREAM4 multifactorial network 4	100	100	100	211
DREAM4 multifactorial network 5	100	100	100	193
DREAM5 in-silico	1643	195	805	4012

2.2   Overview of Methodology

The workflow of PBMarsNet is summarized comprehensively in Fig. 1. The proposed method can be mainly divided into three parts: the pre-weighted processing of candidate transcription factors by PCA-PMI, the multivariate adaptive regression splines based ensemble methods and the importance evaluation for candidate transcription factors. The details of each part will be described in the following subsections.

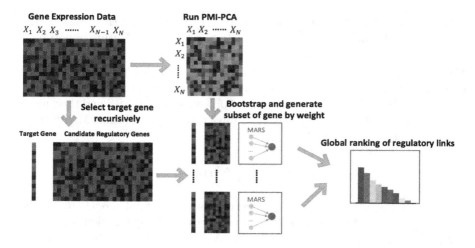

Fig. 1. The workflow of PBMarsNet

Problem Definition. Generally, Gene Regulatory Network is a directed graph where the nodes denote genes and edges denote the regulatory relationships between them. The problem of reconstructing GRNs has been studied for decades. Most of existing methods infer the GRNs from gene expression data, including multifactorial, knock-out, knock-down or time-series gene expression data. In this paper, we focus on inferring the GRNs from multifactorial gene expression data. The expression level of multifactorial data are obtained under different conditions simultaneously. The gene expression data G with N genes M samples is defined as follow:

$$G = \begin{bmatrix} x_{11} & x_{21} & \cdots & x_{N1} \\ x_{12} & x_{22} & \cdots & x_{N2} \\ \vdots & \vdots & \ddots & \vdots \\ x_{1M} & x_{2M} & \cdots & x_{NM} \end{bmatrix} \tag{1}$$

where x_{ij} is expression level of gene X_i at sample j. Reconstruction of GRNs is to infer the direct effects and their direction between genes. Recently, some methods, such as GENIE3, TIGRESS and PLSNET, have formalized the problem of reconstructing GRNs into N independent feature selection problems. These methods recursively select a gene as target gene and other genes as candidate regulatory genes. For a specific target gene X_i, its expression level can be defined by other genes as follow:

$$X_i = f(X_i^-) + \varepsilon \tag{2}$$

where f denotes a regression function, $X_i^- = \{X_1, X_2, \cdots, X_i - 1, X_i + 1, \cdots, X_N\}$ and ε is a random noise. The essential problem is to estimate the importance of genes based on the regression function. Previous studies use tree-based method, partial least squares or other feature selection techniques to solve it. The feature importance $Imp(X_j)$ for target gene X_i is used to indicate the weight of directed regulatory link from gene X_j to gene X_i, then rank all the links globally. In this paper, we apply MARS to estimate the feature importance.

Multivariate Adaptive Regression Splines. The Multivariate Adaptive Regression Splines (MARS) is a no-linear regression model introduced by Friedman firstly [28]. The previous studies [28–30] have shown MARS and their modified versions are flexible and powerful to solve regression problems. MARS takes the general form as follow [31]:

$$f(X) = \sum_{k=1}^{K} \beta_k h_k(X) \tag{3}$$

where K is the number of basis function selected in final model, β_m is the coefficient for each basis function. $h_k(X)$ denotes the basis function, which is a key part in MARS. The basis function $h_k(X)$ includes three types: a constant 1, a hinge function and the product of two or more hinge functions. The form of hinge function is as follow [31]:

$$(x)^+ = \begin{cases} x - t & \text{if } x \\ 0 & \text{otherwise} \end{cases} \quad \text{or} \quad (x)^- = \begin{cases} t - x & \text{if x} < t \\ 0 & \text{otherwise} \end{cases} \tag{4}$$

where t is a constant, called knot. For a variable X, it has a pair of hinge function, and both of them are piecewise linear functions. For a dataset with N variables and M samples, if all the variables' values are distinct, there are $2NM$ candidate hinge functions altogether.

The strategy of model building contains forward pass and backward pass, which is similar with stepwise linear regression. In forward pass, the model selects hinge functions of a variable leading to maximum reduction of residual sum of squares (RSS) in each step and repeats the process until convergence. According to Eq. (3), we use $H = \{h_k(X)\}$ to denote the set of basis functions which are selected in the model, and $C = \{(x_k)^+, (x_k)^-\}$ to denote the candidate set of hinge functions. The model starts with a constant $H_0 = \{1\}$. In iteration i, the set H_{i-1} contains the basis functions in previous $i-1$ iterations. Then the model considers to select a pair of hinge functions $(x_j)^+, (x_j)^-$ from C and to add terms $h_k(X)(x_j)^+, h_k(X)(x_j)^-$ that maximizing the reduction of RSS to the model and updating H_{i-1} by including these terms, where $h_k(X) \in H_{i-1}$. The coefficients $\{\beta_k\}$ are estimated by a least squares fit. The degree of MARS is determined by the maximum number of distinct hinge functions in basis functions. Specially, if the degree is set to 1, that means the model is restricted to be additive.

As the forward pass building the model greedily, which typically leads to a large model and overfitting, the backward pass is applied to avoid the problem. In backward pass, the term which contributes least to reduction of RSS is removed in each iteration. The backward pass of MARS uses generalized cross validation (GCV) to control the balance between number of terms in model and the RSS of model. The GCV is defined as follow:

$$GCV = \frac{RSS}{(1 - E/M)^2} \quad \text{and} \quad E = r + cK \tag{5}$$

where E denotes the effective number of parameters in the model, and M is the number of samples. The value of E includes the number of terms r and the number of knots K with a penalty c. In addition, the importance of each variable can also be estimated by the reduction of GCV [28].

The MARS models can catch the nonlinear relationships by hinge functions and their products. The strategies of MARS and Recursive Partitioning methods (such as decision tree) are similar. Previous studies have shown that MARS tends to achieve better performance than Recursive Partitioning methods on numeric and continuous data [28, 32, 33]. Several modified versions of MARS are proposed to accelerate calculating speed or improve the accuracy of results [30, 34].

Ensemble MARS for GRNs Reconstruction. As described above, reconstruction of GRNs can be decomposed into several independent feature selection problems. In this paper, we also apply MARS to address each feature selection problem. However, previous studies have stated some difficulties in reconstructing GRNs, such as the sufficient number of candidate regulatory genes to a good model for target gene and the noise of gene expression data [18, 19]. Motivated by existing ensemble methods (e.g.,GENIE3 and PLSNET), we propose the ensemble MARS by applying weighted bootstrapping.

Bootstrap is a method to estimate parameters or models by random sampling with replacement. The procedure of bootstrap contains generating multiple sample sets from gene expression data by resampling, then, applying MARS upon each sample set to estimate the importance of each candidate regulatory gene. Generally, the result based on bootstrapping becomes more stable and robust. For the purpose of determining number of candidate regulatory genes and controlling the complexity of MARS model, we also select a subset of all regulatory genes in each bootstrap group. However, selecting genes randomly will need more time to ensure the stability of the result. One of the solution is to weight the candidate regulatory genes and generate the subset by weight. Previous studies have applied correlation based and mutual information (MI) based methods to undirect GRNs reconstruction successfully [7, 9, 35]. In this paper, we use PCA-PMI to weight the candidate regulatory genes. Given a pair of genes X,Y with conditional gene Z, the definition of PMI is as follow:

$$PMI(X;Y|Z) = \sum_{x,y,z} p(x,y,z) log \frac{p(x,y|z)}{p^*(x|z)p^*(y|z)} \tag{6}$$

where $p^*(x|z) = \sum_y p(x|z,y)p(y)$ and $p^*(y|z) = \sum_x p(y|z,x)p(x)$, If Z contains two or more genes, the equation could estimate high order PMI. The proposal of PMI is used to solve the underestimation problem caused by conditional mutual information (CMI) [9].

PCA-PMI combines PMI with path consistency algorithm (PCA). PCA is used to remove the indirect regulatory links recursively. The schema of PCA-PMI is as follow. First, calculate the pairwise mutual information of all genes as the confidence of regulatory links and filter the links with lower confidence by a threshold. Next, for a pair of genes x, y select their common adjacent gene z and calculate first order $PMI(x, y|z)$ and filter the links with low confidence. Then, calculate higher order PMI and do the filter until no links deletion.

In summary, for a specific target gene, the framework of Ensemble MARS is as follow:

Step 1. Use PCA-PMI to estimate the weight of pairwise genes.
Step 2. Generate multi subgroups, which are sampled by bootstrap and selecting subsets from weighted candidate regulatory genes. Genes with higher weight will be selected more frequently.
Step 3. Apply MARS on each subgroup.
Step 4. Estimate the feature importance VIM of candidate regulatory genes. The definition of VIM is as follow:

$$VIM(x) = \frac{\sum_{allgroup} Imp(x)}{\sum_{allsubgroup} I(x)} \tag{7}$$

where $Imp(x)$ is the feature importance gotten by MARS. $I(x)$ is a indicator function, when x in subgroup, $I(x) = 1$, otherwise 0

3   Result

We compare PBMarsNet with six state-of-the-art GRNs inference methods. The GENIE3 [22] is the winner in DREAM4 challenge [25]. TIGRESS [17], PLSNET [19] and NIMEFI [23] are ensemble algorithms based on different feature selection techniques, while CLR [4] and ARACNE [36] are mutual information-based methods. The above methods are used widely in the literatures. Moreover, we also present the result of PBMarsNet without PCA-PMI, called BMaresNet to verify the effectiveness of the feature weights.

To compare with these methods, AUROC (Area under the receiver operating characteristic curve) and AUPR (Area under the precision-recall curve) are used to evaluate performance of the methods. The AUROC is calculated by plotting pairs of true positive rate (TPR) and false positive rate (FPR), and the AUPR is based on pairs of precision and recall. The definitions of TPR, FPR, precision and recall are as follows:

$$TPR = TP/(TP + FN) \tag{8}$$

$$FPR = FP/(FP + TN) \tag{9}$$

$$Precision = TP/(TP + FP) \tag{10}$$

$$Recall = TP/(TP + FN) \tag{11}$$

where TP is the number of true positive, FP is the number of the false positive, TN is the number of true negative and FN is the number of false negative. As the structures of GRNs are generally sparse, the value of AUPR is more meaningful than AUROC.

In experiment, we set the penalty, max terms, bootstrap number of PBMarsNet to 10, 5 and 1000, respectively. Other methods use the default parameters. The results of these methods on DREAM4 dataset are shown in Table 2. From the results, BMarsNet and PBMarsNet gets the comparative results on AUROC and AUPR, especially AURP. BMarsNet and PBMarsNet get the best AUPRs on the five networks except Network 2. And comparing with BMarsNet, PBMarsNet achieves better AUPRs, which means selecting from pre-weighted regulatory genes is better than from random. Average of AUPR on five networks are improved 0.03 by PBMarsNet than the second best method.

AUROC and AUPR values obtained for on DREAM5 in-silico dataset are shown in Table 3. In the case, both AUROC and AUPR are improved significantly by PBMarsNet. The result further verifies the effectiveness of PBMarsNet on inferring large scale networks.

Table 2. The result of different methods on the DREAM4 size 100 multifactorial networks

Method	Network 1		Network 2		Network 3		Network 4		Network 5	
	AUPR	AUROC	AUPR	AUROC	AUPR	AUROC	AUPR	AUROC	AUPR	AUROC
GENIE3	0.161	0.750	0.154	0.734	0.234	0.776	0.211	0.800	0.200	0.795
TIGRESS	0.158	0.747	0.161	0.703	0.233	0.761	0.225	0.774	0.233	0.754
CLR	0.143	0.701	0.117	0.695	0.174	0.744	0.181	0.753	0.175	0.723
ARACNE	0.122	0.605	0.102	0.603	0.201	0.691	0159	0.713	0.167	0.661
NIMEFI	0.157	**0.758**	0.157	0.731	0.248	0.776	0.225	0.806	0.241	**0.801**
PLSNET	0.118	0.713	**0.290**	**0.828**	0.202	**0.794**	0.228	**0.819**	0.206	0.786
BMarsNet	**0.174**	0.722	0.161	0.719	**0.265**	0.765	**0.253**	0.766	**0.255**	0.789
PBMarsNet	**0.192**	0.734	0.162	0.717	**0.287**	0.754	**0.263**	0.776	**0.283**	0.778

Table 3. The result of different methods on the DREAM5 in-silico dataset

	GENIE3	TIGRESS	CLR	ARACNE	NIMEFI	PLSNET	BMarsNet	PBMarsNet
AUPR	0.291	0.302	0.254	0.187	0.298	0.270	**0.317**	**0.382**
AUROC	0.814	0.783	0.771	0.763	0.817	**0.862**	0.827	0.829

4   Conclusion

Reconstruction of GRNs is an essential problem in system biology. In this paper, we propose a multivariate adaptive regression splines based method, called PBMarsNet. The method divides the problem of reconstructing GRNs into multiple feature selection problems. The feature importance is estimated by multivariate adaptive regression splines. Moreover, we also apply bootstrap framework to get more robust results. To avoid overfitting in multivariate adaptive regression splines, we select a small subset of candidate genes in each bootstrap run. The probabilities of genes are pre-calculated by PCA-PMI. We evaluate PBMarsNet on the DREAM4 multifactorial benchmarks and DREAM5 in-silico dataset. The results show the better performance than other state-of-the-art methods on different scales of network.

Fund Sponsored. This work was supported in part by the National Natural Science Foundation of China No. 61622213, No. 61732009, No. 61772552 and No. 61728211.

References

1. Zhu, H., Snyder, M.: Protein chip technology. Curr. Opin. Chem. Biol. **7**(1), 55–63 (2003)
2. Zhao, J., Sun, B.K., Erwin, J.A., Song, J.J., Lee, J.T.: Polycomb proteins targeted by a short repeat RNA to the mouse X chromosome. Science **322**(5902), 750–756 (2008)
3. Frenzel, S., Pompe, B.: Partial mutual information for coupling analysis of multivariate time series. Phys. Rev. Lett. **99**(20), 204101 (2007)
4. Faith, J.J., Hayete, B., Thaden, J.T., Mogno, I., Wierzbowski, J., Cottarel, G., Kasif, S., Collins, J.J., Gardner, T.S.: Large-scale mapping and validation of Escherichia coli transcriptional regulation from a compendium of expression profiles. PLoS Biol. **5**(1), e8 (2007)
5. Lachmann, A., Giorgi, F.M., Lopez, G., Califano, A.: ARACNe-AP: gene network reverse engineering through adaptive partitioning inference of mutual information. Bioinformatics **32**(14), 2233–2235 (2016)
6. Meyer, P.E., Kontos, K., Lafitte, F., Bontempi, G.: Information-theoretic inference of large transcriptional regulatory networks. EURASIP J. Bioinf. Syst. Biol. **2007**, 8 (2007)
7. Zhang, X., Zhao, X.M., He, K., Lu, L., Cao, Y., Liu, J., Hao, J.K., Liu, Z.P., Chen, L.: Inferring gene regulatory networks from gene expression data by path consistency algorithm based on conditional mutual information. Bioinformatics **28**(1), 98–104 (2011)
8. Zhang, X., Zhao, J., Hao, J.K., Zhao, X.M., Chen, L.: Conditional mutual inclusive information enables accurate quantification of associations in gene regulatory networks. Nucleic Acids Res. **43**(5), e31 (2014)
9. Zhao, J., Zhou, Y., Zhang, X., Chen, L.: Part mutual information for quantifying direct associations in networks. Proc. Nat. Acad. Sci. **113**(18), 5130–5135 (2016)
10. Zhou, X., Wang, X., Pal, R., Ivanov, I., Bittner, M., Dougherty, E.R.: A Bayesian connectivity-based approach to constructing probabilistic gene regulatory networks. Bioinformatics **20**(17), 2918–2927 (2004)
11. Werhli, A.V., Husmeier, D.: Reconstructing gene regulatory networks with Bayesian networks by combining expression data with multiple sources of prior knowledge. Stat. Appl. Genet. Mol. Biol. **6**(1) (2007)
12. Shermin, A., Orgun, M.A.: Using dynamic Bayesian networks to infer gene regulatory networks from expression profiles. In: ACM Symposium on Applied Computing, pp. 799–803 (2009)
13. Li, Y., Chen, H., Zheng, J., Ngom, A.: The max-min high-order dynamic Bayesian network for learning gene regulatory networks with time-delayed regulations. IEEE/ACM Trans. Comput. Biol. Bioinform. (TCBB) **13**(4), 792–803 (2016)

14. Zheng, J., Chaturvedi, I., Rajapakse, J.C.: Integration of epigenetic data in Bayesian network modeling of gene regulatory network. In: Loog, M., Wessels, L., Reinders, M.J.T., de Ridder, D. (eds.) PRIB 2011. LNCS, vol. 7036, pp. 87–96. Springer, Heidelberg (2011). https://doi.org/10.1007/978-3-642-24855-9_8
15. Liu, F., Zhang, S.W., Guo, W.F., Wei, Z.G., Chen, L.: Inference of gene regulatory network based on local bayesian networks. PLoS Comput. Biol. **12**(8), e1005024 (2016)
16. Omranian, N., Eloundou-Mbebi, J.M., Mueller-Roeber, B., Nikoloski, Z.: Gene regulatory network inference using fused LASSO on multiple data sets. Sci. Rep. **6**, 20533 (2016)
17. Haury, A.C., Mordelet, F., Vera-Licona, P., Vert, J.P.: TIGRESS: trustful inference of gene regulation using stability selection. BMC Syst. Biol. **6**(1), 145 (2012)
18. Singh, N., Vidyasagar, M.: bLARS: an algorithm to infer gene regulatory networks. IEEE/ACM Trans. Comput. Biol. Bioinf. **13**(2), 301–314 (2016)
19. Guo, S., Jiang, Q., Chen, L., Guo, D.: Gene regulatory network inference using PLS-based methods. BMC Bioinform. **17**(1), 545 (2016)
20. Yao, S., Yoo, S., Yu, D.: Prior knowledge driven Granger causality analysis on gene regulatory network discovery. BMC Bioinform. **16**(1), 273 (2015)
21. Li, M., Zheng, R., Li, Y., Wu, F.X., Wang, J.: MGT-SM: a method for constructing cellular signal transduction networks. IEEE/ACM Trans. Comput. Biol. Bioinform. (2017)
22. Irrthum, A., Wehenkel, L., Geurts, P., et al.: Inferring regulatory networks from expression data using tree-based methods. PLoS One **5**(9), e12776 (2010)
23. Ruyssinck, J., Geurts, P., Dhaene, T., Demeester, P., Saeys, Y., et al.: NIMEFI: gene regulatory network inference using multiple ensemble feature importance algorithms. PLoS One **9**(3), e92709 (2014)
24. Huynh-Thu, V.A., Sanguinetti, G.: Combining tree-based and dynamical systems for the inference of gene regulatory networks. Bioinformatics **31**(10), 1614–1622 (2015)
25. Greenfield, A., Madar, A., Ostrer, H., Bonneau, R.: DREAM4: combining genetic and dynamic information to identify biological networks and dynamical models. PLoS One **5**(10), e13397 (2010)
26. Marbach, D., Costello, J.C., Küffner, R., Vega, N.M., Prill, R.J., Camacho, D.M., Allison, K.R., Aderhold, A., Bonneau, R., Chen, Y., et al.: Wisdom of crowds for robust gene network inference. Nat. Methods **9**(8), 796 (2012)
27. Schaffter, T., Marbach, D., Floreano, D.: GeneNetWeaver: in silico benchmark generation and performance profiling of network inference methods. Bioinformatics **27**(16), 2263–2270 (2011)
28. Friedman, J.H.: Multivariate adaptive regression splines. Ann. Stat. 1–67 (1991)
29. Xu, Q.S., Massart, D., Liang, Y.Z., Fang, K.T.: Two-step multivariate adaptive regression splines for modeling a quantitative relationship between gas chromatography retention indices and molecular descriptors. J. Chromatogr. A **998**(1–2), 155–167 (2003)
30. Weber, G.W., Batmaz, İ., Köksal, G., Taylan, P., Yerlikaya-Özkurt, F.: CMARS: a new contribution to nonparametric regression with multivariate adaptive regression splines supported by continuous optimization. Inverse Prob. Sci. Eng. **20**(3), 371–400 (2012)
31. Friedman, J., Hastie, T., Tibshirani, R.: The Elements of Statistical Learning. Springer Series in Statistics, vol. 1. Springer, New York (2001). https://doi.org/10.1007/978-0-387-21606-5

32. Friedman, J.H.: Estimating functions of mixed ordinal and categorical variables using adaptive splines. Technical report, Stanford Univ., CA, Lab for Computational Statistics (1991)
33. Zhou, Y., Leung, H.: Predicting object-oriented software maintainability using multivariate adaptive regression splines. J. Syst. Softw. **80**(8), 1349–1361 (2007)
34. Friedman, J.H.: Fast MARS. Computational Statistics Laboratory of Stanford University (1993)
35. Yu, H., Luscombe, N.M., Qian, J., Gerstein, M.: Genomic analysis of gene expression relationships in transcriptional regulatory networks. Trends Genet. **19**(8), 422–427 (2003)
36. Margolin, A.A., Wang, K., Lim, W.K., Kustagi, M., Nemenman, I., Califano, A.: Reverse engineering cellular networks. Nat. Protoc. **1**(2), 662 (2006)

Genomic Data Analysis

Bounds on Identification of Genome Evolution Pacemakers

Sagi Snir[(✉)]

Department of Evolutionary Biology, University of Haifa, 3498838 Haifa, Israel
ssagi@research.haifa.ac.il

Abstract. Several works have pointed out that the tight correlation between genes' evolutionary rate is better explained by a model denoted as the *Universal Pacemaker* (UPM) rather than by a simple rate constancy as manifested by the classical hypothesis of Molecular Clock (MC). Under UPM, the relative evolutionary rates of all genes remain nearly constant whereas the absolute rates can change arbitrarily according to the pacemaker ticks. This evolutionary framework was recently adapted to model epigenetic aging where methylated sites are the analogs of evolving genes.

A consequent question to the above finding is the determination of the number of such pacemakers and which gene adheres to which pacemaker. This however turns to be a non trivial task and is affected by the number of variables, their random noise, and the amount of available information. To this end, a clustering heuristic was devised exploiting the correlation between corresponding edge lengths across thousands of gene trees. Nevertheless, no theoretical study linking the relationship between the affecting parameters was done.

We here study this question by providing theoretical bounds, expressed by the system parameters, on probabilities for positive and negative results. We corroborate these results by a simulation study that reveals the critical role of the variances.

Keywords: Phylogenetics · Universal Pacemaker · Gene partitioning
Probabilistic geometrical clustering

1 Introduction

The Molecular Clock (MC) [13,22] model is among the most fundamental concepts of molecular evolution. Under MC genes evolve at roughly constant, albeit different from one another (i.e. gene-specific), rates along all lineages of the tree of life. The above implies constancy of gene-specific relative (WRT some reference gene) evolutionary rates as was observed in several studies [5,9,20]. However, other studies have demonstrated striking differences between fast-evolving and slow-evolving organismal lineages, primarily different groups of mammals [1,2].

Supported in part by the VolkswagenStiftung grant, project VWZN3157.

For example, a genome-wide comparison of the evolutionary rates in the primate vs. rodent lineages shows that the number of fixed mutations per unit time in rodents is about twice that in primates, implying that a lineage-specific, genome-wide change of evolutionary rate occurred after the separation of these lineages [3]. These findings suggest that the MC, if exists, is substantially overdispersed.

Alternatively, a different, more relaxed model, that is compatible with both constant relative evolutionary rates and gene specific rate heterogeneity, has been proposed [15]. Under this model deviations of the gene-specific evolutionary rates from the clock can be arbitrarily large (no MC) but, if these changes occur to all genes in the genome to the same degree, the relative evolutionary rates would remain approximately the same. Thus, the conservation of the evolutionary rate distribution follows from a model of evolution that is more general and less constrained than the MC. It was also shown, through several data sets, that the model is more adequate to real data by providing a better explanation to thousands of gene trees spanning almost all history of life on Earth, i.e the entire tree of life and across 3 billion years [15,19,21]. The model, denoted as the *Universal Pacemaker* (UPM), is thus universal in two respects: first, it applies to all genes in a genome; second, it applies to all branches of the tree of life.

Recently, the evolutionary framework of the UPM was adapted to epigenetic aging [12] where methylated sites in a genome correspond to evolving genes. According to this model methylation rate changes affect simultaneously and proportionally all individual's methylated sites [17,18].

An imperative question implied by the UPM model is the following: while it makes sense that genes vary their evolutionary rate, it is expected that genes of different roles are associated with different pacemakers (PMs[1]). Therefore, a subsequent task to the MC/UPM adequacy question, is the determination of the number of PMs and which genes are associated with what PM. Several first attempts towards this goal were done [6,7,16]. These were based either on the level of evolutionary rate or rate correlation. Gene partitioning is a desirable task in general. Researchers have sought to partition genes based on several criteria such as magnitude of evolutionary rate or the model of evolution [8,11,14]. Such partitioning can shed light on questions of co-functionality and alike.

In this work we focus on multiple pacemakers for gene partitioning [6,16]. In [16] a first step in this direction was done. A heuristic approach exploiting gene pairwise correlation served to collocate correlated genes in Euclidean space. Correlation between gene pairs was inferred by means of the Deming regression [4], and a standard geometrical clustering tool (e.g. *kmeans* [10]) was employed subsequently. Although this later work laid the groundwork for this approach, no theoretical analysis has been done regarding its capabilities. In this work we provide such analysis and we focus only at the aspect of sufficient stochastic signal provided by the random variables. We analyse the extreme case of *perfect reconstruction* in which all genes are *correctly identified*, i.e. associated to their PMs.

---

[1] We use the acronym "UPM" to refer to the model and "PM" to the pacemaker as a natural/combinatorial object.

Therefore, this contribution is divided into four main results: We first give lower bound on the probability for perfect reconstruction as a function of the number of genes, PMs, their corresponding variances, and the number of samples (evolutionary periods along the tree). We make use of the fact that genes tend to evolve according to log normal distribution [20] and assume similarly regarding PMs. The latter implies a chi square distribution on the distances between the objects under investigation and therefore, to allow bounded amount of samples, we use bounds on such distribution to bound the probabilities of genes falling too far from their center (PM) and PMs falling too close to each other. Next, we also give bound, in the same parameters as above, for improbable perfect reconstruction, or more accurately, when a gene falls closer to another PM than to its own one. Then, we show by simulation that the ratio of gene/PM variance is indeed a key feature, as is reflected in our analysis. Finally, we tackle a subsequent, related, problem, of assessing the quality of PM reconstruction in simulation (the *partition distance* problem). We show that under the model studied here, the deterministic heuristic algorithm used in [16] returns the correct result.

Comment: Due to space considerations, several proofs and auxiliary figures are deferred to the journal version.

2   The Evolutionary Model

An evolutionary tree is a tree $T = (V, E)$ where the set of species are mapped to the leaves of T and the edges represent ancestry relationships. Each edge j is associated with a *time period* $\{t_j\}$ that indicates the time between ancestor to the respected descendant (see Fig. 1(a)). All genes evolve along T by acquiring mutations proportionally to the time along the various edges. As all genes evolve on T in an identical manner, and since we are concerned only in the actual time periods, henceforth we will identify these time periods with tree edges and completely ignore the topological information of T.

A gene g_i tends to evolve at an intrinsic rate r_i that is constant along time but deviates randomly along the time periods. Let $r_{i,j}$ be the *actual* (or *observed*) rate of gene i at period j. Then $r_{i,j} = r_i e^{\alpha_{i,j}}$ where $\alpha_{i,j} \in \mathbb{R}$ and $0 < e^{\alpha_{i,j}}$ is a multiplicative error factor. The number of mutations in gene g_i along period j is hence $\ell_{i,j} = r_{i,j} t_j$, commonly denoted as the *branch length* of gene g_i at period j. Therefore gene i, g_i, is a set $[g_i]_j$ indicating the branch lengths (i.e number of mutations) for every time period j (we stress that in reality, these times j correspond to edges in the tree and therefore need not be disjoint, i.e. overlapping time periods). Throughout the text, we will reserve the letters i and j (and their derivatives) to index genes and periods respectively (eg. g_i and t_j).

We now extend this model to include a pacemaker that accelerates or decelerates a gene g_i, relative to its intrinsic rate r_i. Formally, a *pacemaker* (or simply PM) P_k is a set of τ *paces* $\beta_{k,j}$, $1 \leq j \leq \tau$ where $\beta_{k,j} \in \mathbb{R}$ is the relative pace of PM P_k during time period j. We reserve k to index PMs. Under the UPM model, a gene g_i that is *associated* with PM P_k has actual rate at period

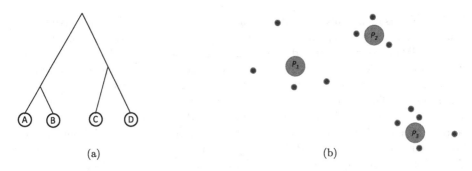

Fig. 1. (a) A phylogenetic (evolutionary) tree over the species $\{A, B, C, D\}$. (b) A scheme of a spatial representation of three pacemakers (red big balls) and their associated genes (small blue balls) centered around them. (Color figure online)

j: $r_{i,j} = r_i e^{\alpha_{i,j}} e^{\beta_{k,j}}$. We assume that every gene is associated with some PM and let $PM(g_i)$ be the PM of gene g_i. Then the latter defines a partition over the set of genes G, where genes g_i and $g_{i'}$ are in the same part if $PM(g_i) = PM(g_{i'})$ (see Fig. 1(b) for illustration). It is important to note that gene rates, as well as pacemakers' paces, are hidden and that we only see for each gene g_i, its set of edge lengths $\ell_{i,j}$. Additionally, the presence of two genes in the same part (PM) does not imply anything about their magnitude of rates, rather on their unison of rate divergence. The above gives rise to the *PM Partition identification Problem:*

Problem 1 (Pacemaker Partition Identification). Given a set of n genes g_i, each with τ branch lengths $\{\ell_{i,j}\}$, find for each gene g_i, its pacemaker $PM(g_i)$. We denote this as the *PMPI* problem.

We say the a gene is *identified* if its PM is (correctly) found. In particular, we are concerned with a particular case of the PMPI problem, the *perfect reconstruction* in which all genes are identified. We first observe the following:

Observation 1. *Assume gene g_i has error factor $\alpha_{i,j} = 0$ for all time periods j, $1 \le j \le \tau$ and let $P_k = PM(g_i)$ be the pacemaker of gene g_i with relative paces $e^{\beta_{k,j}}$. Then at all periods j, $r_{i,j} = r_i e^{\beta_{k,j}}$.*

Literally, Observation 1 states that if the error factor α equals zero, then the rate of gene g_i at all periods is exactly a multiplication of its intrinsic, constant. Rate r_i times the pace of its pacemaker at each specific period, $e^{\beta_{k,j}}$. Observation 1 implies that if genes g_i and $g_{i'}$ belong to the same pacemaker, and both genes have error factor equals zero at all periods, then at all periods, the ratio between the edge lengths at each period is constant and equals to $r_i/r_{i'}$. Indeed, such a strong signal can suffice to classify all genes correctly by simply observing between which pairs that constant ratio holds along all periods. This however is not necessarily true if one of the error factor is not zero or genes g_i and $g_{i'}$ do not belong to the same pacemaker.

For our purpose, we exploit some statistical random structure on the given setting that was observed in nature [20]. This randomness will provide us with signal to distinguish between the objects. Specifically, the goal is to assume that the error factor of each gene is small enough at every period (cf. Observation 1), so that all genes belonging to the same PM, change their actual rate in unison. Similarly, we assume that $[\beta_k]_j$, the paces of a given PM P_k, vary so that genes from different PMs (parts) can be distinguished (otherwise, no difference except their random error factor exists). By [20] we say that for all genes g_i and for all PMs P_k, at all periods j, $\alpha_{i,j}$ and $\beta_{k,j}$ distribute normally around zero with variances σ_P^2 and σ_P^2 respectively. Formally, $\alpha_{i,j} \sim N(0, \sigma_G^2)$, $\beta_{k,j} \sim N(0, \sigma_P^2)$.

3 Sufficient Conditions for Perfect Reconstruction

We here provide a positive result on our capability to obtain perfect reconstruction. The result concerns the amount of information required given the system parameters. Precisely, we give a lower bound on the probability that all genes concentrate around their (unseen) PM in non-intersecting spheres. This forms a sufficient condition for perfect reconstruction. As we operate via expected values as opposed to the probability distributions themselves, the requirement of disjoint spheres poses the restriction $\sigma_P^2 > \sigma_G^2$.

Theorem 1. *Assume $\sigma_P^2 > \sigma_G^2$. Then, given n genes, each with rate error from $N(0, \sigma_G^2)$, and m PMs each with variability $N(0, \sigma_P^2)$. Then if all genes evolve at least τ periods, then the perfect reconstruction is obtained with probability at least $1 - \varepsilon$ where*

$$\varepsilon \le \left(\frac{m^2}{2} + n \right) e^{-\frac{1}{2}\tau \left(\frac{\sigma_P^2 - \sigma_G^2}{\sigma_G^2 + \sigma_P^2} \right)^2} \tag{1}$$

The proof revolves around the attempt to give bounds on the probability that a gene is too "far" from its PM. The notion "far" will be defined in terms of the expected distance between the PMs. The latter will also be bounded, however, from below. Combining all these bounds together, yields the desired result.

Proof. Our model contains the objects of genes and PMs and our sample space is the values of these objects during the τ periods. For an object o, let $v(o)$ be a point (vector) in the τ-dimensional space where each coordinate j in $v(o)$ holds the value of o in period j. Henceforth, an object o will be identified with its corresponding vector in the space. We now can apply geometric tools to measure distances between the objects. Moreover, we are concerned only in the random components of these values. The constant parameters will be ignored as if these are known parameters (in reality these can be inferred given enough samples that is independent of the investigated variables). We work in the logarithmic space, i.e. the log of the values. Hence, for a gene g_i associated with PM P_k, at period j of length t_j, we have

$$\log(\ell_{i,j}) = \log(t_j) + \log(r_i) + \log e^{\alpha_{i,j}} + \log e^{\beta_{k,j}}$$

and since $\log(t_j)$, $\log(r_i)$ are constant, we look only at $\alpha_{i,j} + \beta_{k,j}$. Henceforth we will interchangeably refer by g_i either to the gene object itself or to the vector related to that gene. Therefore we denote by $g_i = [\alpha_i + \beta_k]_j$. Specifically, the j-th coordinate in the vector corresponding to gene g_i will hold the value $\alpha_{i,j} + \beta_{k,j}$. Similarly, for a PM P_k we consider only the values $\beta_{k,j}$ (and the j-th coordinate in the vector β_k corresponding to the paces of PM P_k). By this definition, the random variable holding the difference between a gene and its PM is the gene error factor $\alpha_{i,j}$.

Therefore, the (squared) Euclidean distance, denoted $d(.)$, of a gene g_i from its PM is

$$d^2(g_i - PM(g_i)) = \sum_{1 \le j \le \tau} (g_{i,j} - PM(g_i)_j)^2 = \sum_{1 \le j \le \tau} (\alpha_{i,j})^2. \quad (2)$$

As the random variable $d^2(g_i - PM(g_i))$ distributes identically for all genes i, we simply denote it as $dist_g$. In the sequel, we will make use of versions of the chi square distribution. The following two lemmas are simple extension of the Chernoff bound to handle chi square random variables.

Lemma 1 (Upper Tail, Linear Bound). *Let* $Y = \sum_{1 \le i \le n} X_i^2$ *such that all* X_i *and* $X_{i'}$ *are independent for every* $i \ne i'$, *and* $X_i \sim N(0, \sigma^2)$ *for every* i. *Then:* $\Pr(Y > (1 + \varepsilon)\mathrm{E}[Y]) \le e^{-\frac{1}{8}n\varepsilon^2}$.

Lemma 2 (Lower Tail, Linear Bound). *Let* $Y = \sum_{1 \le i \le n} X_i^2$ *such that all* X_i *and* $X_{i'}$ *are independent for every* $i \ne i'$, *and* $X_i \sim N(0, \sigma^2)$ *for every* i. *Then:* $\Pr(Y < (1 - \varepsilon)\mathrm{E}[Y]) \le e^{-\frac{1}{4}n\varepsilon^2}$.

Now $g_{i,j} - PM(g_i)_j = \alpha_{i,j}$ and since $\alpha_{i,j} \sim N(0, \sigma_G^2)$, it follows that $dist_g$ distributes as non standard chi square.

Our proof hinges around the expected values of the defined distances.

Observation 2

$$\mathrm{E}[dist_g] = \tau \sigma_G^2 \quad (3)$$

Once we set the expected distance of a gene from its PM, we want to control the probability of exceeding this distance and by how much. As this distance distributes as non central chi square RV, we use bounds for deviations from expected values for this type of distribution. Since we are interested in bounding the probability of deviating from the expected value from above (i.e. a gene stays close to its PM), we use a one sided bound for this as is shown below. The following lemma follows directly from Lemma 1 and Observation 2.

Lemma 3. *Set* $\delta_G \ge 0$. *Then, for a gene* g *evolving through* τ *time periods and* $dist_g$ *as defined above,*

$$\Pr(dist_g > (1 + \delta_G)\mathrm{E}[dist_g]) \le e^{-\frac{1}{8}\delta_G^2 \tau} \quad (4)$$

The other distance playing a role in our setting is the distance between the two PMs. Naturally we want this distance to be as large as possible (relative to $dist_g$). This is determined by σ_P^2. For two PMs PM_k and $PM_{k'}$, let $dist_P(k, k')$ be defined as

$$dist_P(k, k') = \sum_{1 \leq j \leq \tau} (\beta_{k,j} - \beta_{k',j})^2. \tag{5}$$

As $dist_P(k, k')$ behaves identically for all pairs of PMs, we simply denote it as $dist_P$.

We now establish $E[dist_P(k, k')]$ that will be useful for bounding the probability of going under that value.

Observation 3

$$E[dist_P] = 2\tau \sigma_P^2 \tag{6}$$

We can now formulate our second central lemma about deviation from the expected distance of two PMs. Recall that we want our PMs to be as far as possible from each other. That means, we are concerned about falling short of their expected distance. Note that we cannot use Lemma 1 since we are bounding the lower tail. Hence, using Lemma 2 and Observation 3 we obtain the following lemma:

Lemma 4. *Set $0 \leq \delta_P \leq 1$. Then, for a pair of PMs P_k and $P_{k'}$ evolving through τ time periods and $dist_P$ as defined above,*

$$Pr(dist_P < (1 - \delta_P)E[dist_p]) \leq e^{-\frac{1}{2}\delta_P^2 \tau}. \tag{7}$$

Now that we obtained upper bounds on probabilities of PMs being too close to one another, and genes too far from their PMs, we can formulate our sufficient condition for PM identification for a single gene:

Observation 4. *Let g_i be a gene and $P_k = PM(g_i)$. Then g_i is identified if $dist(g_i, P_k) < \frac{1}{2}dist(P_k, P_{k'})$ for every $P_k \neq P_{k'}$.*

A special case of the event described in Observation 4 is when the following condition occurs:

Condition 1

1. *For some $\delta_G \geq 0$*

$$dist(g_i, PM(g_i)) < (1 + \delta_G)E[dist_g], \tag{8}$$

2. *And, for some $0 \leq \delta_P \leq 1$ and every PM $P_{k'}$, s.t. $P_{k'} \neq P_k$,*

$$dist(P_k, P_{k'}) \geq (1 - \delta_P)E[dist_P], \tag{9}$$

3. *And*

$$\frac{1}{2}(1 - \delta_P)E[dist_P] \geq (1 + \delta_G)E[dist_g]. \tag{10}$$

In Lemma 3 we bounded the probability of violating Condition 1.1. Let ε_G be that probability. Recall however that we have n genes (with a PM for each such gene). We want to bound the probability that *any* such event occurs (a gene exceeding $(1+\delta_G)\mathrm{E}\,[dist_g]$). Let $\widehat{\varepsilon_G}$ be that probability. Then, by the union bound

$$\widehat{\varepsilon_G} \leq n\varepsilon_G \leq ne^{-\frac{3}{4}\delta_G^2\tau}. \tag{11}$$

Similarly, we can use Lemma 4 to bound the probability that Condition 1.2 is violated. Lemma 4 bounds the probability that a single pair of PMs falls short of $(1 - \delta_P)\mathrm{E}\,[dist_P]$. Let ε_P be that probability. Here however, we have $\binom{m}{2}$ pairs of PMs, so again by the union bound we can bound the probability $\widehat{\varepsilon_P}$ that *any* pair of PMs falls short of this distance:

$$\widehat{\varepsilon_P} \leq \binom{m}{2}\varepsilon_P \leq \frac{m^2}{2}e^{-\frac{1}{2}\delta_P^2\tau}. \tag{12}$$

Finally, Condition 1.3 guarantees that the spheres around any PM do not intersect. That is achieved by requiring that the minimum distance between the PMs, $(1 - \delta_P)\mathrm{E}\,[dist_P]$ is at least twice the sphere of radius $(1 + \delta_G)\mathrm{E}\,[dist_g]$ around every PM. Now since the larger δ_G the tighter the bound (small violation probability) in Lemma 3, we simply require Condition 1.3 to hold in equality. The latter implies an exact relationship between δ_G to δ_P. We start from Condition 1.3 but in equality:

$$\frac{1}{2}(1 - \delta_P)\mathrm{E}\,[dist_P] = (1 + \delta_G)\mathrm{E}\,[dist_g]$$

which implies:

$$\delta_G = \frac{1}{2}\frac{\mathrm{E}\,[dist_p]}{\mathrm{E}\,[dist_g]}(1 - \delta_P) - 1 = ((1 - \delta_P)\sigma_P^2 - \sigma_G^2)/\sigma_G^2, \tag{13}$$

where the last equality stems from Observations 2 and 3.

Now, Condition 1.1 and 1.2 yielded bounds on probabilities that any gene exceeds a specified distance from its PM and any pair of PMs falls too close to one another. We want to guarantee non of these bad events occurs and let ε be this probability, i.e. ε is the probability that either some gene is far from its PM, or some PM pair is close to one another.

$$\varepsilon = \Pr\,[\text{for some gene } g_i, d(g_i, PM(g_i)) \geq (1 + d_G)\mathrm{E}\,[dist_g]$$

$$\bigcup$$

$$\text{for some PM pair } PM_i, PM_j, d(PM_i, PM_j) \leq (1 - d_P)\mathrm{E}\,[dist_P]]$$

Again we can bound ε by using the union bound over $\widehat{\varepsilon_G}$ and $\widehat{\varepsilon_P}$ yielding

$$\varepsilon \leq \widehat{\varepsilon_G} + \widehat{\varepsilon_P}$$

$$\leq ne^{-\frac{3}{4}\delta_G^2\tau} + \frac{m^2}{2}e^{-\frac{1}{2}\delta_P^2\tau}$$

$$= ne^{-\frac{3}{4}\left[\frac{((1-\delta_P)\sigma_P^2 - \sigma_G^2)}{\sigma_G^2}\right]^2\tau} + \frac{m^2}{2}e^{-\frac{1}{2}\delta_P^2\tau} \tag{14}$$

where the last equality holds by plugging in the value of δ_G as implied by (13).

Now, every choice of $0 \leq \delta_P \leq 1$ plugged into (14) leads to a valid bound on the probability to fail in perfect identification. One such particular bound is when $\delta_P = \delta_G$. Then by (13) we obtain that

$$\delta_P = ((1 - \delta_P)\sigma_P^2 - \sigma_G^2)/\sigma_G^2$$

$$\Downarrow$$

$$\delta_P = \frac{\sigma_P^2 - \sigma_G^2}{\sigma_G^2 + \sigma_P^2}. \tag{15}$$

Plugging the latter into (14) yields (1) as requested.

Corollary 1. *If σ_G^2 and σ_P^2 are held constant, then the probability of identifying the PM partition, $1-\varepsilon$, decreases linearly in the number of genes, quadratically in the number of PMs but grows exponentially in the number of tree edges (periods).*

Corollary 2. *If $\tau \leq 2\log\left(\frac{m^2}{2} + n\right)$ then the condition (1) of Theorem 1 yields $\varepsilon > 1$ and hence is irrelevant.*

We conclude this section by noting that the above result only provide us with a guarantee that if the conditions of Eq. (1) hold, we should have enough information to reconstruct the original partition. It does *not* however provide us with tools for this. It also does not imply impossibility conditions for perfect reconstructions if these conditions are not met.

4   Cases of Improbable Perfect Reconstruction

We now analyze the case where perfect reconstruction is improbable. Specifically, we provide a lower bound on the probability that a gene is closer to a PM that is not its own PM. We note that this does not preclude the possibility of reconstruction as the probability of all genes associated with a given PM cluster near another PM still exists. This probability however is negligible and we don't analyze it here. As opposed to the previous section, we do not work through expected values rather by combinatorial arguments and the probability function itself. For sake of clarity, we assume that the gene variance is at least that of the PM's, $\sigma_G^2 \geq \sigma_P^2$. This assumption can be relaxed using a more involved technique that we defer for the journal version. Therefore, the main result of this part is as follows:

Theorem 2. *Assume $\sigma_G^2 \geq \sigma_P^2$. Then some gene falls closer to another PM than its own with probability at least*

$$\left(\frac{1}{e}\frac{1}{m+1}\right)^\tau \left(1 - e^{-(m-1)n/2^\tau}\right). \tag{16}$$

The proof is fairly lengthy and is deferred to the fuller version. We only provide here a very brief intuition. The proof follows by analyzing two independent events: (a) that a gene always (at all periods) falls to the "same side" of some PM, and (b) that the gene always falls far enough from its PM. By analyzing the appropriate probabilities and over the relevant objects (genes, PMs), the theorem follows.

5   Practical Study

In order to study the validity of our bounds we performed a simulation study similar to [16] where we simulated a system of PMs and associated genes (only that here we embedded the values obtained instead of working through gene pairs). Numbers of genes n and PMs m were held constant per experiment. We ran four experiments where n was set to 100 in all experiments and m was set to $2, 4, 6, 8$. Number of samples (edges) τ was set to 25 in all experiments. Each gene i was associated randomly with some intrinsic rate r_i. A very influential factor in our system is the ratio between the two variances: σ_G and σ_P. Therefore we set $\sigma_P = 1$ along the whole experiment and varied σ_G, $0.01 \le \sigma_G \le 100$. Every PM P_i was associated with an intrinsic variance σ_P^2 that sets its relative pace to $e^{\beta_{i,j}}$ where $\beta_{i,j} \sim N(0, \sigma_P^2)$. Similarly, every gene sets its rate at period j to $r_{i,j} = r_i e^{\alpha_{i,j}} e^{\beta_{i,j}}$ where $\alpha_{i,j} \sim N(0, \sigma_G^2)$.

Every gene was associated with a PM, same number of genes for each PM. Number of PMs varied from 2 to 8 (i.e. 12 to 50 genes per PM). We also checked in Corollary 2 that the necessary conditions for reconstruction hold (this does not imply reconstruction yet) and we find that indeed $\tau = 25 > 2\log(2 + 100) = 2\log(m^2/2 + n)$. We mapped our genes (i.e. g_is) to the τ-dimension space and applied standard *kmeans* [10] to it. The inferred partition was compared to the original partition by the greedy algorithm that we used in [16]. The measured quantity is the partition distance where zero signifies perfect reconstruction. As our theoretical analyses suggest we used as the independent variable the ratio between σ_G and σ_P. Indeed our results demonstrate that for two and four PMs, for any ratio of $\sigma_G/\sigma_P \le 1$ a quite accurate reconstruction was achieved and so as to six clusters but for ratio a little less than 1. This is in accordance with our analytic results presented above where for $\sigma_G/\sigma_P < 1$ with logarithmic (in n) number of samples a perfect reconstruction is obtained with high probability. The prominent and interesting result stemming from this study is that for all numbers of PMs, the point $\sigma_G/\sigma_P = 1$ is a critical point, as alluded by our analysis, and after this point, i.e. $\sigma_G/\sigma_P > 1$, quality of reconstruction drops sharply and rapidly reaches saturation that tends to the random similarity of $\frac{1}{m}$.

6   Concluding Remarks and Further Research

The pacemaker paradigm was proved useful in both sequence evolution and epigenetic aging as it accounts for correlation among changing objects (genes, methylation sites) and reducing the overall variance. According to the paradigm, each object adheres to a single pacemaker. This axiom induces a partition over the set of evolving objects - the pacemaker partition. Due to its complexity, until now, all practical works were done assuming a single pacemaker, i.e. the trivial partition. In this work we have provided the first analytical study of the pacemaker partition identification problem. Although the origin of the problem comes from evolutionary biology, it provides challenging statistical problems that may be of separate, independent, interest. Our analysis relies on the randomness of the variables and makes use of it in order to distinguish between them. There is a contrast between the pacemaker's variance and the gene's variance in which the first facilitates identification while the latter blurs the signal. Our analytic results are reinforced by our simulation study and in particular, the existence of the critical point of $\sigma_G/\sigma_P = 1$ that appears to be independent of the other system parameters. A positive, but perhaps unsurprising, conclusion, is that both the probabilities for possibility and impossibility of perfect reconstruction (see Eqs. (1, 16)), are affected exponentially, however in opposite direction, by the number of samples.

The natural and immediate next objective to tackle is the extension to imperfect reconstruction in which some genes are incorrectly identified. Here, we would like to give bounds, again in the system parameters, on the number of genes identified/misidentified. These will provide more realistic results and allow the relaxation of the constraints on the variances.

Acknowledgments. We would like to thank Eugene Koonin and Yuri Wolf for inspiring the question, and Ilan Newman and Nick Harvey for helpful discussions. We also thank helpful and meticulous comments of the anonymous reviewers, used to clarify exposition. Part of this work was done while the author was visiting the NIH, USA, supported by Intramural funds of the US Department of Health and Human Services.

References

1. Bromham, L.: Why do species vary in their rate of molecular evolution? Biol. Lett. **5**(3), 401–404 (2009)
2. Bromham, L.: The genome as a life-history character: why rate of molecular evolution varies between mammal species. Philos. Trans. Roy. Soc. B: Biol. Sci. **366**(1577), 2503–2513 (2011)
3. Mouse Genome Sequencing Consortium: Initial sequencing and comparative analysis of the mouse genome. Nature **20**, 520–562 (2002)
4. Deming, W.: Statistical Adjustment of Data. Wiley, Hoboken (1943)
5. Drummond, D.A., Wilke, C.O.: Mistranslation-induced protein misfolding as a dominant constraint on coding-sequence evolution. Cell **134**(2), 341–352 (2008)
6. Duchne, S., Ho, S.Y.W.: Mammalian genome evolution is governed by multiple pacemakers. Bioinformatics **31**, 2061–2065 (2015)
7. Duchne, S., Ho, S.Y.: Using multiple relaxed-clock models to estimate evolutionary timescales from DNA sequence data. Mol. Phylogenet. Evol. **77**, 65–70 (2014)
8. Duchne, S., Molak, M., Ho, S.Y.W.: ClockstaR: choosing the number of relaxed-clock models in molecular phylogenetic analysis. Bioinformatics **30**(7), 1017–1019 (2014). https://doi.org/10.1093/bioinformatics/btt665

9. Grishin, N.V., Wolf, Y.I., Koonin, E.V.: From complete genomes to measures of substitution rate variability within and between proteins. Genome Res. **10**(7), 991–1000 (2000). https://doi.org/10.1101/gr.10.7.991. http://genome.cshlp.org/content/10/7/991.abstract

10. Hartigan, J.A., Wong, M.A.: A k-means clustering algorithm. Appl. Stat. **28**, 100–108 (1979)

11. Ho, S.Y.W., Lanfear, R.: Improved characterisation of among-lineage rate variation in cetacean mitogenomes using codon-partitioned relaxed clocks. Mitochondrial DNA **21**(3–4), 138–146 (2010)

12. Horvath, S.: DNA methylation age of human tissues and cell types. Genome Biol. **14**(10), 1–20 (2013). https://doi.org/10.1186/gb-2013-14-10-r115

13. Kimura, M.: Molecular evolutionary clock and the neutral theory. J. Mol. Evol. **26**, 24–33 (1987)

14. Lanfear, R., Calcott, B., Ho, S.Y.W., Guindon, S.: PartitionFinder: combined selection of partitioning schemes and substitution models for phylogenetic analyses. Mol. Biol. Evol. **29**(6), 1695–1701 (2012). https://doi.org/10.1093/molbev/mss020

15. Snir, S., Wolf, Y., Koonin, E.: Universal pacemaker of genome evolution. PLoS Comput. Biol. (in Press)

16. Snir, S.: On the number of genomic pacemakers: a geometric approach. Algorithm. Mol. Biol. **9**, 26 (2014). Extended abstract appeared in WABI 2014

17. Snir, S., Pellegrini, M.: An epigenetic PaceMaker is detected via a fast conditional EM algorithm. Epigenomics (2018, accepted)

18. Snir, S., vonHoldt, B.M., Pellegrini, M.: A statistical framework to identify deviation from time linearity in epigenetic aging. PLoS Comput. Biol. **12**(11), 1–15 (2016). https://doi.org/10.1371/journal.pcbi.1005183

19. Snir, S., Wolf, Y.I., Koonin, E.V.: Universal pacemaker of genome evolution in animals and fungi and variation of evolutionary rates in diverse organisms. Genome Biol. Evol. **6**(6), 1268–1278 (2014)

20. Wolf, Y.I., Novichkov, P.S., Karev, G.P., Koonin, E.V., Lipman, D.J.: The universal distribution of evolutionary rates of genes and distinct characteristics of eukaryotic genes of different apparent ages. Proc. Nat. Acad. Sci. **106**(18), 7273–7280 (2009)

21. Wolf, Y.I., Snir, S., Koonin, E.V.: Stability along with extreme variability in core genome evolution. Genome Biol. Evol. **5**(7), 1393–1402 (2013)

22. Zuckerkandl, E., Pauling, L.: Molecules as documents of evolutionary history. J. Theoret. Biol. **8**(2), 357–366 (1965)

REXTAL: Regional Extension
of Assemblies Using Linked-Reads

Tunazzina Islam[1]([✉]), Desh Ranjan[1], Eleanor Young[2], Ming Xiao[2,3],
Mohammad Zubair[1], and Harold Riethman[4]

[1] Department of Computer Science, Old Dominion University, Norfolk, VA, USA
{tislam,dranjan,zubair}@cs.odu.edu
[2] School of Biomedical Engineering, Drexel University, Philadelphia, PA, USA
eay25@glink.drexel.edu
[3] Institute of Molecular Medicine and Infectious Disease,
School of Medicine, Drexel University, Philadelphia, USA
mx44@drexel.edu
[4] School of Medical Diagnostic and Translational Sciences,
Old Dominion University, Norfolk, VA, USA
hriethma@odu.edu

Abstract. It is currently impossible to get complete de novo assembly of segmentally duplicated genome regions using genome-wide short-read datasets. Here, we devise a new computational method called Regional Extension of Assemblies Using Linked-Reads (REXTAL) for improved region-specific assembly of segmental duplication-containing DNA, leveraging genomic short-read datasets generated from large DNA molecules partitioned and barcoded using the Gel Bead in Emulsion (GEM) microfluidic method [1]. We show that using REXTAL, it is possible to extend assembly of single-copy diploid DNA into adjacent, otherwise inaccessible subtelomere segmental duplication regions and other subtelomeric gap regions. Moreover, REXTAL is computationally more efficient for the directed assembly of such regions from multiple genomes (e.g., for the comparison of structural variation) than genome-wide assembly approaches.

Keywords: 10X sequencing · Linked-read sequencing · Subtelomere
Assembly · Segmental duplication · Structural variation · Genome gaps

1 Introduction

Massively parallel short-read DNA sequencing has dramatically reduced the cost and increased the throughput of DNA sequence acquisition; it is now cheap and straightforward to do a variety of whole- genome analyses by comparing datasets of newly sequenced genomes with the human reference sequence. However, even with the use of paired-end read approaches using input molecules of various lengths, de novo assembly of human genomes has remained problematic because

© Springer International Publishing AG, part of Springer Nature 2018
F. Zhang et al. (Eds.): ISBRA 2018, LNBI 10847, pp. 63–78, 2018.
https://doi.org/10.1007/978-3-319-94968-0_6

of abundant interspersed repeats. A recently developed approach pioneered by 10X Genomics generates short-read datasets from large genomic DNA molecules first partitioned and barcoded using the Gel Bead in Emulsion (GEM) microfluidic method [1]. The bioinformatic pipeline for assembly of these reads (Supernova; [2]) takes advantage of the very large number of sets of linked reads. Each set of linked reads is comprised of low-read coverage of a small number of large genomic DNA molecules (roughly 10) and is associated with a unique barcode. This approach enables efficient de novo assembly of much of the human genome, with large segments separable into haplotypes [2]. However, even with these new methods, evolutionarily recent segmentally duplicated DNA such as that found in subtelomere regions remain inaccessible to de novo assembly due to the long stretches of highly similar (>95% identity) DNA. The problem for subtelomere DNA analysis is amplified by the relative lack of high-quality reference assemblies and abundance of structural variation in these regions. To address this problem and attempt to better assemble human subtelomere regions, we have developed a computational approach designed to leverage linked-reads from genomic GEM datasets to extend de novo assemblies from subtelomeric 1-copy DNA regions into adjacent segmentally duplicated and gap regions of human subtelomeres.

Conceptually, what the Gel Bead in Emulsion (GEM) [1] microfluidic method enables us to do is illustrated in Fig. 1. There are approximately one million partitions, each with a unique barcode. Each partition receives approximately 10 molecules of length approximately 50 kb-100 kb. Short reads of length 150 bases are obtained from these molecules with the barcode for the partition attached at the beginning of the first read in a pair [2]. Sets of these read pairs having same barcodes attached to them are called linked-reads.

Supernova assembly [2] takes advantage of linked reads to separate haplotypes over long distances, and these separated haplotypes are represented as megabubbles in the assembly. The chain of megabubbles generates scaffolds [2]. Supernova uses the barcode information after initial whole-genome assembly for bridging long gaps. It finds all the reads of corresponding barcodes that are present in sequence adjacent to the left and right sides of the assembly gap. Then it assembles this set of reads and tries to fill the gap [2]. We refer to this method as genome-wide assembly method. As in all genome-wide assemblies, reads from evolutionarily recent segmental duplications such as those near subtelomeres are collapsed into artifactual DNA segment assemblies; these assembly artifacts are typically either located at a single genomic locus or excluded entirely from the initially assembled genome [3]. REXTAL differs from the genome-wide assembly method in that we use the barcode information for selection of reads from anticipated segmental duplication or gap regions adjacent to a specified 1-copy DNA segment before doing the assembly. We initially find reads matching the 1-copy DNA segment (bait DNA segment) based upon the reference human genome (HG38), then select all reads for barcodes represented in these initial matching reads. This set of reads should represent a very limited subset of all genomic reads, and approximately 10% of the barcode-selected reads should be derived specifically from the selected 1-copy DNA and 50 kb-100 kb segments of

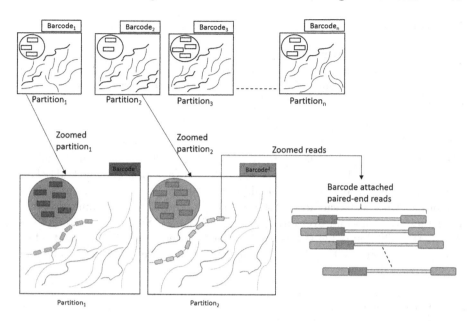

Fig. 1. Conceptual description of GEM microfluidic method. Circle (blue, magenta) represents gel beads. Each bead contains many copies of a 16-base barcode (Rectangles inside the circle) unique to that bead. Each partition gets one gel bead. The 10 curve lines inside the large square (represents partition) represent molecules of length approximately 50 kb-100 kb. The green and orange ovals represent short reads of length 150 bases which are obtained from these molecules (curve lines). (Color figure online)

flanking DNA. We show here that this is indeed the case, enabling the extension of existing assemblies into adjacent segmental duplication and gap regions.

While the primary motivation of our work is to improve the assembly of subtelomeric gap regions and extend the assembly to inaccessible subtelomere segmental duplication regions of genomes of human individuals from their 10X genomic data, REXTAL can be applied more generally for enriching region-specific linked reads and improving the assembly of any specified 1-copy genome region of an individual from any species for which a reference genome exists. For targeted region-specific assemblies from many individuals for which 10X datasets are available (e.g., analysis of structural variation at specific loci), REXTAL is faster and more accurate than genome-wide assembly method. In this scenario, for genome-wide assembly, we need to assemble the whole genome of the individuals and then extract the assembled portion of the specific region. But in our case, we first extract the specific region from the 10X dataset by aligning with a 1-copy segment of the reference genome and then use our bioinformatic pipeline to do the assembly.

2   Method

In Subsect. 2.1, we present the input data description. Subsection 2.2 presents processing of raw data to get our key input data. In Subsects. 2.3, 2.4 and 2.5 we show our assembly pipeline step by step. Subsection 2.6 shows further analysis after assembly.

2.1   Data

The key input data is 10X Genomics linked-reads from individual human genomes, in our case from the genome of a publically available cell line GM19440. Our dataset has approximately 1.49 billion 10X Genomics linked-reads in paired-end format, with each read about 150 bp. The Supernova whole genome assembly using these data had an overall coverage of 103 and a Supernova N50 scaffold of 19.1 Mb. The loupe file shows a mean depth coverage of 67.4. Human reference genome assembly HG38 was used to select test subtelomere regions for the targeted assemblies.

2.2   Data Processing

We processed the raw 10X Genomics data using Long Ranger Basic software developed by 10X Genomics (and freely available to any researcher) to generate barcode-filtered 10XG linked-reads. The Long Ranger basic pipe-line performs basic read and barcode processing including read trimming, barcode error correction, barcode whitelisting, and attaching barcodes to reads. We used the UCSC browser [4] to access HG38 and selected subtelomere DNA segments for analysis.

2.3   Alignment of Subtelomeric Region with Linked-Reads

Masking Out Repeats. We used RepeatMasker [5] and Tandem Repeats Finder [6] to screen bait DNA segment sequences for interspersed repeats, low complexity DNA sequences, and tandem repeats in order to minimize the possibility of false-positive contaminant read identification in the initial selection of reads matching specified 1-copy DNA segments.

Alignment Using BLAT. We used BLAT (BLAST-like alignment tool) [7] with default parameter to do the alignment of the masked subtelomeric region with genome-wide reads from GM19440.

Reads Selection. The output of BLAT gives reads which have a good match with a given subtelomeric bait region. However, it is possible that many reads that would have originated within this given subtelomeric region could have been missed because of masking out repeat regions done previously. More importantly, we were especially interested in capturing reads from the large source DNA molecules extending from the flanks of the targeted 1-copy bait segment. We therefore initially collected all reads that shared a barcode with any read matching the 1-copy segment.

2.4 Barcode Frequency Range and Clustering Pattern Selection

We further reduced this subset of selected reads based on the frequency of occurrence and the clustering pattern of reads from each barcode identified as matching within the specified 1-copy segment. We estimated that each barcode should have approximately 800 reads based on the following calculation: we assumed there are 1 million partitions in the genome with each partition containing 10 molecules of 50 kb each [2]. With the length of each read 150 bp and 0.25X coverage of each single molecule in the partition, we should have approximately (0.25 x 500000 bp)/150 = 833 reads with each barcode. For each barcode, approximately 1/10 of these reads (about 80) should originate from a single locus, and since about 50% of the bait locus (the specified 1-copy region used for BLAT) is masked, about 40 reads/partition should be matched if the entire 50 kb is within the bait locus. If the source DNA molecule partially overlaps the bait locus and extends into the adjacent region, then this number would be smaller and dependent on the extent of the overlap. So, a key challenge was to identify the range of matching reads for each barcode that would minimize inclusion of false positive barcodes while maximizing inclusion of true positive barcodes that would permit extension of the assembly into adjacent DNA. Histogram analysis to check the frequency of the occurrence of each barcode revealed vast over-representation of barcodes with one or two reads, so we required a minimum of three reads per barcode in order to include that barcode for read selection. In addition, we required all matching reads from a single barcode to originate within less than the estimated maximum input molecule size of 100 kb within a given bait region in order to qualify for inclusion. We then empirically tested a variety of barcode frequency ranges meeting both of the above requirements for final read selection, using the ability of the selected reads to assemble the original bait region and extend into flanking DNA as the metric for optimization as described below.

2.5 Assembly of Subset of Reads

To get the assembly of the selected paired-end barcode reads Supernova [2] was used. It can generate assembled scaffolds in four styles named: raw, megabubbles, pseudohap, and pseudohap2. We used pseudohap2 style here. An overview of our assembly strategy is shown in Fig. 2.

2.6 Alignment of Assembled Scaffolds with Reference

To measure the quality of the assembly, we aligned specified subtelomeric regions of the HG38 reference sequence corresponding to our unmasked single-copy bait segments along with their flanking reference DNA segments as query with our generated assembled scaffolds as subject using NCBI BLAST [8], requiring high identity matches (≥98%) for retention of each local alignment. The resulting output hit table of these local alignments lists the sequence identifier, the start and stop points for each local stretch of sequence similarity, and the percent identity of the match. From this information one can map high-similarity alignments

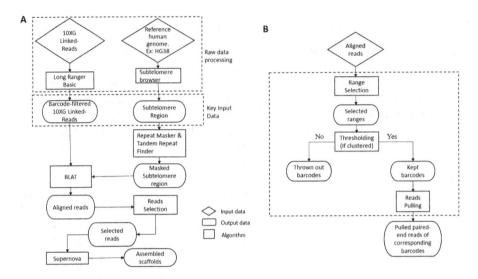

Fig. 2. A: Flowchart of REXTAL. B: Details of Reads Selection algorithm is shown inside dotted box.

of our regional assembly (prepared using barcode-selected linked reads) across the query reference sequence and, by merging the high-quality local alignments, evaluate assembly coverage relative to regions of the reference sequence using a parameter we define as the Lengthwise Assembled Fraction (LAF; see Fig. 5). Intuitively, LAF is defined as the fraction of a targeted reference sequence that is accurately assembled by the regional sequence assembly. Regions of the reference query sequence with highest LAF have the best coverage of assembled sequence, and the limit of assembly extension regions corresponding to flanking reference sequence can be ascertained by a sudden decrease in LAF. Details of LAF calculation are presented in 3.4.

3   Results and Discussions

We tested our read selection and regional assembly strategy (Fig. 2) on four human subtelomere regions with representative patterns of sequence organization (base pair coordinates listed are from HG38; Fig. 3). The 2p subtelomere is a 500 kb sized segment of 1-copy DNA (10,001 to 500,000); 19p subtelomere has a very large segmental duplication region next to the telomere (10,001–259,447) followed by a 300 Kb-sized 1-copy region (259,448–559,447), 10p has a smaller segmental duplication region near the telomere (10,001–88,570) followed by a 300 kb 1-copy region (88,571–388,571); 5p has multiple segmental duplication regions (10,001–49,495 and 210,596–305,378) separated and flanked by two 1-copy regions (49,496–210,595 and 305,379–510,000).

We processed the raw input data from GM19440 as described in Subsect. 2.2. Table 1 presents some characteristics of the output obtained after processing

Fig. 3. Four different chromosomes with different characteristics. The blue rectangle represents single copy region and the magenta rectangle represents segmental duplication region. (Color figure online)

the raw data with Long Ranger Basic software. Interspersed repeats and tandem repeats from the 1-copy regions of these subtelomeres were masked and used as bait segments to select matching reads from the GM19440 linked-read dataset using BLAT. Barcodes for matching reads were identified and characterized according to occurrence frequency and clustering within the bait DNA segments.

Table 1. Some characteristics of the obtained data.

Number of reads	$1.49 * 10^9$	Number of reads without barcode	$9.8 * 10^7$
Number of paired-end reads	$0.75 * 10^9$	Barcode whitelist	0.933959
Number of barcoded reads	$1.39 * 10^9$	Barcode diversity	743369.62

3.1 Barcode Range and Clustering Analysis

We tested a wide variety of Barcode ranges empirically for their ability to select read sets capable of generating high-quality regional assemblies corresponding to the bait segment itself (Fig. 4) as well as extending assemblies of the bait segment into adjacent DNA (Fig. 6). In all cases, a secondary filter was applied requiring that barcodes used for reads selection contained only reads mapping to a single 100 kb segment of the bait DNA (cluster) as anticipated from linked-read library preparation (Table 2). Initial experiments with 2p focused on selection of reads from barcode ranges that produced high-quality assemblies of the 500 kb bait segment, and follow-up work with all four subtelomeres fine-tuned these parameters to optimize both high-quality assembly of bait segments as well as maximal extension into adjacent segmental duplication regions and single-copy regions. Table 2 shows the selected number of barcodes and number of reads after thresholding for 2p and 19p 1-copy region for our chosen ranges.

3.2 Generation of Assembled Scaffolds

After pulling out reads according to our selected range and clustering parameters, we used Supernova assembler for the assembly of the collected paired-end reads. We analyzed assembled scaffolds in pseudohap2 style and calculated the length of each assembled scaffolds.

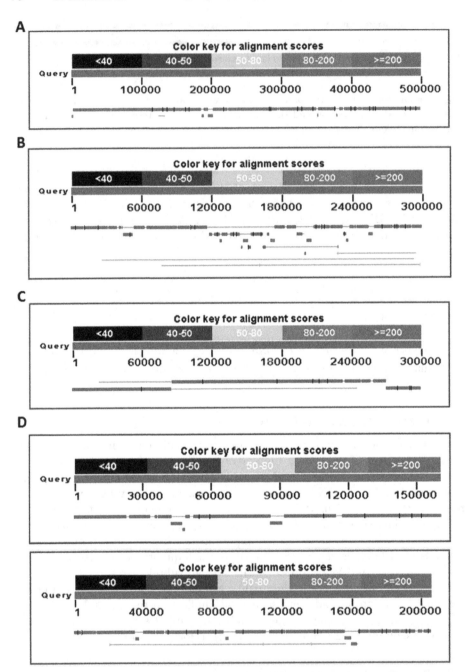

Fig. 4. A: Alignment of 2p 500 kb as query with assembled scaffolds of 2p for range 10–60 as subject in BLAST. B: Alignment of 19p 1-copy 300 kb as query with assembled scaffolds of 19p 1-copy for range 3–70 as subject in BLAST. C: Alignment of 10p 1-copy 300 kb as query with assembled scaffolds of 10p 1-copy for range 3–70 as subject in BLAST. D: Alignments of two 1-copy regions of 5p as query with assembled scaffolds of 5p 1-copy regions for range 3–70 as subject in BLAST.

Table 2. Results after range selection and clustering step

chr^a	$freq^b$	bc^c	bc^d	$read^e$	chr^a	$freq^b$	bc^c	bc^d	$read^e$
2p	10–50	1639	1223	2074096	19p	3–70	1493	1378	2482446
	10–60	1726	1281	2177142		5–70	1142	1026	1870206
	10–70	1807	1330	2265538		10–70	770	662	1265324

a: Chromosomal region.
b: Barcode frequency ranges.
c: Number of selected barcode after range selection.
d: Number of selected barcode after clustering.
e: Number of collected reads of corresponding barcodes.

3.3 Alignment of the Scaffolds with Reference

We aligned the 2p 1-copy region, 19p 1-copy region, 10p 1-copy region and 5p 1-copy regions as the query with corresponding generated assembled scaffolds of 2p, 19p, 10p and 5p as the subject using BLAST with default parameters and retaining only local alignments with ≥98% identity. Figure 4 shows a graphical representation (using the NCBI BLAST output visualization tool) of these BLAST alignments with near-optimal barcode frequencies for retention of linked-reads prior to assembly. While the respective assemblies cover most of each of the 1-copy bait regions, the extent of coverage as well as the number of scaffolds contributing substantially to coverage vary according to subtelomere. We, therefore, developed a more quantitative metric for assembly coverage in order to better quantify the assembly quality and compare them with the assemblies generated de novo from the whole-genome dataset using Supernova.

3.4 Assembly Quality Measurement

Standard assembly quality measurements (QUAST [9]) are not suitable to our case as we are doing region specific assemblies rather than genome-wide assemblies. We are focused on coverage and accuracy of our assembly over the targeted region and have developed a metric called Length-wise Assembled Fraction (LAF) for quality measurement of our regional assemblies. As mentioned previously, LAF measures the fraction of a targeted reference sequence that is accurately assembled by the regional sequence assembly.

Quality in Single Copy Region. We extracted reference sequences of 2p, 19p, 10p, and 5p from HG38 and then aligned them with corresponding assembled scaffolds using BLAST, requiring ≥98% of identity for retention of each local alignment. This generates positions of each local alignment including query start positions and query end positions. The starting positions of the query were sorted in increasing order. Local alignments were merged by (1) deleting local alignments located entirely within other higher-quality alignments; and (2) Local alignments with partial overlap, the overlap regions were merged by selecting the alignment with equivalent or higher % identity in the overlap region. The regions of the query sequence not aligned with sequences in the assembly scaffold are designated as gaps.

For LAF calculation, we considered a number of subsequences of the assembly. More precisely we considered subsequences of the assembly whose end points are start and end positions of n contigs (Fig. 5).

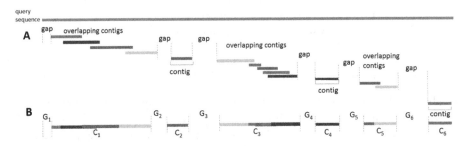

Fig. 5. Top magenta rectangle represents the query sequence. A: Partially overlapped local alignment regions and gaps in coverage of the query sequence. B: Considering partially overlapped local alignment regions as sequence contigs and each sequence contig region (C) is followed by one sequence gap (G). Dotted blue lines represent starting position and ending position of gap. (Color figure online)

We present an algorithm (Algorithm 1) to compute the LAF of given contig and gap lengths. The input to the algorithm are two arrays C and G each of size n. C[i] is the length of ith contig and G[i] is the length of gap before the ith contig. The algorithm computes LAF and outputs an array S of size $2n$. The values in this S array correspond to LAF for $2n$ different subsequences of the assembled sequence, all starting at the reference start position and ending at the end of each contig and gap. To see the accuracy of REXTAL in subtelomeric region, we calculated the LAF with regular intervals. For example: for all ranges of 2p, we took the intervals as the distance from coordinate 1 of the reference query sequence to the starting positions of the 1st gap after 200 kb, 300 kb, 400 kb, and

---

Algorithm 1. CALCULATE LAF (C, G)

1: **construct** C', $C' \leftarrow [C_1, C_1 + C_2, ..., (C_1 + C_2 + ... + C_n)]$
2: **construct** G', $G' \leftarrow [G_1, G_1 + G_2, ..., (G_1 + G_2 + ... + G_n)]$
3: $S[1] \leftarrow 0$
4: $S[2] \leftarrow C'[1]/(C'[1] + G'[1])$
5: **for** $i = 1$ *to* $n - 1$ **do**
6:      $S[2i + 1] \leftarrow C'[i]/(C'[i] + G'[i + 1])$
7:      $S[2i + 2] \leftarrow C'[i + 1]/(C'[i + 1] + G'[i + 1])$
8: **return** S

500 kb respectively. For range 10–60 of 2p subtelomeric region we achieve good LAF (Table 3).

For the calculation of LAF, for all ranges of 19p 1-copy, we calculated the LAF from coordinate 1 of the reference query sequence up to the starting positions of 1^{st}t gap after 50 kb, 100 kb, 150 kb, 200 kb, 250 kb, and 300 kb respectively. We achieve good LAF for range 3–70 of 19p 1-copy (Table 3). We fixed the range 3–70 for 10p and 5p. Table 3 shows the LAF of 10p 1-copy with same intervals taken for 19p 1-copy.

The 5p has multiple segmental duplication regions as well as multiple single copy regions. 1st segmental duplication region is 10,001–49,495 bp, 1st 1-copy region is 49,496–210,595 bp, 2nd segmental duplication region is 210,596–305,378 bp, and 2nd 1-copy region is 305,379–677,959 bp. We applied our assembly pipeline both for 1st 1-copy and 2nd 1-copy (305,379–510,000 bp) region. Because of the length variation of 1-copy region we chose different set of intervals for 1st 1-copy and 2nd 1-copy. We calculated the LAF from coordinate 1 of the reference query sequence up to the starting positions of 1st gap after 30 kb, 60 kb, 90 kb, 120 kb, and 150 kb respectively for 1st 1-copy region and for the 2nd 1-copy region we chose the intervals from coordinate 1 of the reference query sequence to the starting position of 1st gap after 30 kb, 60 kb, 90 kb, 120 kb, 150 kb, 180 kb, and 210 kb (Table 3).

Quality in Extended Region. We can extend our assembly of single-copy diploid DNA into adjacent and other subtelomeric gap regions. To see the extension of our assembly to extended single copy region, we extracted the reference 2p (10,001–700,000 bp) with length 700 kb, 19p (259,448–759,447 bp) with length 500 kb, 10p (88,571–588,571 bp) with length 500 kb, and 5p 2nd 1-copy (305,379–677,959 bp) with length 372,580 bp from HG38. Following BLAST analysis using the extended reference sequence as the query and the assembled scaffolds as subject, we used Algorithm 1 to measure the LAF only for the extended region i.e. >500k for 2p, >300k for 19p and 10p 1-copy, >204,621 bp for 5p 2nd 1-copy.

We calculated the LAF with regular intervals only from the edge of the bait segment into the extended region. We took the intervals as from the end of the bait segment to the starting positions of 1st gap after 10 kb, 20 kb, 30 kb, 40 kb, and 50 kb respectively. To decide the cut-off point for the extended region, we checked all LAFs of the extended region and we stopped where we noticed a sharp drop of the LAF. The reason for this sharp drop is after this contig there is a big gap and after that, there is no significant length of assembled contig to increase the LAF (Table 4).

Quality in Segmental Duplication Region. As segmental duplication region contains segments of DNA with near-identical duplicated subtelomere sequence, this region is hard to assemble de novo with whole genome reads. We can extend our assembly into subtelomere segmental duplication regions. Following BLAST analysis using the HG38 reference subtelomere assembly including the segmental duplication region along with the adjacent bait region, we used Algorithm 1, to

Table 3. Quality comparison for 1-copy region

Chromosomal region	Interval size[a]	LAF[b]	LAF[c]	Chromosomal region	Interval size[a]	LAF[b]	LAF[c]
19p	50 kb	0.9	0.91	5p (1st 1-copy)	30 kb	0.97	0.98
	100 kb	0.91	0.91		60 kb	0.94	0.9
	150 kb	0.89	0.87		90 kb	0.94	0.91
	200 kb	0.88	0.86		120 kb	0.94	0.92
	250 kb	0.88	0.86		150 kb	0.95	0.93
	300 kb	0.89	0.87				
10p	50 kb	0.99	0.99	5p (2nd 1-copy)	30 kb	0.99	0.99
	100 kb	0.99	0.99		60 kb	0.96	0.96
	150 kb	0.99	0.99		90 kb	0.93	0.96
	200 kb	0.99	0.99		120 kb	0.92	0.95
	250 kb	0.98	0.97		150 kb	0.93	0.95
	300 kb	0.97	0.68		180 kb	0.93	0.94
					210 kb	0.93	0.93
2p	200 kb	0.99	0.98				
	300 kb	0.98	0.98				
	400 kb	0.97	0.97				
	500 kb	0.97	0.97				

a: Starting position of 1st gap after the given interval size.
b: LAF for REXTAL. For 2p the range is 10–60 and for 19p, 10, 5p the range is 3–70.
c: LAF for genome-wide assembly method.

measure the LAF only for the segmental duplication region of 19p, 10p, and 5p and then chose the cut-off point. Table 5 shows the analysis of segmental duplication region with extension length as well as LAF.

3.5　Comparison with Genome-Wide Assembly

For a fair comparison with genome-wide assembly method, we need to extract all contigs in the genome-wide assembly that overlap (including potential extensions into flanking DNA) with the reference sequence. To do so we use BWA index [10] of the reference genome (hg38). We have the genome-wide assembly of our input data using Supernova. For alignment using BWA-MEM [10], we aligned the genome-wide assembled reads against the indexed reference genome and generated a .sam file. Using SAMtools [11] we converted the .sam file into a .bam file, sort, and index the results. We extracted specific region of specific chromosomes (here 2p, 19p, 10p, and 5p) from that indexed results using SAMtools and aligned them with the same reference queries used for analysis of the barcode-selected read assemblies using BLAST with ≥98% of identity (see Fig. 6B, D, F and H).

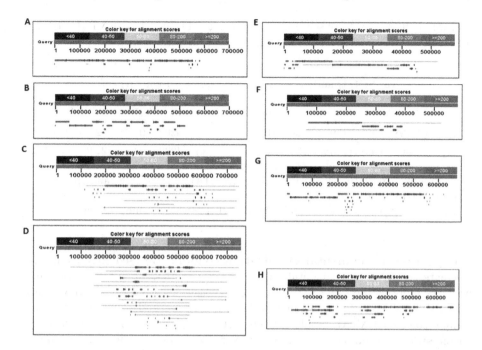

Fig. 6. A: Alignment of 2p with assembled scaffolds of 2p for range 10–60 of REXTAL. B: Alignment of 2p as query with assembled scaffolds of 2p extracted from genome-wide assembly. C: Alignment of 19p with assembled scaffolds of 19p 1-copy for range 3–70 of REXTAL. D: Alignment of 19p with assembled scaffolds of 19p 1-copy region extracted from genome-wide assembly. E: Alignment of 10p with assembled scaffolds of 10p 1-copy for range 3–70 of REXTAL. F: Alignment of 10p with assembled scaffolds of 10p 1-copy region extracted from genome-wide assembly. G: Alignment of 5p with assembled scaffolds of 5p 1-copy regions for range 3–70 of REXTAL. H: Alignment of 5p with assembled scaffolds of 5p 1-copy regions extracted from genome-wide assembly.

Comparison in Single Copy Region. To measure the quality of subtelomeric region assembly of extracted 2p, 19p, 10p, and 5p 1-copy region from the genome-wide assembly, we followed the same steps that mentioned previously for REXTAL to measure quality in the 1-copy region (see 3.4). We calculated the LAF with regular intervals using Algorithm 1. Table 3 shows the comparison of LAF between REXTAL and genome-wide assembly method. For 2p and 5p 2nd 1-copy we get similar LAF with genome-wide method (Table 3). We get better LAF using REXTAL for 19p, 10p 1-copy, and 5p 1st 1-copy than genome-wide method (Table 3).

Comparison in Extended Region. To show the extension of single copy region in genome-wide assembly method, we followed the same steps that we discussed for REXTAL to measure quality in the extended 1-copy region. We calculated the LAF using Algorithm 1. Then we decided the cut-off point. We compared our result for the extended 1-copy region with the genome-wide method in Table 4. It is easy to observe that the results obtained by REXTAL are significantly better than the genome-wide method for these four loci.

Table 4. Quality comparison for extended 1-copy region

chr[a]	(EL, LAF)[b]	(EL, LAF)[c]	chr[a]	(EL, LAF)[b]	(EL, LAF)[c]
2p	(33798, 0.99)	(16954, 1.00)	10p	(52022, 0.93)	(12437, 1.00)
19p	(43666, 0.93)	(6738, 0.99)	5p (2nd 1-copy)	(42326, 0.98)	(22485, 0.97)

a: Chromosomal region.
b: Extension length (in bases) and LAF for REXTAL. For 19p, 10p, and 5p 1-copy region the range is 3–70.
c: Extension length (in bases) and LAF for genome-wide assembly method.

Comparison in Segmental Duplication Region. We used Algorithm 1 to calculate the LAF for segmental duplication region that we got from genome-wide assembly method and compared the extension achieved by REXTAL with the extension achieved by genome-wide assembly method. Table 5 shows the comparison of REXTAL result for the segmental duplication region with the genome-wide method. Once again note that for segmental duplication region the results obtained by REXTAL are notably superior to the genome-wide method for all loci that have been tested. In particular, extensions from the 5p 1st 1-copy and the 2nd 1-copy region together (94,950 bp) cover the entire 2nd segmental duplication region (Table 5).

Table 5. Quality comparison for segmental duplication region

Chromosomal region	SD_L[a]	(EL, LAF)[b]	(EL, LAF)[c]
19p	249446	(67099, 0.98)	(5549, 1.00)
10p	78569	(40089, 0.98)	(4606, 1.00)
5p (1st 1-copy extends to 1st SD)	39495	(36477, 0.98)	(23129, 0.99)
5p (1st 1-copy extends to 2nd SD)	94782	(51860, 0.96)	(65, 1.00)
5p (2nd 1-copy extends to 2nd SD)	94782	(43090, 0.92)	(1307, 1.00)

a: Length of segmental duplication region (in bases) of corresponding chromosomal region.
b: Extension length (in bases) and LAF of 19p, 10p, and 5p for REXTAL.
c: Extension length (in bases) and LAF for genome-wide assembly method.

Figure 7 shows the comparison of extended segmental duplication region for 19p and 10p using REXTAL and genome-wide assembly method.

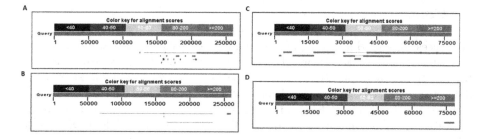

Fig. 7. A: Alignment of 19p segmental duplication region with assembled scaffolds of 19p 1-copy for range 3–70 of REXTAL. B: Alignment of 19p segmental duplication region with assembled scaffolds of 19p 1-copy region extracted from genome-wide assembly. C: Alignment of 10p segmental duplication region with assembled scaffolds of 10p 1-copy for range 3–70 of REXTAL. D: Alignment of 10p segmental duplication region with assembled scaffolds of 10p 1-copy region extracted from genome-wide assembly.

3.6   Efficiency Considerations

For targeted region-specific assemblies from multiple individuals for which 10X datasets are available, REXTAL is faster and more accurate than genome-wide assembly method. For genome-wide assembly, we need to assemble the whole genome of the individuals first and then extract the assembled portion of the specific region. To do whole genome assembly using Supernova takes approximately 36–48 h [2]. Before extraction of the specific region, we need to align the genome-wide assembled reads against the indexed reference genome (hg38) using BWA-MEM. This takes approximately 25 h. Then we can extract the specific region of the specific chromosome. For multiple individuals, although we want to do region-specific assembly, the genome-wide assembly method assembles the whole genome for each individual first and does the alignment – these two steps are time-consuming.

In contrast, REXTAL extracts the reads relevant for assembling the targeted region from the 10X dataset by aligning the targeted region with a 1-copy segment of the reference genome (hg38) using BLAT. This step takes 2–5 h. Reads selection step mentioned in 3rd paragraph of Subsect. 2.3 and Barcode frequency range and clustering pattern selection step described in Subsect. 2.4 together take around 2–3 h. Assembly of the subset of selected reads using Supernova takes approximately 5–15 min. So in total, the region-specific assembly using REXTAL takes approximately 4–8 h. In case of targeted region-specific assembly for multiple individuals, our method REXTAL is approximately 9 times faster than genome-wide assembly method. The configuration of the machine where we ran REXTAL is CPU: 32 cores (2, 16 core processors → Intel(R) Xeon(R) CPU E5-2683 v4/Broadwell @ 2.10 GHz), Memory: 128 GB RAM, Network: FDR IB (56 Gbps fabric).

4   Conclusion

We successfully used a new computational method called Regional Extension of Assemblies Using Linked-Reads (REXTAL) for improved region-specific assembly of segmental duplication-containing DNA, leveraging genomic short-read datasets generated from large DNA molecules partitioned and barcoded using the Gel Bead in Emulsion (GEM) microfluidic method [1]. We showed that using REXTAL, it is possible to extend assembly of single-copy diploid DNA into adjacent, otherwise inaccessible subtelomere segmental duplication regions. In future experiments, using larger source DNA molecules for barcode sequencing approaches could further extend assemblies into and through segmental duplications, and optical maps of large single molecules extending from the 1-copy regions through segmental duplications and gaps could be used to optimally guide and validate these assemblies.

Acknowledgement. The work in this paper is supported in part by NIH R21CA177395 (HR and MX), and Modeling and Simulation Scholarship (to TI) from Old Dominion University.

References

1. Zheng, G.X.-L.-P., et al.: Haplotyping germline and cancer genomes with high-throughput linked-read sequencing. Nature Biotechnol. **34**, 303–311 (2016)
2. Weisenfeld, N.I., Kumar, V., Shah, P., Church, D.M., Jaffe, D.B.: Direct determination of diploid genome sequences. Genome Res. **27**, 757–767 (2017)
3. Alkan, C., Sajjadian, S., Eichler, E.E.: Limitations of next-generation genome sequence assembly. Nature Methods **8**, 61 (2011)
4. Kent, W.J., Sugnet, C.W., Furey, T.S., Roskin, K.M., Pringle, T.H., Zahler, A.M., Haussler, D.: The human genome browser at UCSC. Genome Res. **12**, 996–1006 (2002)
5. Smit, A.F.: 2010 RepeatMasker Open-3.0 (1996). http://www.repeatmasker.org/
6. Benson, G.: Tandem repeats finder: a program to analyze DNA sequences. Nucleic Acids Res. **27**, 573 (1999)
7. Kent, W.J.: BLAT the BLAST-like alignment tool. Genome Res. **12**, 656–664 (2002)
8. Altschul, S.F., Madden, T.L., Schffer, A.A., Zhang, J., Zhang, Z., Miller, W., Lipman, D.J.: Gapped BLAST and PSI-BLAST: a new generation of protein database search programs. Nucleic Acids Res. **25**, 3389–3402 (1997)
9. Gurevich, A., Saveliev, V., Vyahhi, N., Tesler, G.: QUAST: quality assessment tool for genome assemblies. Bioinformatics **29**, 1072–1075 (2013)
10. Li, H., Durbin, R.: Fast and accurate short read alignment with Burrows-Wheeler transform. Bioinformatics **25**, 1754–1760 (2009)
11. Li, H., Handsaker, B., Wysoker, A., Fennell, T., Ruan, J., Homer, N., Marth, G., Abecasis, G., Durbin, R.: The sequence alignment/map format and SAMtools. Bioinformatics **25**, 2078–2079 (2009). 1000 Genome Project Data Processing Subgroup

A Scalable Reference-Free Metagenomic Binning Pipeline

Terry Ma[1] and Xin Xing[2(✉)]

[1] Lambert High School, Suwanee, GA, USA
151232@forsythk12.org
[2] University of Georgia, Athens, GA, USA
xinxing@uga.edu

Abstract. Metagenomics studies microbial genomes in an ecosystem such as the gastrointestinal tract of a human through sequencing thousands of organism in parallel. The sheer number of genomic fragments are challenging for current metagenomic binning software to process. Here we present a scalable reference-free metagenomic binning pipeline designed to handle large scale metagenomic data. It allows users to input several tera base pairs (TB) of reads and produces highly accurate binning results, even at a species level. The pipeline outputs all binned species in multiple metagenomic samples and their estimated relative abundance. We integrate the pipeline into an open-source software, MetaMat, which is freely available at: https://github.com/BioAlgs/MetaMat.

Keywords: Metagenomic binning · Parallel computing
Disease diagnosis

1 Introduction

Extensive evidence suggests that microbial ecosystems especially the gut ecosystem plays a crucial role in human health. A disruption of these ecosystems can cause some perplexing diseases such as autism, diabetes, and coronary heart disease. Thus, our understanding of the biodiversity and composition of these ecosystems is crucial in developing potential diagnostic methods for diseases.

A recent development in next generation sequencing technology (NGS) allows us to sequence mass genomes extracted directly from the environment. NGS technology allows us to bypass several difficulties in the study of microbial ecosystems, such as a need for highly trained personnel and a need for high cost high maintenance equipment. This type of research is referred to as a metagenomic study. Due to technological limitations, NGS cannot sequence the entire genome and can only sequence short fragments of these genomes. After sequencing, the sequenced fragmented genomes must be reassembled back into whole genomes. The process of reassembly is known as metagenomic binning.

Several current binning methods include reference-based binning methods such as MAGEN [1], MetaPhyler [2], Kraken [3] and CLARK [4] that require us

© Springer International Publishing AG, part of Springer Nature 2018
F. Zhang et al. (Eds.): ISBRA 2018, LNBI 10847, pp. 79–83, 2018.
https://doi.org/10.1007/978-3-319-94968-0_7

to know the reference genomes, which can be a serious limitation since a very limited number of existing genomes have references. In contrast, the k-mer or the oligonucleotide-frequency based methods [5] are reference-free. However, the binning accuracy of k-mer based methods can be significantly compromised for short contigs (e.g., <10kb) or genetically similar species. An improvement of k-mer based methods, coverage-based methods such as CONCOCT [6], MaxBin [7], MetaBAT [8], Groopm [9] and VizBin [10] have higher accuracy but have a high computational cost involved in the binning and feature extraction steps, impeding these binning methods in processing large metagenomic data sets. All of these methods also fail in generating an accurate estimate of the relative abundance of the species which can hinder identification of the pathogen and diagnosis for those affected.

The pipeline that we propose here integrates pooled genome assembly, fast binning algorithm, and disease-microbial species association studies. A novel and fast computational method of which a linear computational cost is achieved is proposed for large scale metagenomic datasets that has hundreds of samples and thousands of giga base pairs of reads. In addition, parallel computing is used on multiple metagenomic samples which can bin millions of contigs in a couple of hours, a speed that cannot be achieved by other existing methods. Furthermore, MetaMat pipeline not only outputs the binning results but also outputs the relative abundance as an input for downstream biological analysis such as disease classification and disease-microbial species association. In summary, MetaMat can greatly benefit the biologists' study of microbial-environment.

2   The MetaMat Pipeline

MetaMat is developed using shell script, C++, and R programing languages. To ease the usage of the software, we integrate the whole pipeline into one command with arguments to allow users to customize the parameters. We release the source code of the software on the public repository. The release of the source code allows users to add features and extend the functionality of the code which makes it more versatile for advanced users. We test the software on tens of simulated and real metagenomic data sets. For the simulated data, we consider the different sequence depths, number of samples and number of species. MetaMat shows robust performance under these different conditions. We also test MetaMat on real datasets with hundreds of metagenomic samples including several tera base pairs (TB) of reads. For real data sets, we verify the binning results by comparing our bins with the ones obtained from BLAST search. MetaMat can achieve a very high accuracy when compared with existing methods.

MetaMat targets on large scale biomedical studies where species in each biomedical sample or human subject have different distributions. This assumption can be satisfied in most of scientific studies. For example, we can safely assume that the microbial distributions in human gut are different for different subjects. Under this assumption, we can further assume that the cross-sample relative abundance of one species is different from that of another given species.

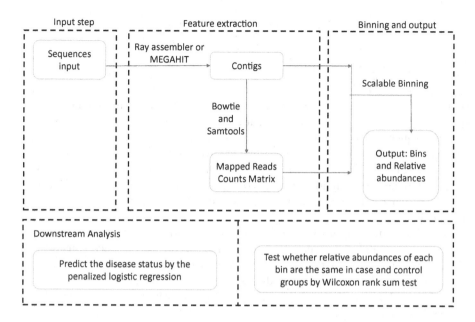

Fig. 1. Software architecture of the MetaMat system.

Because the relative abundance of a contig equals the relative abundance of the species that contains it, we can make use of the relative abundance of each species across samples to sort contigs based on the fact that contigs from different species have different relative abundances across samples.

MetaMat accepts sequenced reads outputted from the sequencing machine as an input. We require that the reads are in FASTA or FASTQ format and are stored in different files for different subjects. If the input files are FASTQ files, a quality control will automatically be applied. In the MetaMat pipeline, we mainly have three steps as shown in Fig. 1. The first step is to specify the input data which includes multiple FASTA or FASTQ files. The customized parameters inputed, such as the number of CPU cores is used to run the pipeline. The second step aims at extracting useful features used for binning. Particularly, we use overlap information to assemble the short reads into longer sequences which are called contigs. Several well developed open source software are used such as Ray assembler [11] and MEGAHIT assembler [12]. Next, the sequenced reads are aligned to the assembled contigs by Bowtie 2 [13] and the counts of the mapped reads from different samples on each contig are extracted. The assembled contigs and extracted mapped reads counts matrix are passed to the third step in which we apply a scalable binning algorithm based on searching for the matched angle among the mapped read counts vector of contigs. The computational complexity is of the order $O(NKP)$. The computational complexity of our algorithm is linear and much faster than any other available software, thus, it is able to bin millions of contigs within several hours. In addition, we implement silhouette statistics [14] to estimate the number of species. We define the silhouette statistics as

$$C(K) = \frac{1}{n} \sum_{i=1}^{n} \frac{b_i(K) - a_i(K)}{\max(a_i(K), b_i(K))}, \qquad (1)$$

where $a_i(K) = \frac{1}{n_k} \sum_{i' \in \mathcal{S}_k} d_{ii'}$, $b_i(K) = \min_{k' \neq k} \frac{1}{n_{k'}} \sum_{i' \in \S_{k'}} d_{ii'}$, and $d_{ii'}$ is the angular distance between the mapped reads counts vectors of the ith and i'th contig. We choose $K_{opt} = \arg\max_{1 \leq K \leq D} C(K)$ as our estimated number of species. We test this criteria on several simulated metagenomic data sets with varying sequence depths, number of samples and number of species. This criteria can accurately select the number of species with error less than 5%.

The MetaMat output consists of two files. One of the files contains the binning information i.e. the bin id for each contig, while the other file contains the $K_{opt} \times P$ matrix in which each entry is the relative abundance of each bin in each sample.

MetaMat can be applied to case-control metagenomic studies where the outputted relative abundance can be used in identifying the species causing the disease and diagnosing patients. In our online instruction, we include two downstream pipelines for predicting the disease status using penalized logistic regression [15], and for identifying the species with significantly different relative abundances between the case and control groups by using the Wilcoxon rank sum test [16].

3   Implementation

MetaMat is an integrated pipeline utilizing a parallel computing architecture to divide computational resources among multicore computers or computer clusters with multiple nodes. The shell script is used to combine the power of different software. In the feature extraction steps, we use GNU parallel [17] to process multiple samples at the same time. In the binning step, we use the R packages 'doParallel' [18] and 'foreach' [19] to simultaneously run the scalable binning algorithm with different candidate number of bins K.

As a demonstration, we applied MetaMat on three simulated data sets and ran the pipeline on a computer server with 12 CPU cores. The binning accuracy and performance of the silhouette statistic are reported in the Github repository.

4   Discussion

In this paper, we present a pipeline that addresses some emerging issues in metagenomic research based on high-throughput sequencing technologies. The proposed pipeline integrates parallel feature extraction, scalable binning, and downstream analysis, which will lead to a deeper understanding of how microbial ecosystems affect our health and help design new probiotics for disease prevention and intervention.

Acknowledgement. This research was supported in part by the National Institutes of Health grant R01 GM113242-01 and the National Science Foundation grants DMS-1440038 and DMS-1440037.

References

1. Huson, D.H., Auch, A.F., Qi, J., Schuster, S.C.: Megan analysis of metagenomic data. Genome Res. **17**(3), 377–386 (2007)
2. Liu, B., Gibbons, T., Ghodsi, M., Pop, M.: MetaPhyler: taxonomic profiling for metagenomic sequences. In: 2010 IEEE International Conference on Bioinformatics and Biomedicine (BIBM), pp. 95–100. IEEE (2010)
3. Wood, D.E., Salzberg, S.L.: Kraken: ultrafast metagenomic sequence classification using exact alignments. Genome Biol. **15**(3), R46 (2014)
4. Ounit, R., Wanamaker, S., Close, T.J., Lonardi, S.: CLARK: fast and accurate classification of metagenomic and genomic sequences using discriminative k-mers. BMC Genom. **16**(1), 236 (2015)
5. Abe, T., Sugawara, H., Kinouchi, M., Kanaya, S., Ikemura, T.: Novel phylogenetic studies of genomic sequence fragments derived from uncultured microbe mixtures in environmental and clinical samples. DNA Res. **12**(5), 281–290 (2005)
6. Alneberg, J., Bjarnason, B.S., Bruijn, I.D., Schirmer, M., Quick, J., Ijaz, U.Z., Lahti, L., Loman, N.J., Andersson, A.F., Quince, C.: Binning metagenomic contigs by coverage and composition. Nature Methods **11**(11), 1144 (2014)
7. Yu-Wei, W., Simmons, B.A., Singer, S.W.: MaxBin 2.0: an automated binning algorithm to recover genomes from multiple metagenomic datasets. Bioinformatics **32**(4), 605–607 (2015)
8. Kang, D.D., Froula, J., Egan, R., Wang, Z.: MetaBat, an efficient tool for accurately reconstructing single genomes from complex microbial communities. PeerJ **3**, e1165 (2015)
9. Imelfort, M., Parks, D., Woodcroft, B.J., Dennis, P., Hugenholtz, P., Tyson, G.W.: GroopM: an automated tool for the recovery of population genomes from related metagenomes. PeerJ **2**, e603 (2014)
10. Laczny, C.C., Sternal, T., Plugaru, V., Gawron, P., Atashpendar, A., Margossian, H.H., Coronado, S., Van der Maaten, L., Vlassis, N., Wilmes, P.: VizBin-an application for reference-independent visualization and human-augmented binning of metagenomic data. Microbiome **3**(1), 1 (2015)
11. Boisvert, S., Raymond, F., Godzaridis, É., Laviolette, F., Corbeil, J.: Ray meta: scalable de novo metagenome assembly and profiling. Genome Biol. **13**(12), R122 (2012)
12. Li, D., Liu, C.-M., Luo, R., Sadakane, K., Lam, T.-W.: MEGAHIT: an ultra-fast single-node solution for large and complex metagenomics assembly via succinct de Bruijn graph. Bioinformatics **31**(10), 1674–1676 (2015)
13. Langmead, B., Salzberg, S.L.: Fast gapped-read alignment with Bowtie 2. Nature Methods **9**(4), 357 (2012)
14. Rousseeuw, P.: Silhouettes: a graphical aid to the interpretation and validation of cluster analysis. J. Comput. Appl. Math. **20**(1), 53–65 (1987)
15. Tibshirani, R.: Regression shrinkage and selection via the lasso. J. Roy. Stat. Soc. Ser. B (Methodol.) **58**, 267–288 (1996)
16. Haynes, W.: Wilcoxon rank sum test. In: Dubitzky, W., Wolkenhauer, O., Cho, K.H., Yokota, H. (eds.) Encyclopedia of Systems Biology, pp. 2354–2355. Springer, New York (2013). https://doi.org/10.1007/978-1-4419-9863-7
17. Tange, O., et al.: GNU parallel-the command-line power tool. USENIX Mag. **36**(1), 42–47 (2011)
18. Analytics Revolution, Weston, S.: doParallel: foreach parallel adaptor for the parallel package. R package version, vol. 1, no. 8 (2014)
19. Analytics Revolution, Weston, S.: Foreach: foreach looping construct for R. R package version, vol. 1, no. 1 (2013)

Cancer Data Analysis

The Review of the Major Entropy Methods and Applications in Biomedical Signal Research

Guangdi Liu[1,2], Yuan Xia[1], Chuanwei Yang[3(✉)], and Le Zhang[1,4(✉)]

[1] College of Computer and Information Science, Southwest University,
Chongqing 400715, China
[2] Library of Chengdu University, Chengdu University, Chengdu City, Sichuan
Province 610106, China
[3] Systems Biology, The University of Texas MD Anderson Cancer Center,
Houston, USA
johnyang80@hotmail.com
[4] College of Computer Science, Sichuan University, Chengdu 610065, China
zhangle06@scu.edu.cn

Abstract. Since biomedical signals are high dimensional data sets with a lot of noise signal, the results processed by the classical signal processing method are subjected to the impact of the noise and interference. Entropy as a measure of disorder or uncertainty in the data has been applied in signal processing research areas. This review is to introduce the application of entropy in the analysis of biomedical signals and discuss the advantages and shortcomings of various entropies. Especially, the utilization and application of entropy concept in cancer research are highlighted.

Keywords: Shannon entropy · Biomedical signal · Cancer research
Approximate entropy · Correntropy

1 Introduction

Biomedical signals are the concentrated expression of human life information and also the window to peep into the phenomenon of life [1]. Biomedical signals play an important role in life science research, health care, disease prevention and treatment. Therefore, the research on the detection, treatment and application of biomedical signals will not only help us to understand the law of life activities, but also to explore the methods of disease prevention and treatment, and the development of the medical equipment [2, 3].

The term 'Entropy' may be appreciated both from the viewpoint of thermodynamics and from the information theory as a display of system uncertainty or disorder or heterogeneity. The concept of "information entropy" is proposed by C. E. Shannon, who is the father of information theory in 1948 [4]. Currently, the applications of entropy in biomedical signals have received extensive research attentions [5–7]. For example, Pincus [8] proposed the concept of approximate entropy to investigate the changes of heart rate for infantile sudden illness. Approximate entropy is good at

© Springer International Publishing AG, part of Springer Nature 2018
F. Zhang et al. (Eds.): ISBRA 2018, LNBI 10847, pp. 87–100, 2018.
https://doi.org/10.1007/978-3-319-94968-0_8

solving the problem of common signal with short noise in biomedical signals. Based on the previous research [9–11], Richman and Moorman [12] developed the sample entropy. Sample entropy, compared with approximate entropy, does not count self-matches and shows better relative consistency and less dependence on data length [12, 13]. Based on the correntropy function definition, Liu et al. [14] developed a generalized correlation function as correntropy, which extends the definition to the general case of two arbitrary random variables and provides its probabilistic and geometric meaning. This theoretical framework will help us understand and apply correntropy judiciously to the nonlinear, non-Gaussian signal processing [14]. Correntropy has received increasing attention in domains of machine learning and signal processing [15]. Therefore, it is necessary for this study to review the current entropy studies in biomedical signals research area.

2 Entropy Methods

The notion of entropy is introduced by Rudolph Clausius in the context of thermodynamics [16]. It abides by the second law of thermodynamics, which states that the change of the entropy in the volume element is equal to the ratio of the heat state changes in temperature [17]. Boltzmann [18] explores the entropy from the perspective of molecular motion theory. The relationship that between the entropy change at the macroscopic level and the heat absorbed by the system is extended to a certain relationship between the entropy and the distribution of subsystems at the microscopic level of the system. The statistical interpretation of thermodynamic entropy [19] is described by Eq. 1.

$$S = k \ln W \qquad (1)$$

Here, S is entropy, $k = 1.38 \times 10^{-23} J/K$ is Boltzmann's constant, and W is probability, determined by the configurationally properties of a statistical system.

The microscopic interpretation of entropy by the Boltzmann formula not only creates an opportunity for the generalization of entropy concept, but also supports and favors the use of entropy in biomedical signal analysis. Then, current entropy(Kolmogorov entropy, approximate entropy [8], dynamic approximation entropy, sample entropy, mode entropy, multi-scale entropy [20], basic scale entropy, joint entropy [21], fuzzy approximation entropy and fuzzy measure entropy Algorithm [22, 23]) greatly enhance the level of biomedical signal analysis. Thus, this study not only reviews the role of entropy in biomedical signal study from the four major entropy research fields (Fig. 1), but also discusses their advantages and shortcomings.

2.1 Shannon Entropy

Shannon [24–26] propose the Shannon entropy as Eq. 2.

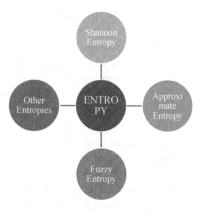

Fig. 1. Major entropy research fields

$$H(X) = -\sum\nolimits_{i=1}^{n} (P(x_i) \log_a P(x_i)) \tag{2}$$

Here a is the base of the logarithm used [27]. $H(X)$ indicates the information entropy of a random variable X. $X = [x_1, x_2, \ldots, x_n]$ indicates n possible independent events in a message. $P = [P_1, P_2, \ldots, P_n]$ indicates the probability of occurrence of these n events, which satisfies $\sum_{i=1}^{n} P_i = 1$. The index i is from 1 to n.

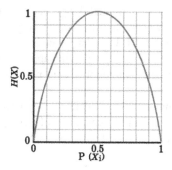

Fig. 2. Shannon entropy with different probabilities

The uncertainty of the signal source at different ratios is shown in the Fig. 2.

If X can assume values 0 and 1, entropy of X is defined as $H(X) = -P(x_i = 0)\log_2 P(x_i = 0) - P(x_i = 1)\log_2 P(x_i = 1)$. It has value if $P(x_i = 0) = 1$ or $P(x_i = 1) = 1$. The entropy reaches maximum when $P(x_i = 0) = P(x_i = 1) = 1/2$ (the value of entropy is then 1).

Shannon entropy is characterized by a degree of uncertainty associated with the occurrence of the result [17]. The greater value of the entropy, the more uncertain the results are. In addition, the Shannon entropy also reflects the average uncertainty of information sources. The more uncertain the information source is, the more disordered the information source is [25].

Since Shannon entropy is better adapted to the normal distributions [28, 29], it is wildly used for biomedical research. For example, indicated by Bruhn et al. [30], Shannon entropy is calculated directly from the digitized voltage signal, without any data transformation when it is applied to the measurement of the electroencephalographic effects of desflurane. Also, Stefanie Fuhrman et al. [31] consider that Shannon entropy, unlike these traditional measures, quantifies the information content of gene expression patterns over entire time courses or anatomical areas, which provides a more complete measure of each gene's participation in a disease process, and permits a rank ordering by physiological relevance. Moreover, De Araujo et al. [32] considers that Shannon entropy is particularly useful for the location of the indicative patterns of event-related fMRI signals. In addition, typical applications for the Shannon entropy include estimating potency and relative potency of drugs and their combinations [30]. For example, Fuhrman et al. [31] propose that the use of Shannon entropy as a method for selecting the most likely drug target candidates from among thousands of genes assayed in parallel. However, Shannon's entropy also does have several shortcomings [29], which are (i) the possibility of losing more information due to aggregation; (ii) the possibility of over-estimation of entropy level if too many zones are used [33], and (iii) this method fails to explain temporal relationships between different values extracted from a time series signal [34].

2.2   Approximate Entropy

Nonlinear dynamical analysis is a powerful approach to understanding biological systems. The calculations, however, usually require very long data sets that can be difficult or impossible to obtain [12]. In order to analyze the short and noisy clinical time series data, Pincus et al. [8, 12, 35] develop the approximate entropy for a measure of regularity closely related to the Kolmogorov entropy, the rate of generation of new information.

This family of statistics, named approximate entropy (ApEn), is rooted in the work of Grassberger and Procaccia [10] and Eckmann and Ruelle [36]. And has been widely applied in clinical cardiovascular studies, [37–53]). Mathematically, ApEn is calculated by

$$ApEn = \ln\left(\frac{C_m(\mathrm{r})}{C_{m+1}(\mathrm{r})}\right) \tag{3}$$

Here $ApEn$ is the Approximate Entropy, r is the similarity coefficient of the samples and $C_m(\mathrm{r})$ is the pattern mean of length m and $C_{m+1}(\mathrm{r})$ is the pattern mean of length $m+1$. The pattern mean is calculated by computing the count of similar patterns of length, m and length, $m+1$ [29].

The advantages of this entropy are: (i) it can be calculated for a relatively short series of noisy data [54]; (ii) it can potentially differentiate a variety of systems such as periodic and multiple periodic systems, chaotic systems, and stochastic systems [8]; and (iii) it can have statistically accurate results compared to the Kolmogorov–Sinai entropy [8, 29]. The drawbacks of approximate entropy are described as following [29]: (i) ApEn is heavily dependent on the record length and is uniformly lower than expected for short records [12]; (ii) significant noise compromises meaningful interpretation of this entropy [29, 54]; (iii) this entropy is a biased statistic as it depends on length of the time series, and counts self-matches [35]; and (iv) it lacks relative consistency.

Approximate entropy has been successfully applied to the analysis of physiological time-series. For example, Akareddy and Kulkarni [55] improve approximate entropy by employing it in EEG signal classification for epileptic seizure detection. Their results show that the accuracy of the proposed method is better than the existing method for epileptic seizure identification [55]. In addition, ApEn also has been applied to the analysis of the heart rate signal [12], blood pressure signal [56], male sex hormone secretion curve [57] and other time-series [20] in the complexity of the study.

2.3   Fuzzy Entropy

Fuzzy entropy, a measure of time series regularity, is proposed by Chen et al. [22]. Compared with approximate entropy, it is insensitive to noise and is highly sensitive to changes in the information content [58]. The main limitation is that Fuzzy entropy (FuzzyEn) only focuses on the local similarity of the signal vector and ignores the global similarity, so it is not sensitive to the small change of signal complexity.

Therefore, Liu and Zhao [23] introduce a new entropy measure - Fuzzy Measure Entropy (FuzzyMEn), which utilizes the fuzzy local and fuzzy global measure entropy to reflect the whole complexity implied in physiological signals and improves the limitation of FuzzyEn, which only focus on the local complexity.

The recently introduced Fuzzy entropy has been successfully used to process biological signal. For example, Ahmed et al. [59] propose the multivariate multiscale fuzzy entropy (MMFE) algorithm and demonstrate its superiority over the multivariate multiscale entropy (MMSE) on both synthetic and real-world uterine electromyography (EMG) short duration signals. A Q-based multivariate sub-band fuzzy entropy has been proposed by Bhattacharyya et al. [60]. It not only analyses the complexity of multivariate electroencephalogram (EEG) signals in different frequency scales, but also proposes such a multivariate sub-band fuzzy entropy that can measure the complexity of the multivariate biomedical signals [60].

2.4   Other Entropies

The previous three type of entropies discussed in Sects. 2.1–2.3 is a form of conditional probability. Here, we introduce another type of entropy measure, which does not use the conditional probability form [61, 62], such as permutation entropy [63] and Correntropy [14].

2.4.1   Permutation Entropy

In order to quantify the complexity of time series by using symbolic dynamics, the permutation entropy(PE) is proposed by Bandt and Pompe [63] to map a continuous time series onto a symbolic sequence [17]. PE is given by

$$PE = -\sum_{j=1}^{n} p_j \log_2 p_j \tag{4}$$

where $j = 1, 2, \ldots, n$, p_j represents relative frequencies of the possible sequence patterns and n implies permutation order [29].

The advantages of this entropy are [29]: (i) it is simplicity, robustness and low computational complexity [64, 65], (ii) it is applicable to real and noisy data, (iii) it does not require any model assumption and is suitable for the analysis of nonlinear processes [66], and (iv) it is useful to analyze huge data sets and requires less pre-processing time and fine-tuning of parameters [66]. However, the main limitation is unable to distinguish well defined patterns of a particular design [67].

2.4.2 Correntropy

Kolmogorov et al. [68] further refine the concept of the Shannon entropy [69] by proposing the concept of Kolmogorov entropy to measure the random or disorder of the system movement. Because Kolmogorov entropy costs a lot of computing resource [70], Crassbreger and Procaccia [10] develop the concept of correlation dimensions and employ the correntropy to approximate Kolmogorov entropy. In recent years, correntropy [14, 71, 72] has been successfully applied in non-Gaussian signal processing [73–77]. In particular, the correntropy function is a generalized correlation that quantifies sums of higher order moments of the signal, which opens the door for new spectral definitions that quantify more precisely the signal structure.

According to Santamaria et al. [71], correntropy is a generalized similarity measure between two random variables x and y, described by Eq. 5.

$$V_\sigma(x, y) = E[k_\sigma(x - y)] \tag{5}$$

Here $k_\sigma(\bullet)$ denotes a positive definite kernel. $V_\sigma(x, y)$ denotes the correntropy. $E[\bullet]$ denotes the expectation operator. The most widely used kernel is the Gaussian kernel (Eq. 6).

$$k_\sigma(x - y) = \exp\left(-\frac{e^2}{2\sigma^2}\right) \tag{6}$$

where $e = x - y$, and $\sigma > 0$ denotes the kernel width [6]. Another interesting property of the correntropy function is its robustness against impulsive noise. This additional advantage is due to the fact that when an outlier is present, the inner product in the feature space is computed via the Gaussian kernel tends to be zero [71]. In other words, the Gaussian kernel is robust to impulsive noises since it approaches zero when the error is very large. It increases as the error decreases especially when the error is closing to the origin, and reaches its maximum at the origin [6, 71].

The advantages of correntropy are illustrated as following. First, the kernel size controls all its properties. Due to the close relationship between m-estimation and methods of information theoretic learning (ITL), choosing an appropriate kernel size [14] in the correntropy criterion becomes practical. Second, the correntropy measure has good property of outlier rejection. Correntropy has been successfully employed to construct different cost functions for biomedical signal processing and machine learning. The major drawback of correntropy is the limitation of its kernel function (Gaussian kernel [14]). For this reason, it is hot for entropy scientists to [5, 6, 73, 77, 78] develop the new Gaussian function for different application scenarios' requirements.

3 The Application of Entropy in Bioinformatics Research

3.1 Cardiovascular Research

Entropy has already been wildly used in cardiovascular related signal analysis [15, 73, 78–81]. For example, Wang et al. [82] investigate that the pulses' approximate entropies of patients with cardiovascular disease preferred to smaller value and less irregularity by applying the approximate entropy. Christopher et al. [83] illustrate the influence of parameter selection on entropy measures' potential for cardiovascular risk stratification and support the potential use of entropy measures in future studies. In addition, Chicote et al. [84] introduce, adapt and fully characterize six entropy indices for VF shock outcome prediction, based on the classical definitions of entropy to measure the regularity and predictability of a time series. Indicated by Hagmair et al. [85] the potential of entropy measures for cardiovascular risk stratification in cohorts the parameters were not optimized for, and it provides additional insights into the parameter choice.

Ferlazzo et al. [86] use permutation entropy (PE) to disclose abnormalities of cerebral activity in patients with typical absences (TAs). And PE seems to be a useful tool to disclose abnormalities of cerebral electric activity not revealed by conventional EEG recordings, opening interesting prospective for future studies. In addition, Li et al. [87] propose to use permutation entropy to explore whether the changes in electroencephalogram (EEG) data can effectively distinguish different phases in human absence epilepsy, i.e., the seizure-free, the pre-seizure and seizure phases. The experimental results show the mean value of PE gradually decreases from the seizure-free to the seizure phase and provides evidence that these three different seizure phases in absence epilepsy can be effectively distinguished. Indicated by Melia et al. [88], a novel approach to this issue based on correntropy function analysis of EEG signals was proposed in order to detect patients suffering from Excessive daytime sleepiness(EDS).

3.2 Cancer Research

Information dynamics are tightly associated with tumorigenesis and cancer progression. Chromosomal aberrations, genomic mutations and epigenetic perturbations are common occurrences during cancer development [76]. These genomic, genetic and epigenetic events decrease the information content by producing the disorder and increasing the entropy inside the cells [89]. The concept of entropy has been applied to diverse subjects in cancer research [90]. Shannon entropy is often used for computing the signaling entropy and structural entropy in cancer. Signaling entropy, a measure of signaling promiscuity or degree of uncertainty, increases in cancer as a result of increased signaling disorder from clonal evolution and clonal diversity of cancer cells and has a negative impact on patient outcomes [91]. Computation of single-cell signaling entropy based on a cell's transcriptome has been used to estimate the differentiation potency and plasticity of a single normal stem cell and to identify cancer stem cell populations in diverse cancer types such as acute myeloid leukemia (AML) and melanoma [92]. Shannon entropy has also been successfully employed for cancer biomarker discovery based on data from high-throughput technologies such as RNA-seq in both prostate cancer and melanoma [93]. Based on the metastatic tumor distribution data from autopsy of 3827 untreated cadavers collected in New England from 1914 to 1943, one research team categorized 12 common cancer types into high entropy (skin cancer, breast cancer, kidney and lung cancer), medium-level entropy (stomach cancer, uterine cancer, pancreatic and ovarian cancer), and low entropy cancers (colorectal cancer, cervical cancer, bladder cancer, and prostate cancers). The rank of entropy values correlates well with the complexity and metastatic potential of each cancer type [94]. Structural entropy is especially useful in the pathological study of cancers by describing entropy information from tissue sections, nuclear texture and karyotypes etc. [95] Castro et al. applied Shannon's entropy to quantify the karyotypic diversity of 14 epithelial tumor types (n = 1232) with scores of their aneuploidy status [96]. Nielsen et al. reported that entropy based nuclear texture features could be independent prognostic markers in cancer [97]. Entropy concept has been used to evaluate results from laboratory tests such as serum biomarker tests, histological and cytological images for grading and diagnosis purposes respectively [98]. Judging from the recent progress, we may witness a dramatic increase in entropy based research and applications in biological sciences and medicine in the next decade as more and more data become available.

4   Conclusions

This comprehensive review presents the characteristics of major entropies and their applications in biomedical signal research, and briefly illustrates the development trend and existing problems of entropy theory in recent years.

Recently, entropy has been used in analyzing short, sparse, and noisy medical time series data. Furthermore, it is widely employed in the multi-variety and multi-scale medical systems [99]. Especially, more and more analytical methods and indicators for entropy research have been applied to clinical research especially cancer research with encouraging results. Here, we summarize the past research and looking forward to the future of entropy research trend. The following aspects will be of far-reaching significance:

(i)   Integrating more clinical diagnosis results into biomedical signal research, it could improve the performance of entropy algorithm. For example, Liu et al. [100] show that Entropy-based Consensus Clustering (ECC) displays superior performance on the pan-omics data by integrating clinical data, such as somatic mutations and DNA methylations in cancer samples.

(ii) It could be another future research direction to develop such a novel entropy measure algorithms that are suitable for short-term data analysis with less computation. For example, Huang et al. [101] propose that correlation dimension and second-order entropy at the minimum embedding dimension could not only analyze pathologic human voices, but also remarkably decrease computation time for the clinical applications.

(iii) It will be the research hotspots to improve the sensitivity and specificity of the measure index of entropy in biomedical research. For example, Ma et al. [102] propose that Poincaré SD2 analysis [103] and complexity index (derived from multiscale entropy) are more sensitive in distinguishing the alterations caused by Metabolic syndrome (MetS) [104]. So large-scale screening to detect early stage cardiac dysfunction may help to prevent or alleviate various late cardiovascular complications.

We already introduce that entropy measures have been used to analyze short, sparse, and noisy medical time series, but they still have many limitations. For example, how to use entropy measures for the classification of pathological and non-pathological data [17] is still unsolved. At present, there is little knowledge concerning how to solve the problems of classification and to select appropriate data ranges to use. Therefore, more work is needed to overcome these shortcomings in the distant future.

Acknowledgments. This research was supported by the National Natural Science Foundation of China (No. 61372138) and the National Science and Technology Major Project (No. 2018ZX10201002).

References

1. Chang, T., Dai, M., Xu, C., Fu, F., You, F., Dong, X.: Research on eit boundary measured voltage data denoising based on a subspace method. Biotechnol. Biotechnol. Equip. **27**, 4157–4161 (2013)
2. Enderle, J., Blanchard, S., Bronzino, J.: Introduction to Biomedical Engineering. Xi'an Jiaotong University Press (2010)
3. Goette, J.: Review of Various Biomedical Signals
4. Clausius, R.: Ueber die Wärmeleitung gasförmiger Körper. Annalen Der Physik **191**, 1–56 (2010)
5. Bao, R., Rong, H., Angelov, P.P., Chen, B., Wong, P.K.: Correntropy-based evolving fuzzy neural system. IEEE Trans. Fuzzy Syst. **PP**, 1 (2017)
6. Wang, W., Zhao, J., Qu, H., Chen, B.: A correntropy inspired variable step-size sign algorithm against impulsive noises. Sig. Process. **141**, 168–175 (2017)
7. Peng, S., Chen, B., Sun, L., Ser, W., Lin, Z.: Constrained maximum correntropy adaptive filtering. Sig. Process. **140**, 116–126 (2017)
8. Pincus, S.M.: Approximate entropy as a measure of system complexity. Proc. Natl. Acad. Sci. U.S.A. **88**, 2297–2301 (1991)
9. Grassberger, P.: Finite sample corrections to entropy and dimension estimates. Phys. Lett. A **128**, 369–373 (1988)
10. Grassberger, P., Procaccia, I.: Estimation of the Kolmogorov entropy from a chaotic signal. Phys. Rev. A **28**, 2591–2593 (1983)

11. Grassberger, P., Schreiber, T., Schaffrath, C.: Nonlinear time sequence analysis. Int. J. Bifurcat. Chaos **01**, 9100040 (1991)

12. Richman, J.S., Moorman, J.R.: Physiological time-series analysis using approximate entropy and sample entropy. Am. J. Physiol. Heart Circ. Physiol. **278**, H2039 (2000)

13. Yan, J.-J., Wang, Y.-Q., Guo, R., Zhou, J.-Z., Yan, H.-X., Xia, C.-M., Shen, Y.: Nonlinear analysis of auscultation signals in TCM using the combination of wavelet packet transform and sample entropy. Evid.-Based Complement. Altern. Med. **2012** (2012)

14. Liu, W., Pokharel, P.P., Principe, J.C.: Correntropy: properties and applications in non-gaussian signal processing. IEEE Trans. Sig. Process. **55**, 5286–5298 (2007)

15. Chen, B., Xing, L., Zhao, H., Zheng, N., Principe, J.C.: Generalized correntropy for robust adaptive filtering. IEEE Trans. Sig. Process. **64**, 3376–3387 (2015)

16. Clausius, R.: On the Motive Power of Heat, and on the Laws Which Can Be Deduced From It for the Theory of Heat (1960)

17. Borowska, M.: Entropy-based algorithms in the analysis of biomedical signals. Stud. Log. Gramm. Rhetor. **43**, 21–32 (2015)

18. Goldstein, S., Lebowitz, J.L.: On the (Boltzmann) entropy of non-equilibrium systems. Physica D **193**, 53–66 (2004)

19. Stephen, G.: Vorlesungen über Gastheorie. University of California Press (1964)

20. Costa, M., Goldberger, A.L., Peng, C.K.: Multiscale entropy analysis of complex physiologic time series. Phys. Rev. Lett. **89**, 068102 (2002)

21. Li, J., Ning, X., Ma, O.: Nonlinear dynamical complexity analysis of short-term heartbeat series using joint entropy. J. Biomed. Eng. **24**, 285–289 (2007)

22. Chen, W., Wang, Z., Xie, H., Yu, W.: Characterization of surface EMG signal based on fuzzy entropy. IEEE Trans. Neural Syst. Rehabil. Eng. **15**, 266 (2007)

23. Liu, C., Zhao, L.: Using fuzzy measure entropy to improve the stability of traditional entropy measures. In: Computing in Cardiology, pp. 681–684. IEEE (2011)

24. Shannon, C.E.: The mathematical theory of communication. 1963. MD Comput. Med. Pract. **14**, 306 (1997)

25. Shannon, C.E.: A mathematical theory of communication: the bell system technical journal. J. Franklin Inst. **196**, 519–520 (1938)

26. Shannon, C.: A Mathematical Theory of Communication (1948)

27. Schneider, T.D.: Information Theory Primer (1995)

28. Tellenbach, B., Burkhart, M., Sornette, D., Maillart, T.: Beyond shannon: characterizing internet traffic with generalized entropy metrics. In: Moon, S.B., Teixeira, R., Uhlig, S. (eds.) PAM 2009. LNCS, vol. 5448, pp. 239–248. Springer, Heidelberg (2009). https://doi.org/10.1007/978-3-642-00975-4_24

29. Acharya, U.R., Fujita, H., Sudarshan, V.K., Bhat, S., Koh, J.E.W.: Application of entropies for automated diagnosis of epilepsy using EEG signals. Knowl.-Based Syst. **88**, 85–96 (2015)

30. Bruhn, J., Lehmann, L.E., Röpcke, H., Bouillon, T.W., Hoeft, A.: Shannon entropy applied to the measurement of the electroencephalographic effects of desflurane. Anesthesiol.: J. Am. Soc. Anesthesiol. **95**, 30–35 (2001)

31. Fuhrman, S., Cunningham, M.J., Wen, X., Zweiger, G., Seilhamer, J.J., Somogyi, R.: The application of Shannon entropy in the identification of putative drug targets. Biosystems **55**, 5–14 (2000)

32. De Araujo, D., Tedeschi, W., Santos, A., Elias, J., Neves, U., Baffa, O.: Shannon entropy applied to the analysis of event-related fMRI time series. NeuroImage **20**, 311–317 (2003)

33. Heikkila, E.J., Hu, L.: Adjusting spatial-entropy measures for scale and resolution effects. Environ. Plan. **33**, 845–861 (2006)

34. Feldman, D.P., Crutchfield, J.P.: Measures of statistical complexity: why? Phys. Lett. A **238**, 244–252 (1997)

35. Pincus, S.: Approximate entropy (ApEn) as a complexity measure. Chaos: An Interdisciplinary. J. Nonlinear Sci. **5**, 110–117 (1995)

36. Eckmann, J.-P., Ruelle, D.: Ergodic theory of chaos and strange attractors. Rev. Mod. Phys. **57**, 617 (1985)

37. Fleisher, L.A., Dipietro, J.A., Johnson, T.R., Pincus, S.: Complementary and noncoincident increases in heart rate variability and irregularity during fetal development. Clin. Sci. **92**, 345–349 (1997)

38. Fleisher, L.A., Pincus, S.M., Rosenbaum, S.H.: Approximate entropy of heart rate as a correlate of postoperative ventricular dysfunction. Anesthesiology **78**, 683–692 (1993)

39. Goldberger, A.L., Mietus, J.E., Rigney, D.R., Wood, M.L., Fortney, S.M.: Effects of head-down bed rest on complex heart rate variability: response to LBNP testing. J. Appl. Physiol. **77**, 2863–2869 (1994)

40. Mäkikallio, T.H., Seppänen, T., Niemelä, M., Airaksinen, K.E.J., Tulppo, M., Huikuri, H. V.: Abnormalities in beat to beat complexity of heart rate dynamics in patients with a previous myocardial infarction 1. J. Am. Coll. Cardiol. **28**, 1005–1011 (1996)

41. Korpelainen, J.T., Sotaniemi, K.A., Mäkikallio, A., Huikuri, H.V., Myllylä, V.V.: Dynamic behavior of heart rate in ischemic stroke. Stroke **30**, 1008–1013 (1999)

42. Palazzolo, J.A., Estafanous, F.G., Murray, P.A.: Entropy measures of heart rate variation in conscious dogs. Am. J. Physiol. **274**, H1099 (1998)

43. Mäkikallio, T.H., Ristimäe, T., Airaksinen, K.E.J., Peng, C.K., Goldberger, A.L., Huikuri, H.V.: Heart rate dynamics in patients with stable angina pectoris and utility of fractal and complexity measures. Am. J. Cardiol. **81**, 27–31 (1998)

44. Ho, K.K., Moody, G.B., Peng, C.K., Mietus, J.E., Larson, M.G., Levy, D., Goldberger, A. L.: Predicting survival in heart failure case and control subjects by use of fully automated methods for deriving nonlinear and conventional indices of heart rate dynamics. Circulation **96**, 842 (1997)

45. Lipsitz, L.A., Pincus, S.M., Morin, R.J., Tong, S., Eberle, L.P., Gootman, P.M.: Preliminary evidence for the evolution in complexity of heart rate dynamics during autonomic maturation in neonatal swine. J. Auton. Nerv. Syst. **65**, 1 (1997)

46. Nelson, J.C., Rizwan-uddin, Griffin, M.P., Moorman, J.R.: Probing the order within neonatal heart rate variability. Pediatr. Res. **43**, 823–831 (1998)

47. Hogue, C.W., Domitrovich, P.P., Stein, P.K., Despotis, G.D., Re, L., Schuessler, R.B., Kleiger, R.E., Rottman, J.N.: RR interval dynamics before atrial fibrillation in patients after coronary artery bypass graft surgery. Circulation **98**, 429–434 (1998)

48. Pincus, S.M., Viscarello, R.R.: Approximate entropy: a regularity measure for fetal heart rate analysis. Obstet. Gynecol. **79**, 249–255 (1992)

49. Ryan, S.M., Goldberger, A.L., Pincus, S.M., Mietus, J., Lipsitz, L.A.: Gender- and age-related differences in heart rate dynamics: are women more complex than men? J. Am. Coll. Cardiol. **24**, 1700 (1994)

50. Pincus, S.M., Cummins, T.R., Haddad, G.G.: Heart rate control in normal and aborted-SIDS infants. Am. J. Physiol. **264**, R638 (1993)

51. Tulppo, M.P., Makikallio, T.H., Takala, T.E., Seppanen, T., Huikuri, H.V.: Quantitative beat-to-beat analysis of heart rate dynamics during exercise. Am. J. Physiol. **271**, 244–252 (1996)

52. Pincus, S.M., Gladstone, I.M., Ehrenkranz, R.A.: A regularity statistic for medical data analysis. J. Clin. Monit. **7**, 335–345 (1991)

53. Schuckers, S.A.C.: Use of approximate entropy measurements to classify ventricular tachycardia and fibrillation. J. Electrocardiol. **31**, 101–105 (1998)

54. Seely Andrew, J.E., Macklem, P.T.: Complex systems and the technology of variability analysis. Crit. Care **8**, R367 (2004)
55. Akareddy, S.M., Kulkarni, P.K.: EEG signal classification for epilepsy seizure detection using improved approximate entropy. Int. J. Public Health Sci. **2**, 23–32 (2013)
56. Hornero, R., Aboy, M., Abasolo, D., Mcnames, J., Goldstein, B.: Interpretation of approximate entropy: analysis of intracranial pressure approximate entropy during acute intracranial hypertension. IEEE Trans. Biomed. Eng. **52**, 1671–1680 (2005)
57. Veldhuis, J.D., Liem, A.Y., South, S., Weltman, A., Weltman, J., Clemmons, D.A., Abbott, R., Mulligan, T., Johnson, M.L., Pincus, S.: Differential impact of age, sex steroid hormones, and obesity on basal versus pulsatile growth hormone secretion in men as assessed in an ultrasensitive chemiluminescence assay. J. Clin. Endocrinol. Metab. **80**, 3209–3222 (1995)
58. Chakraborty, B., Bhattacharyya, S., Dutta, P.: An unsupervised video shot boundary detection technique using fuzzy entropy estimation of video content (2011)
59. Ahmed, M., Chanwimalueang, T., Thayyil, S., Mandic, D.: A multivariate multiscale fuzzy entropy algorithm with application to uterine EMG complexity analysis. Entropy **19**, 2 (2016)
60. Bhattacharyya, A., Pachori, R.B., Acharya, U.R.: Tunable-Q wavelet transform based multivariate sub-band fuzzy entropy with application to focal EEG signal analysis. Entropy **19**(3), 01–14 (2017)
61. Porta, A., Guzzetti, S., Montano, N., Furlan, R., Pagani, M., Malliani, A., Cerutti, S.: Entropy, entropy rate, and pattern classification as tools to typify complexity in short heart period variability series. IEEE Trans. Bio-med. Eng. **48**, 1282–1291 (2001)
62. Porta, A., Castiglioni, P., Bari, V., Bassani, T., Marchi, A., Cividjian, A., Quintin, L., Di, R. M.: K-nearest-neighbor conditional entropy approach for the assessment of the short-term complexity of cardiovascular control. Physiol. Meas. **34**, 17 (2013)
63. Bandt, C., Pompe, B.: Permutation entropy: a natural complexity measure for time series. Phys. Rev. Lett. **88**, 174102 (2002)
64. Li, X., Ouyang, G., Richards, D.A.: Predictability analysis of absence seizures with permutation entropy. Epilepsy Res. **77**, 70–74 (2007)
65. Liang, Z., Wang, Y., Sun, X., Li, D., Voss, L.J., Sleigh, J.W., Satoshi, H., Li, X.: EEG entropy measures in anesthesia. Front. Comput. Neurosci. **9**, 16 (2015)
66. Zanin, M., Zunino, L., Rosso, O.A., Papo, D.: Permutation entropy and its main biomedical and econophysics applications: a review. Entropy **14**, 1553–1577 (2012)
67. Fadlallah, B., Chen, B., Keil, A., Príncipe, J.: Weighted-permutation entropy: a complexity measure for time series incorporating amplitude information. Phys. Rev. E: Stat. Nonlin. Soft Matter Phys. **87**, 022911 (2013)
68. Shiryayev, A.N.: New metric invariant of transitive dynamical systems and automorphisms of Lebesgue spaces. In: Shiryayev, A.N. (ed.) Selected Works of A. N. Kolmogorov. Mathematics and Its Applications, vol. 27, pp. 57–61. Springer, Dordrecht (1993). https://doi.org/10.1007/978-94-017-2973-4_5
69. Xu, Q.: Measuring information content from observations for data assimilation: relative entropy versus Shannon entropy difference. Tellus A **59**, 198–209 (2007)
70. Abel, M., Biferale, L., Cencini, M., Falcioni, M., Vergni, D., Vulpiani, A.: Exit-time approach to ε-entropy. Phys. Rev. Lett. **84**, 6002 (2000)
71. Santamaria, I., Pokharel, P.P., Principe, J.C.: Generalized correlation function: definition, properties, and application to blind equalization. IEEE Trans. Sig. Process. **54**, 2187–2197 (2006)
72. Li, R., Liu, W., Principe, J.C.: A unifying criterion for instantaneous blind source separation based on correntropy. Sig. Process. **87**, 1872–1881 (2009)

73. Wu, Z., Peng, S., Chen, B., Zhao, H.: Robust hammerstein adaptive filtering under maximum correntropy criterion. Entropy **17**, 7149–7166 (2015)
74. Jeong, K.H.: The correntropy mace filter for image recognition. In: Proceedings of the 2006 IEEE Signal Processing Society Workshop on Machine Learning for Signal Processing, pp. 9–14 (2006)
75. Ma, W., Qu, H., Gui, G., Xu, L., Zhao, J., Chen, B.: Maximum correntropy criterion based sparse adaptive filtering algorithms for robust channel estimation under non-Gaussian environments. J. Franklin Inst. **352**, 2708–2727 (2015)
76. Zhang, L., Liu, Y., Wang, M., Wu, Z., Li, N., Zhang, J., Yang, C.: EZH2-, CHD4-, and IDH-linked epigenetic perturbation and its association with survival in glioma patients. J. Mol. Cell Biol. **9**, 477–488 (2017). https://doi.org/10.1093/jmcb/mjx056
77. Singh, A., Principe, J.C.: Using Correntropy as a cost function in linear adaptive filters. In: International Joint Conference on Neural Networks, pp. 2950–2955 (2009)
78. Chen, B., Principe, J.C.: Maximum correntropy estimation is a smoothed MAP estimation. IEEE Sig. Process. Lett. **19**, 491–494 (2012)
79. Chen, B., Liu, X., Zhao, H., Principe, J.C.: Maximum correntropy Kalman filter ✭. Automatica **76**, 70–77 (2017)
80. Chen, B., Xing, L., Liang, J., Zheng, N., Principe, J.C.: Steady-state mean-square error analysis for adaptive filtering under the maximum correntropy criterion. IEEE Sig. Process. Lett. **21**, 880–884 (2014)
81. Lu, L., Zhao, H.: Active impulsive noise control using maximum correntropy with adaptive kernel size. Mech. Syst. Sig. Process. **87**, 180–191 (2017)
82. Wang, K., Xu, L., Li, Z., Zhang, D.: Approximate entropy based pulse variability analysis. In: Proceedings of the IEEE Symposium on Computer-Based Medical Systems, pp. 236–241 (2003)
83. Mayer, C., Bachler, M., Holzinger, A., Stein, P., Wassertheurer, S.: The effect of threshold values and weighting factors on the association between entropy measures and mortality after myocardial infarction in the cardiac arrhythmia suppression trial (CAST). Entropy **18**, 129 (2016)
84. Chicote, B., Irusta, U., Alcaraz, R., Rieta, J.J., Aramendi, E., Isasi, I., Alonso, D., Ibarguren, K.: Application of entropy-based features to predict defibrillation outcome in cardiac arrest. Entropy **18**, 313 (2016)
85. Hagmair, S., Bachler, M., Braunisch, M.C., Lorenz, G., Schmaderer, C., Hasenau, A.L., Stülpnagel, L.V., Bauer, A., Rizas, K.D., Wassertheurer, S.: Challenging recently published parameter sets for entropy measures in risk prediction for end-stage renal disease patients. Entropy **19**, 582 (2017)
86. Ferlazzo, E., Mammone, N., Cianci, V., Gasparini, S., Gambardella, A., Labate, A., Latella, M.A., Sofia, V., Elia, M., Morabito, F.C.: Permutation entropy of scalp EEG: a tool to investigate epilepsies: suggestions from absence epilepsies. Clin. Neurophysiol. Official J. Int. Fed. Clin. Neurophysiol. **125**, 13–20 (2014)
87. Li, J., Yan, J., Liu, X., Ouyang, G.: Using permutation entropy to measure the changes in EEG signals during absence seizures. Entropy **16**, 3049–3061 (2014)
88. Melia, U., Guaita, M., Vallverdú, M., Montserrat, J.M., Vilaseca, I., Salamero, M., Gaig, C., Caminal, P., Santamaria, J.: Correntropy measures to detect daytime sleepiness from EEG signals. Physiol. Meas. **35**, 2067 (2014)
89. Gatenby, R.A., Frieden, B.R.: Information dynamics in carcinogenesis and tumor growth. Mutat. Res. **568**, 259 (2004)
90. Ritchie, W., Granjeaud, S., Puthier, D., Gautheret, D.: Entropy measures quantify global splicing disorders in cancer. PLoS Comput. Biol. **4**, e1000011 (2008)

91. Banerji, C.R.S., Severini, S., Caldas, C., Teschendorff, A.E.: Intra-tumour signalling entropy determines clinical outcome in breast and lung cancer. PLoS Comput. Biol. **11**, e1004115 (2015)

92. Teschendorff, A.E., Enver, T.: Single-cell entropy for accurate estimation of differentiation potency from a cell's transcriptome. Nat. Commun. **8**, 15599 (2017)

93. Regina Berretta, P.M.: Cancer biomarker discovery: the entropic hallmark. PLoS ONE **5**, e12262 (2010)

94. Newton, P.K., Jeremy, M., Brian, H., Kelly, B., Lyudmila, B., Jorge, N., Peter, K.: Entropy, complexity, and Markov diagrams for random walk cancer models. Sci. Rep. **4**, 7558 (2014)

95. Kayser, K., Kayser, G., Metze, K.: The concept of structural entropy in tissue-based diagnosis. Anal. Quant. Cytol. Histol. **29**, 296 (2007)

96. Castro, M.A., Onsten, T.T., de Almeida, R.M., Moreira, J.C.: Profiling cytogenetic diversity with entropy-based karyotypic analysis. J. Theor. Biol. **234**, 487 (2005)

97. Nielsen, B., Hveem, T.S., Kildal, W., Abeler, V.M., Kristensen, G.B., Albregtsen, F., Danielsen, H.E., Rohde, G.K.: Entropy-based adaptive nuclear texture features are independent prognostic markers in a total population of uterine sarcomas. Cytometry Part A: J. Int. Soc. Anal. Cytol. **87**, 315–325 (2015)

98. Vollmer, R.T.: Entropy and information content of laboratory test results. Am. J. Clin. Pathol. **127**, 60–65 (2007)

99. Humeauheurtier, A.: The multiscale entropy algorithm and its variants: a review. Entropy **17**, 3110–3123 (2016)

100. Liu, H., Zhao, R., Fang, H., Cheng, F., Fu, Y., Liu, Y.Y.: A novel clustering method for patient stratification (2016)

101. Huang, N., Zhang, Y., Calawerts, W., Jiang, J.J.: Optimized nonlinear dynamic analysis of pathologic voices with laryngeal paralysis based on the minimum embedding dimension. J. Voice Official J. Voice Found. **31**, 249.e1–249.e7 (2017)

102. Ma, Y., Tseng, P.H., Ahn, A., Wu, M.S., Ho, Y.L., Chen, M.F., Peng, C.K.: Cardiac autonomic alteration and metabolic syndrome: an ambulatory ECG-based study in a general population. Sci. Rep. **7**, 44363 (2017)

103. Mourot, L., Bouhaddi, M., Perrey, S., Rouillon, J.D., Regnard, J.: Quantitative Poincaré plot analysis of heart rate variability: effect of endurance training. Eur. J. Appl. Physiol. **91**, 79–87 (2004)

104. Malik, S., Wong, N.D., Franklin, S.S., Kamath, T.V., Gilbert, J., Pio, J.R., Williams, G.R.: Impact of the metabolic syndrome on mortality from coronary heart disease, cardiovascular disease, and all causes in United States adults. Circulation **110**, 1245–1250 (2004)

Inferring Dysregulated Pathways of Driving Cancer Subtypes Through Multi-omics Integration

Kai Shi[1,2], Lin Gao[1(✉)], and Bingbo Wang[1]

[1] School of Computer Science and Technology, Xidian University,
Xian 710126, Shaanxi, China
mail_shikai@foxmail.com, lgao@mail.xidian.edu.cn
[2] College of Science, Guilin University of Technology,
Guilin 541004, Guangxi, China

Abstract. The rapid accumulation of multi-omics cancer data has created the opportunity for biological discovery and biomedical applications. In this study, we propose an approach that integrates multi-omics data to identify dysregulated pathways driving cancer subtypes, which simultaneously considers DNA methylation, DNA copy number, somatic mutation and gene expression profiles. After applying it to Breast Invasive Carcinoma (BRCA) in TCGA, we identify distinct top 30 dysregulated pathways for each breast cancer subtypes. The result suggests that dysregulated pathways of different subtypes display common and specific patterns. Furthermore, 44 differentially expressed genes with corresponding genetic and epigenetic dysregulation are retrieved from the subtype-specific pathways. Literature validation and functional enrichment analysis indicate that these genes are function associated with BRCA. Our method provides a new insight for identifying the driver of cancer subtypes through multi-omics data integration.

Keywords: Dysregulated pathways · BRCA · Data integration
Maximum relevance minimum redundancy · Disease gene

1 Introduction

In the last few years, the rapid accumulation of multi-omics cancer data has created the opportunity for biological discovery and biomedical applications. It is a known fact that carcinogenesis is associated with multiple levels. Genetic and epigenetic mechanisms affect the gene expression signatures. So, different scale data pose a new challenge: how to integrate them and improve the understanding of disease mechanism and pathology.

Electronic supplementary material The online version of this chapter (https://doi.org/10.1007/978-3-319-94968-0_9) contains supplementary material, which is available to authorized users.

In cancer genomics, a key problem is to identify "driver genes" or "driver pathways", which is causally implicated in oncogenesis [6,7,32]. During the last years, a lot of methods mainly focus on deriving dysregulated genes from gene expression data. Recently, the network or pathway-based methods are further considered to identify robust biomarkers [8,14,16,20,22]. The "pathway driven"analysis has achieved more robust biomarkers to the disease of interest. Some of them regarded inferring dysregulated pathways as a feature selection problem. Lee *et al.* [20] used a gene subset of the pathway to infer the pathway activity. Han *et al.* [14] used Gaussian Bayesian networks to create individualized features that reflect pathway activity, they incorporated gene interactions using probabilistic graphical models to more accurately represent the underlying biology and achieve better performance. Furthermore, some methods considered the structure information of the network to infer more active pathways [16,22]. Hung *et al.* [16] weighted genes based on the neighbor genes to improve the pathway enrichment analysis. Liu *et al.* [22] used directed random walk to identify disease genes and nominated active pathways by mining the structure information of pathway. But relatively few methods utilize different scale information to infer the driver of cancer subtypes.

Here, we shift the focus to infer dysregulated pathways that drive cancer subtypes and propose an approach for integrating DNA methylation, DNA copy number, somatic mutation and gene expression data in the context of pathway information. Because of the cross-talk among various biological pathways contributes to the complexity and redundancy of molecular mechanisms [23,31]. Inferring dysregulated pathways is translated to an optimization problem that how to infer the most relevance pathways to disease, at the same time, with the least redundancy among each other.

Breast cancer is a complex disease that it is not only diverse with many subtypes, but also widely multi-level dysregulated [5,9,25]. By application of our method to BRCA dataset from TCGA, we identify subtype-specific dysregulated pathways by investigating the disease driver factors on four functional levels. The further analysis reveals common and specific patterns among these pathways. Our results verify the previous suggestion that distinct biological pathways drive the pathobiology of different breast cancer subtypes [2,4].

2   Materials and Methods

2.1   Resources and Datasets

Our method integrating multi-omic data requires gene expression, DNA copy number, somatic mutation and DNA methylation data from the same breast cancer subtypes. All the BRCA associated datasets are downloaded during April 2016 in TCGA by an R/Bioconductor package TCGAbiolinks. For our multiple level analysis, BRCA datasets include DNA copy number level 3 data(Affymetrix SNP 6.0 platform), DNA methylation level 3 data(Illumina HumanMethylation27), gene expression data(AgilentG4502A_07_3), and somatic mutation level

2 data. We focus on the intrinsic molecular subtypes of breast tumors: luminal A (LumA), luminal B (LumB), human epidermal growth factor receptor 2 (HER2)-enriched, and Basal-like. More details of samples are shown in Table 1.

Table 1. TCGA breast cancer datasets with normal and tumor samples

Subtypes	Type of samples	Data categories		
		DNA copy number data	DNA methylation data	Gene expression data
Basal-like	Normal	133	27	63
	Tumor	96	58	96
HER2-enriched	Normal	133	27	63
	Tumor	56	44	58
Luminal A	Normal	133	27	63
	Tumor	228	119	231
Luminal B	Normal	133	27	63
	Tumor	127	80	127

All of the raw data are dealt with the instruction of TCGAbiolinks. For DNA copy number data, segmentation data is processed and transformed into a gene-level matrix based on genomic location of genes by Bioconductor package "CNTools".

The pathway information is obtained from the KEGG database by an R/Bioconductor package KEGGgraph. After removing the disease pathways, 276 pathways are analyzed in our research.

2.2 Method for Inferring Dysregulated Pathways

The key idea of our method is to model the disease on the pathway level. In our method, we formalize it as a multi-objective optimization problem to infer dysregulated pathways with maximum relevance and minimum redundancy. Our method involves two steps: seed genes selection and dysregulated pathways inference. The framework of our method is shown in Fig. 1.

Seed Genes Selection. The first step of our method is to identify aberrant genes regarded as seed genes at multi-level events. For methylation analysis, differentially methylated CpG sits are searched using beta-values by comparing control and disease samples. Then, genes with differential methylation (hypomethylation and hypermethylation) are identified in disease group compared to normal group ($p - values < 0.01, diffmean < 0.25$). In DNA copy number analysis, we use t statistic test ($FDR < 0.001$) to identify altered genes with genomic amplification or deletion. As for somatic mutation data, single nucleotide variants (SNVs) and small insertions and deletions (Indels) are considered. For gene

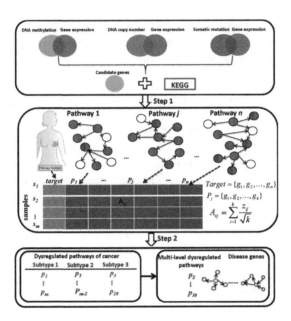

Fig. 1. Schematic flowchart of data integration and dysregulated pathways identification. Step 1: seed genes selection. Genes with differential expression and corresponding differential methylation, or differential copy number alteration or somatic mutation are selected. Step 2: dysregulated pathways inference. A feature matrix is constructed based on pathway level and the optimal approach is applied to screen dysregulated pathways with maximum relevance and minimum redundancy. Finally, the dysregulated pathways are performed the literature validation and functional enrichment analysis.

expression data, differentially expressed genes (DEGs) are nominated between two groups($FDR < 0.001$). Furthermore, genes with significantly altered on three levels (DNA copy number, DNA methylation, somatic mutation) are intersected with the differential expression genes, separately. Finally, the seed genes are extracted by merging the genes of the following three groups:

(i) genes with differential expression and corresponding differential methylation;
(ii) genes with differential expression and corresponding differential copy number alteration;
(iii) genes with differential expression and corresponding somatic mutation.

Dysregulated Pathways Inference. Recently, incorporating pathway information into the disease dysfunciton analysis has been paid more attentions. Motivated by Lee *et al.* [20], we propose a new definition of dysfunctional pathways to precisely formulate disease system. In our definition, the activity of a dysfunctional pathway is represented by seed genes rather than the entire signaling pathway. First, for each gene i in sample j, the expression value g_{ij} is normalized to z-transformed score $z_{ij} = \frac{g_{ij} - \mu_i}{\delta_i}$ with $\mu_i = 0$ and $\delta_i = 1$ over all samples. For pathway j composed of k seed genes, $p_j = \{g_1, g_2, ..., g_k\}$, the activity of pathway j on sample s is designated a combined z-score representing the averaged normalized score in the seed gene set.

$$a_{sj} = \sum_{i=1}^{k} \frac{z_{ij}}{\sqrt{k}}, \quad s = 1, 2, \ldots, m \tag{1}$$

Finally, after differential analysis ($p - value < 0.01$), the feature matrix A with differential pathways is constructed for next analysis. There are a lot of modeling methods to analyze disease, but understanding disease remains a challenge. The key of our method is to model the disease state based on the pathway with two aspects. On the one hand, the pathways need to be highly related with disease; at the same time, there is the least redundancy among them. Inspired by [29], we formalize it as a multi-objective optimization problem with maximum relevance and minimum redundancy for pathways (mRMR_P). The final-selected pathways are regarded as the core features to screen disease genes.

In our work, mutual information is used as the measure of the pathway relevance. If a pathway is significant dysregulation in different classes, it should has larger mutual information. The mutual information is defined as below:

$$I(x, y) = \sum_{i,j} p(x_i, y_j) log \frac{p(x_i, y_j)}{p(x_i)p(y_j)}, \tag{2}$$

where x and y are two variables, $p(x, y)$ is their join probabilistic distribution, and $p(x)$ and $p(y)$ are the marginal probabilities, respectively. Further, Ω is defined as the whole pathway set ($|\Omega| = n$), S is as the selected pathway set ($|S| = m$) and Ω_R is as the remaining pathway set ($\Omega_R = \Omega - S$). The relevance D between the pathway p_i in Ω_R and the target class (disease state) c is defined as:

$$D = I(c, p_i), \tag{3}$$

The redundancy R between the pathway p_i in Ω_R and all the pathways in S is defined as:

$$R = \frac{1}{m} \sum_{p_j \in S} I(p_j, p_i), \tag{4}$$

The mRMR_P is defined as a combination criteria optimizing the minimum redundancy (Eq. 3) and the maximum relevance (Eq. 4) simultaneously.

$$\max_{p_i \in \Omega_R} [I(c, p_i) - \frac{1}{m} \sum_{p_j \in S} I(p_j, p_i)], \tag{5}$$

To resolve the optimization problem, a heuristic algorithm based on incremental way is adopted to rank pathways. After $N(= n + m)$ rounds evaluations, the final pathway set W can be written as

$$W = \{p'_1, p'_2, \ldots, p'_h, \ldots, p'_N\} \tag{6}$$

In Eq. 6, the smaller the index value, the more significant the pathway will be.

3   Results

3.1   Screening Significan Seed Genes

For each breast cancer PAM50 subtype, we respectively identify 5667, 4742, 3903 and 4857 seed gene set in the step 1. Genes in each seed gene set are differential expression, and aggregating the other three differential components: methylation, copy number and mutation.

3.2   Inferring Dysregulated Pathways

Then, genes from seed gene set are mapped into 276 non-disease KEGG pathways. After differential analysis, we retrieve 205, 185, 200 and 194 significant pathways with t statistic test for Basal-like, HER2-enriched, Luminal A and Luminal B, respectively. Based on these significant pathways, a feature matrix is constructed following Eq. 1 as the input of mRMR_P. After performing the mRMR_P, we get a top 30 pathways list (Supplement table S1) which represent that the dysregulated pathways with the maximum relevance and minimum redundancy according to the subtype. The venn diagram (Fig. 2) shows the number of common and specific dysregulated pathways in the subtypes. The Fig. 3 shows the heat map of top 30 dysregulated pathways for each subtype of breast cancer. The result indicates that the screened pathways can significantly discriminate disease and normal samples, especially in Basal-like and HER2-enriched subtypes. In other word, these dysregulated pathways are highly associated with the target disease subtype.

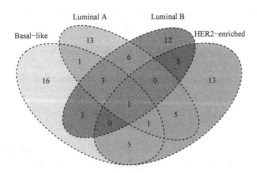

Fig. 2. The venn graph of dysregulated pathways for each subtype of breast cancer

3.3   Extracting Common Dysregulation Genes from the Top Pathways

44 common dysregulated genes are extracted from the top 30 pathways for four subtypes of breast cancer (Table 2). Among those dysregulated genes, *PIK3CA*, *FGFR1*, *YWHAZ* and *CCND2* are considered as known cancer genes from the

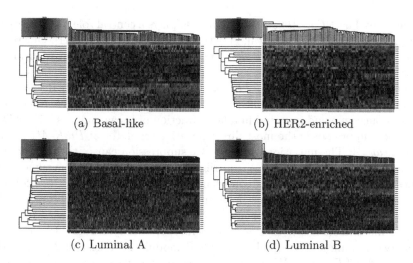

(a) Basal-like   (b) HER2-enriched

(c) Luminal A   (d) Luminal B

Fig. 3. Active levels of the highest ranking pathways. (a-c) The top 30 dysregulated pathways of each breast cancer subtype samples are clustered by complete linkage hierarchical clustering

Cancer Gene Census(CCG)[12]. They show different dysregulated patterns in the four subtypes. Some of the rest genes show overexpression such as *FGF2*, *LAMB3*, *CCNE2*, and so on. In addition, literature validation shows that the rest genes present highly association with breast cancer.

Table 2. 44 common dysregulated genes of four subtypes on top 30 pathways

FGFR1*	LAMB3	IRS1	EFNA3	C4BPA	IGF1	JAM3	CRTC2
FGF2	LAMC1	PCK1	EFNA4	ITGAX	ANGPT1	ADRA1A	CCNE2
HSPA8	PTK2	AP2M1	EPHA2	MYL9	EFNA1	SERPING1	EIF4E2
PLA2G4A	RELN	ARF1	FLT4	MYLK2	**YWHAZ***	F11R	**CCND2***
PIK3CA*	ATP6V0B	PIP5K1A	RILP	NPR1	LEP	MYLK	PPP1R12B
PIK3R3	ATP6V0D2	ADCY4	DYNC2H1				

3.4 Exploring Subtype-Common and Specific Pattern of Pathways

Breast cancer is one of the most common types of cancer in women, and it is comprised of distinct subtypes. This means that it may respond differently to drug and pathways targeted therapies.

(1) Analysis of subtype-common dysregulated pathways

Supplement table S2 shows the common pathways among the subtypes. Our result displays an interest phenomenon that all subtypes involve the *PI3K-Akt signaling pathway*. The common dysregulated genes for subtypes in this pathway are shown on Fig. 4. These dysfunction genes mainly focus on the upstream of this pathway. The *PI3K-Akt signaling pathway* plays a central role in cellular processes, responsible for cell survival, growth, division and motility [11]. Our result further confirms that *PI3K-Akt signaling pathway* may be related with multiple breast cancer subtypes [1,5]. Furthermore, we find that all subtypes exhibit dysregulated on *PIK3CA*, and *PTEN* is dysregulated except in HER2-enriched subtype. They are very important genes for breast cancer. *PIK3CA* is discovered as a viral oncogene and mutated in a large fraction of breast tumors [17]. *PTEN* as a tumor suppressor is the most direct negative regulator of *PI3K-Akt signaling pathway* [1]. The mutation or epigenetic suppression causes the inactivation of *PTEN*. The mutation in *PTEN* contributes to *PTEN* hamartoma tumor syndromes (*PHTS*) which is a high incidence of breast cancer [15]. In addition, there are specifically distinctions between the major molecular subtypes in the *PI3K-Akt signaling pathway*. For all subtypes, this pathway is divided into two or three modules (Fig. 5). In Basal-like subtype, the dysregulated genes constitute two modules. The small dysregulted module is consist of *IL2*, *IL2RB*, *IL2RA*, *IL3*, *CSF3* and *GH2*. The bigger module involves core genes such as *EGFR*, *PIK3CA*, *PIK3CB*, *PIK3R1*, *PIK3R3*, *PIK3AP1* and as on. In HER2-enriched subtype, one of module is consist of *EPO*, *GH2*, *Il2*, *IL3*, *Il7*, *GHR* and *PRLR*. At the same time, it includes a new module in which *AKT3* as the central node. This module also appears in Luminal A and Luminal B subtypes, but the upstream and downstream genes of *AKT3* are different. *AKT3* is one of the AKT kinase, which is responsible to metabolism, proliferation, cell survival, growth and angiogenesis. The mutations in AKT kinase are frequently observed in breast tumors [30].

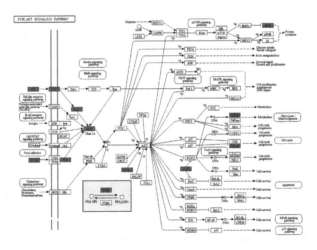

Fig. 4. The common dysregulated genes of the four breast cancer subtypes in PI3K-Akt signaling pathway

(2) Analysis of subtype-specific dysregulated pathways

Supplement table S3 shows the subtype-specific dysregualted pathways. For Basal-like subtype, there are 16 unique dysregulated pathways. The number of

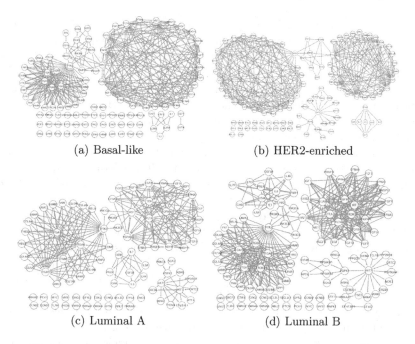

(a) Basal-like

(b) HER2-enriched

(c) Luminal A

(d) Luminal B

Fig. 5. The different modules of seed genes in PI3K pathway for breast cancer subtypes

subtype-specific dysregulated pathways for Basal-like is the largest in the four subtypes. The MAPK signaling pathway is the highest relevance pathway in the Basal-like subtype. Researches support that the MAPK signaling pathway plays an important role in Basal-like breast cancer [3]. *DUDP4* regulates the MAP-ERK kinase(MEK) and c-jun-NH2-kinase(JNK), and the loss of DUSP4 actives the MAPK signaling pathways [3]. We find that Epstein-Barr virus infection is the second relevance pathway with this subtype. However, there is a controversy about the association of *EBV* with breast cancer [24]. Some researches supported the presence of *EBV* in breast cancer [19], other groups reported negative results [13]. This pathway also is the third relevance pathway in Luminal B subtype. Our interesting result indicates that the Epstein-Barr virus, commonly known as the kissing disease, may raise the risk of breast cancer.

For HER2-enriched subtype, we find that it shares 7, 7 and 6 pathways with Basal-like, Luminal A and Luminal B subtypes, respectively. Excepted the PI3K-Akt signaling pathway, the Ras signaling pathway is shared by the Basal-like, HER2-enriched and Luminal A subtypes, especially Basal-like and HER2-enriched: ranked on the top relevance pathway list. It is considered deregulated in breast cancer, many of components of which control cell growth and differentiation. The dysfunction of any one component of this pathway may trigger the same events as the dysfunction of ras itself, and considering the aberrations of upstream or downstream of the pathway is important to provide additional targets for drug design [10]. More important is that *RASAL2* shows low expression, which is reported that it plays a causal role in breast cancer development and metastasis, especially Luminal B subtype [26]. Another 5 pathways which HER2-enriched shared with Luminal A subtype include "Vascular smooth muscle contraction", "Jak-STAT signaling pathway", "Phospholipase D signaling pathway", "Ovarian steroidogenesis" and "Vibrio cholerae infection". These pathways are different with the pathways which HER2-enriched shared with Luminal B, including "AMPK signaling pathway", "p53 signaling pathway", "Phagosome", "Hematopoietic cell lineage", and "Porphyrin and chlorophyll metabolism". Most of them are associated with two genes: *HER2* and *TP53* genes. AMP-activated protein kinase (*AMPK*) is an important mediator in maintaining cellular energy homeostasis. *AMPK* regulates activity of *HER2* and *EGFR* in breast cancer [18].

13 and 12 exclusive pathways have been identified respectively in the Luminal A, Luminal B subtypes. For Luminal A subtype, the unique pathways are "ErbB signaling pathway", "Influenza A", "Glycerophospholipid metabolism", "Bacterial invasion of epithelial cells" and so on. For Luminal B subtype, the unique pathways are "Rap1 signaling pathway", "Purine metabolism", "FoxO signaling pathway" and so on. In addition to the common pathway "PI3K-Akt signaling pathway" that is present in all subtypes, there are 9 pathways shared by Luminal A and Luminal B, including Cell adhesion molecules (CAMs), Endocytosis, Inflammatory mediator regulation of TRP channels, Regulation of actin cytoskeleton, Toxoplasmosis, HIF-1 signaling pathway and so on. Cell adhesion molecules (CAMs) are known as membrane receptors that mediate cell-cell and cell-matrix interactions, and are essential for breast cancer metastasis [21]. Defective endocytosis contributes multiple oncogenic alterations, such as altered ubiquitylation, altered cytoskeletal interactions and alterations to Rab family members [27]. Researches demonstrated that the activity of the transient receptor potential (TRP) channels are altered in breast cancer and they may be new marker [28].

4   Discussion

In this work, we try to model the disease state on the pathway level. An optimal strategy algorithm is applied to the subtypes of breast cancer, considering somatic mutation, copy number, methylation and gene expression data. The result shows that the top 30 key pathways of different subtypes display common and differ dysregulation patterns. 44 differentially expressed genes with corresponding genetic and epigenetic dysregulation are retrieved from the subtype-specific pathways. Literature validation shows that these dysregulated pathways and genes have significant function associated with BRCA. Our method provides a new strategy for the identification of dysregulated pathways and disease genes.

Acknowledgements. This work was supported by the National NSFC (Grant No. 61532014 & No. 61432010 & No. 61672407 & No. 61772395), the Fundamental Research Funds for the Central Universities (No. JB150303) and the Fundamental Research Funds for young teacher (2017KY0264).

References

1. Adams, J.R., Schachter, N.F., Liu, J.C., et al.: Elevated PI3K signaling drives multiple breast cancer subtypes. Oncotarget **2**(6), 435–447 (2011)
2. Andre, F., Job, B., Dessen, P., et al.: Molecular characterization of breast cancer with high-resolution oligonucleotide comparative genomic hybridization array. Clin. Cancer Res. Off. J. Am. Assoc. Cancer Res. **15**(2), 441–451 (2009)
3. Balko, J.M., Schwarz, L.J., Bhola, N.E., et al.: Activation of MAPK pathways due to DUSP4 loss promotes cancer stem cell-like phenotypes in basal-like breast cancer. Cancer Res. **73**(20), 6346–6358 (2013)
4. Bergamaschi, A., Kim, Y.H., Wang, P., et al.: Distinct patterns of dna copy number alteration are associated with different clinicopathological features and gene-expression subtypes of breast cancer. Genes. Chromosom. Cancer **45**(11), 1033–1040 (2006)
5. Cancer Genome Atlas Network: Comprehensive molecular portraits of human breast tumours. Nature **490**(7418), 61–70 (2012)
6. Cheng, F., Zhao, J., Zhao, Z.: Advances in computational approaches for prioritizing driver mutations and significantly mutated genes in cancer genomes. Brief. Bioinform. **17**(4), 642–656 (2015)
7. Creixell, P., Reimand, J., Haider, S., et al.: Pathway and network analysis of cancer genomes. Nat. Methods **12**(7), 615 (2015)
8. Dimitrakopoulos, C.M., Beerenwinkel, N.: Computational approaches for the identification of cancer genes and pathways. Wiley Interdiscip. Rev. Syst. Biol. Med. **9**(1) (2017)
9. Dutta, B., Pusztai, L., Qi, Y., et al.: A network-based, integrative study to identify core biological pathways that drive breast cancer clinical subtypes. Br. J. Cancer **106**(6), 1107–1116 (2012)
10. Eckert, L.B., Repasky, G.A., Ulkü, A.S., et al.: Involvement of ras activation in human breast cancer cell signaling, invasion, and anoikis. Cancer Res. **64**(13), 4585–4592 (2004)
11. Fruman, D.A., Chiu, H., Hopkins, B.D., Bagrodia, S., et al.: The PI3K pathway in human disease. Cell **170**(4), 605–635 (2017)
12. Futreal, P.A., Coin, L., Marshall, M., et al.: A census of human cancer genes. Nat. Rev. Cancer **4**(3), 177–183 (2004)
13. Glaser, S.L., Ambinder, R.F., DiGiuseppe, J.A., et al.: Absence of Epstein-Barr virus EBER-1 transcripts in an epidemiologically diverse group of breast cancers. Int. J. Cancer **75**(4), 555–558 (1998)
14. Han, L., Maciejewski, M., Brockel, C., et al.: A probabilistic pathway score (PROPS) for classification with applications to inflammatory bowel disease. Bioinformatics **1**, 9 (2017)
15. Hollander, M.C., Blumenthal, G.M., Dennis, P.A.: PTEN loss in the continuum of common cancers, rare syndromes and mouse models. Nat. Rev. Cancer **11**(4), 289–301 (2011)
16. Hung, J.H., Whitfield, T.W., Yang, T.H., et al.: Identification of functional modules that correlate with phenotypic difference: the influence of network topology. Genome Biol. **11**(2), 1 (2010)
17. Isakoff, S.J., Engelman, J.A., Irie, H.Y., et al.: Breast cancer-associated PIK3CA mutations are oncogenic in mammary epithelial cells. Cancer Res. **65**(23), 10992–11000 (2005)

18. Jhaveri, T.Z., Woo, J., Shang, X., et al.: AMP-activated kinase (AMPK) regulates activity of HER2 and EGFR in breast cancer. Oncotarget **6**(17), 14754–14765 (2015)
19. Khabaz, M.N.: Association of Epstein-Barr virus infection and breast carcinoma. Arch. Med. Sci. **9**(4), 745–751 (2013)
20. Lee, E., Chuang, H.Y., Kim, J.W., et al.: Inferring pathway activity toward precise disease classification. PLoS Comput. Biol. **4**(11), e1000217 (2008)
21. Li, D.M., Feng, Y.M.: Signaling mechanism of cell adhesion molecules in breast cancer metastasis: potential therapeutic targets. Breast Cancer Res. Treat. **128**(1), 7–21 (2011)
22. Liu, W., Bai, X., Liu, Y., et al.: Topologically inferring pathway activity toward precise cancer classification via integrating genomic and metabolomic data: prostate cancer as a case. Sci. Rep. **5**, 13192 (2015)
23. Logue, J.S., Morrison, D.K.: Complexity in the signaling network: insights from the use of targeted inhibitors in cancer therapy. Genes Dev. **26**(7), 641–650 (2012)
24. Magrath, I., Bhatia, K.: Breast cancer: a new Epstein-Barr virus-associated disease? J. Natl. Cancer Inst. **91**(16), 1349–1350 (1999)
25. Marcotte, R., Sayad, A., Brown, K.R., et al.: Functional genomic landscape of human breast cancer drivers, vulnerabilities, and resistance. Cell **164**(1–2), 293–309 (2016)
26. McLaughlin, S.K., Olsen, S.N., et al.: The RasGAP gene, RASAL2, is a tumor and metastasis suppressor. Cancer Cell **24**(3), 365–378 (2013)
27. Mosesson, Y., Mills, G.B., Yarden, Y.: Derailed endocytosis: an emerging feature of cancer. Nat. Rev. Cancer **8**(11), 835–850 (2008)
28. Ouadid-Ahidouch, H., Dhennin-Duthille, I., Gautier, M., et al.: TRP calcium channel and breast cancer: expression, role and correlation with clinical parameters. Bull. Cancer **99**(6), 655–664 (2012)
29. Peng, H., Long, F., Ding, C.: Feature selection based on mutual information criteria of max-dependency, max-relevance, and min-redundancy. IEEE Trans. Pattern Anal. Mach. Intell. **27**(8), 1226–1238 (2005)
30. Troxell, M.L., Levine, J., Beadling, C., et al.: High prevalence of PIK3CA/AKT pathway mutations in papillary neoplasms of the breast. Mod. Pathol. Off. J. U.S. Can. Acad. Pathol. Inc. **23**(1), 27–37 (2010)
31. Ulitsky, I., Shamir, R.: Pathway redundancy and protein essentiality revealed in the Saccharomyces cerevisiae interaction networks. Mol. Syst. Biol. **3**, 104 (2007)
32. Zhang, J., Zhang, S.: Discovery of cancer common and specific driver gene sets. Nucleic Acids Res. **45**(10), e86 (2017)

An Extension of Deep Pathway Analysis: A Pathway Route Analysis Framework Incorporating Multi-dimensional Cancer Genomics Data

Yue Zhao[✉]

Computer Science and Engineering Department,
University of Connecticut, Storrs, CT 06269, USA
yue.2.zhao@uconn.edu

Abstract. Recent breakthroughs in cancer research have happened via the up-and-coming field of pathway analysis. By applying statistical methods to previously known gene and protein regulatory information, pathway analysis provides a meaningful way to interpret genomic data. In this paper we propose systematic methodology framework for studying biological pathways; one that cross-analyzes mutation information, transcriptome and proteomics data. Each pathway route is encoded as a bayesian network which is initialized with a sequence of conditional probabilities specifically designed to encode directionality of regulatory relationships defined by the pathways. Proteomics regulations, such as phosphorylation, is modeled by dynamically generated bayesian network through combining certain type of proteomics data to the regulated target. The entire pipeline is automated in R. The effectiveness of our model is demonstrated through its ability to distinguish real pathways from decoy pathways on TCGA mRNA-seq, mutation, copy number variation and phosphorylation data for both breast cancer and ovarian cancer study.

Keywords: Pathway analysis · Bayesian network · Data integration

1 Introduction

Pathway analysis has been a crucial player in recent cancer research. By combining previously defined gene and protein regulatory information with computational methods, large sets of genomic data can be interpreted. Ample gene/protein regulatory relationships are summarized in the literature that is organized into various forms of gene/protein regulatory networks and pathways. However, pathway analysis research is still a novel area, particularly when it comes to practical problems. Therefore, developing a more comprehensive way to analyze pathways by combining multiple genomic data sets, which are now readily available through various high-throughput sequencing technologies

© Springer International Publishing AG, part of Springer Nature 2018
F. Zhang et al. (Eds.): ISBRA 2018, LNBI 10847, pp. 113–124, 2018.
https://doi.org/10.1007/978-3-319-94968-0_10

(e.g., RNA-Seq, DNA-Seq, ChIP-Seq), is of great significance. [1,2] proposed an approach by modeling the pathway route as an analysis unit. The goal of this paper is to extend that pathway analysis framework to give it the ability to include proteomics and CNV data along with the specific types of regulation mentioned above. Together with existing transcriptome and mutation data, we aim to pinpoint precise pathway routes perturbed. This analysis approach tends to provide deeper insight of biological mechanisms behind cancer development. The rest of the paper is outlined as follows: Sect. 2 reviews existing pathway analysis methods, Sect. 3 describes the model settings and assumptions in detail, Sect. 4 presents a significance study similar to that of [3] using TCGA Breast Cancer (BRCA) and Ovarian Cancer data (OV). Finally we conclude with Sect. 5.

2   Related Work

Great efforts have been made to incorporate pathway information into genomic data analysis. One of the first popular methods of analyzing genome-wide experimental data was using gene set enrichment analysis methods [4]. [5] encoded the pathway network into a penalty function and performed model selection by optimizing the function to pick meaningful genes and subnetworks. [6] proposed SPIA which measures pathway significance by statistical testing against random permutation. [3] presented PARADIGM, a novel method modeling the pathway as a factor graph to do patient specific inference. [7] introduce a computational framework for de novo identification of subnetworks in a large gene interaction network that are mutated in a significant number of patients. [8] ranked the pathways by p-value obtained from encoding pathway logic into a global network. The p-value was calculated based on a hypothesis test where the null hypothesis was that the pathway is picked randomly. [9,10] encoded the pathway as a Bayesian network. After removing cycles in the graph, they trained the model with expression data. [1] dynamically encoded pathway routes as a Bayesian network by incorporating expression and mutation data to perform the pathway analysis.

3   Methods

3.1   Model and Terms

Figure 1(a) illustrates the pipeline of this approach. Figure 1(a.A) shows an example pathway, ErbB, which has been adapted from the KEGG pathway database [11]. A pathway is a graph with biological molecules as nodes and regulation interactions as edges. The edges in the pathway can be categorized as two subtypes: Protein activation and inhibition or gene activation and inhibition which are defined as expression and repression in KEGG pathway database. The framework for KEGG pathway database can be generalized to other pathway databases, (Reactome, for instance) since the category of

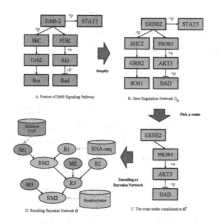

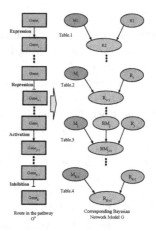

(a) Conversion Pipeline. Part of the ErbB pathway shown in part A. The pathway is simplified by keeping only specific interactions and genes, resulting in a gene regulation network G_B in Part B. A route G^* shown in part C, starting from ERBB2 to BAD, is extracted from G_B and converted to a Bayesian Network G in part D.

(b) Converting the route in the pathway to Bayesian Network. The pathway route G^* on the left is converted to Bayesian Network G on the right. Especially, for $Gene_j$ activating $Gene_{j+1}$, the RM_{j+1} has an extra parent RM_j. This illustrates the special case when $Gene_{j-1}$ is activating or inhibiting $Gene_j$.

Fig. 1. Bayesian network construction

pathway edges still holds. The former type of edges do not affect the expression level of the target gene, however, they can affect the corresponding protein product structure and function. Thus we may define them as (Protein) **Functional Interactions**. Moreover, these edges have the following subtypes, namely, phosphorylation(+p), ubiquitination(+u), glycosylation(+g), methylation(+m), dephosphorylation(−p), deubiquitination(−u), deglycosylation(−g), or demethylation(−m), where tags in the parentheses represent this information in the KEGG database as shown in Fig. 1(a.A). Those tags may be defined as **Evidence Tag** for unambiguousness because they determine the type of data that will be associated with the interactions in this approach.

On the other hand, the Expression and Repression subtype edges, only affect the expression level of its target gene, while the protein function is unaffected. Thus, we categorize these two types of interactions as **Expression Interactions**. Functional Interactions and Expression Interactions will be handled in distinct ways in the model.

The pipeline is initialized with the simplification of the pathway. Figure 1(a.B) presents the result of this process: a gene regulation network G_B. Unlike most existing approaches that merely keep activation and inhibition interactions after the simplification, the Evidence Tag can be furthermore kept in G_B. The next step is to identify all possible "routes" available from the given G_B. As an example, Fig. 1(a.C) shows a route, G^*, which starts from ERBB2 and ends at BAD. The selected route is then converted into a discrete Bayesian Network, G, shown by Fig. 1(a.D). The Bayesian Network encoding the biological logic in pathway route G^* is integrated with its corresponding omics data to measure the perturbance of G^*.

Before proceeding to introduce the measure integrating G and omics data, it is necessary to describe the conversion process from a pathway route G^* to a Bayesian network G in detail. As illustrated in Fig. 1(b), for a gene regulation network G_B, a path G^* is simply a subgraph of G_B, $G^* \subseteq G_B$, $G^* = (V^*, E^*)$ where $V^* = \{g_1, \ldots, g_{k_{G^*}}\}$, g_i represents the ith gene and k_{G^*} is the number of genes contained in path G^*, $E^* = \{e_{ij} | 1 \leq i < k_{G^*}$ and $j = i + 1\}$. For each edge in G^*, $e_{i-1,i}, 1 < i \leq k_{G^*}$, if $i < k_{G^*}$ and $e_{i-1,i}$ is one of the Functional Interactions in G^*, three nodes are created in the corresponding Bayesian Network G: R_i, M_i and RM_i for g_i. On the other hand, if $i < k_{G^*}$ and $e_{i-1,i}$ is one of the Expression Interactions, only two nodes R_i and M_i will be created. The first gene, g_1 will always have two nodes created R_1 and M_1 while $g_{k_{G^*}}$ will only have one node, either $R_{k_{G^*}}$ (if $e_{k_{G^*}-1,k_{G^*}}$ is Expression Interaction) or $RM_{k_{G^*}}$ (if $e_{k_{G^*}-1,k_{G^*}}$ is Functional Interaction). In this way, there will usually be three nodes for target genes of Functional interactions and two nodes for that of Expression interactions.

After creating nodes for each gene in the path G^*, the edges in the Bayesian network G will be added dynamically according to the edges in pathway route G^*. For $g_i \in V^*, 1 < i \leq k_{G^*}$, if $e_{i-1,i} \in E^*$ is one of the Functional Interactions, edges will be created pointing from all the nodes for parent gene g_{i-1} to the RM node for the child gene g_i. Namely, we will add edges from $R_{i-1}, M_{i-1}, (RM_{i-1})$, to RM_i. On the other hand if $e_{i-1,i}$ is an Expression Interaction, edges from the nodes of g_{i-1} $(R_{i-1}, M_{i-1}, (RM_{i-1}))$ to R_i will be created instead. The conditional probability table corresponding to edges in Bayesian Network G is determined by the type of the edge in G^* as shown in details from Tables 1, 2, 3 and 4. The assumption is that, given the edge $e_{i-1,i}$, the expression level (R_i) (or functional status RM_i), of the gene g_i is affected by its parent's expression status R_{i-1}, the DNA functional status M_{i-1} (and the Protein functional status RM_{i-1} if exists). After conversion, the resulting Bayesian Network G is formally defined as follows: $G = (V, E)$, where $V = RR \bigcup MM \bigcup RMS$, $RR = \{R_i, i \in \{1, \ldots, k_{G^*}\}\}$ where R_i is a random variable representing expression level status on gene g_i. $MM = \{M_i, i \in \{1, \ldots, k_{G^*} - 1\}\}$ where M_i is a random variable representing DNA functional status on gene g_i. $RMS = \{RM_i, i \in \{j : e_{j-1,j}$ is one of Functional Interactions in $G^*\}\}$ where RM_i is a random variable representing protein functional status on gene g_i. This setting is motivated from central dogma shown in Fig. 2(a). M_i, R_i and RM_i are now defined in detail. Since the DNA information is not affected by any interactions in the pathway route and M_i doesn't have a parent node in G, the random variable M_i follows a Bernoulli distribution as shown in (1). The Bernoulli random variable M_i has two possible values: $+1$ represents that g_i functions normally on DNA level, and -1 represents function loss, i.e. g_i's DNA original biological function is disrupted. The probability distribution indicates the prior has no specific preference on these two levels:

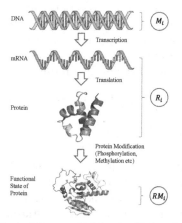 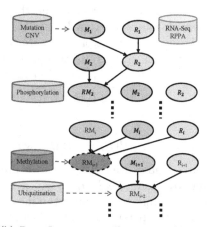

(a) Central Dogma of Molecular Biology for gene g_i where different nodes for g_i corresponds to a certain stage

(b) Data Intergration illustration. All the nodes with same color will have the same data resource.

Fig. 2. Linking data with bayesian network nodes based on central dogma (Color figure online)

$$M_i = \begin{cases} +1 & p = 0.5 \\ -1 & p = 0.5 \end{cases} \qquad (1)$$

Random variable R_i follows a different probability distribution based on the location of gene g_i in path G^*: Suppose g_i is the starting node in G^*, R_i's distribution is shown in (2).

$$R_i = \begin{cases} +1 & p = 1/3 \\ 0 & p = 1/3 \\ -1 & p = 1/3 \end{cases} \qquad (2)$$

where $+1$ represents gene g_i is not down regulated in expression level, -1 represents gene g_i is down regulated in test cases and $R_i = 0$ for missing values. For cancer research, test case is equivalent to the tumor cell. On the other hand, if g_i ($i > 1$) has a parent, gene g_{i-1} in G^*, R_i will follow the conditional probability table in Table 1 (Table 2) if $e_{i-1,i}$ is expression (repression) in G^*. In order to illustrate our model more clearly, we define the following operator &, which is really similar to the AND operator, in (3).

$$A_1 \& A_2 \&, \dots, A_{n-1} \& A_n = \begin{cases} -1 & \exists i \in [1,n] \ s.t. \ A_i = -1 \\ +1 & otherwise \end{cases} \qquad (3)$$

Next we show the biological logic behind the conditional probability table for R_i. Here we focus on the expression table (Table 1); the repression table (Table 2) is built in a similar way. If the parent gene of g_i, g_{i-1}, has no function loss in DNA, overexpression and the functional status of g_{i-1}'s protein is fully activated, namely $M_{i-1}\&R_{i-1}\&RM_{i-1} = +1$, then the target g_i will also be highly likely to overexpress, i.e. $R_i = +1$, given the edge between them in G^* is 'expression'. If there is no Functional Interaction $e_{i-2,i-1}$ targeting at g_{i-1}, there will be just R_{i-1} and M_{i-1} in the conditional table. As a result,

$$P(R_i = +1|M_{i-1}\&R_{i-1}(\&RM_{i-1}) = +1) = 1 - \epsilon_1 - \epsilon_2$$

while

$$P(R_i = -1|M_{i-1}\&R_{i-1}(\&RM_{i-1}) = +1) = \epsilon_1$$

where ϵ_1 and ϵ_2 are respectively the probability of observing $R_i = -1$ and $R_i = 0$ given $M_{i-1}\&R_{i-1}(\&RM_{i-1}) = +1$. $1 - \epsilon_1 - \epsilon_2$ should be close to 1. Here $\epsilon_1 < \epsilon_2$ indicating that we penalize the inconsistency more than the uncertainty. Similarly, if the parent gene of g_i has DNA function loss, caused by mutation for instance, or its expression level is down-regulated in test case, or the protein of g_{i-1} is not activated successfully ($M_{i-1}\&R_{i-1}\&RM_{i-1} = -1$), then the downstream regulation process towards g_i is likely not to be functioning. Therefore, g_i would tend to be down-regulated, namely $R_i = -1$, and hence the corresponding probability would be flipped.

Similar to R_i, Random variable RM_i has three possible values: $\{+1, 0, -1\}$, where $+1$ represents gene g_i has its protein activated by its parent gene g_{i-1} through $e_{i-1,i}$, -1 represents gene g_i is inhibited and otherwise $RM_i = 0$. Recall that RM_i will be attached only when interaction $e_{i-1,i}$ in G^* is Functional Interaction, g_i will always have a parent, gene g_{i-1} in G^*. RM_i will follow the conditional probability table from Tables 3 and 4.

The biological logic behind the conditional probability table for RM_i is built based on the central dogma, as shown in Fig. 2(a). Here we focus on the functional activation table (Table 3); the functional inhibition table (Table 4) is built in a similar way. If the parent gene of g_i, g_{i-1}, has no function loss in DNA, it is overexpressed and g_{i-1}'s protein is successfully activated (if RM_{i-1} exists) ($M_{i-1}\&R_{i-1}\&RM_{i-1} = +1$), then interaction $e_{i-1,i}$ will take effect, thus the target g_i protein will also be highly likely to be regulated successfully, namely, $RM_i = +1$ given the edge between them in G^* is functional activation. As a result,

$$P(RM_i = +1|M_{i-1}\&R_{i-1}\&RM_{i-1} = +1) = 1 - \epsilon_1 - \epsilon_2$$

while

$$P(RM_i = -1|M_{i-1}\&R_{i-1}\&RM_{i-1} = +1) = \epsilon_1$$

where ϵ_1 and ϵ_2 are respectively the probability of observing $RM_i = -1$ and $RM_i = 0$. Similarly, if the parent gene of g_i has DNA function loss, caused by mutation for instance, or its expression level is down-regulated, or the protein of g_{i-1} is not activated successfully ($M_{i-1}\&R_{i-1}\&RM_{i-1} = -1$), then the downstream regulation process towards g_i likely not to be functioning. Therefore, g_i would tend to be not activated, namely $RM_i = -1$, and hence the corresponding probability would be flipped.

Table 1. The regulation process $e_{i-1,i}$ in G^* is expression

$(RM_{i-1}\&)M_{i-1}\&R_{i-1}$	$R_i = +1$	$R_i = 0$	$R_i = -1$
$+1$	$1 - \epsilon_1 - \epsilon_2$	ϵ_2	ϵ_1
-1	ϵ_1	ϵ_2	$1 - \epsilon_1 - \epsilon_2$

$$^*0 < \epsilon_1 < \epsilon_2 << 1 - \epsilon_1 - \epsilon_2$$

Table 2. The regulation process $e_{i-1,i}$ in G^* is repression

$(RM_{i-1}\&)M_{i-1}\&R_{i-1}$	$R_i = +1$	$R_i = 0$	$R_i = -1$
$+1$	ϵ_1	ϵ_2	$1 - \epsilon_1 - \epsilon_2$
-1	$1 - \epsilon_1 - \epsilon_2$	ϵ_2	ϵ_1

Table 3. The regulation process $e_{i-1,i}$ in G^* is functional activation

$(RM_{i-1}\&)M_{i-1}\&R_{i-1}$	$RM_i = +1$	$RM_i = 0$	$RM_i = -1$
$+1$	$1 - \epsilon_1 - \epsilon_2$	ϵ_2	ϵ_1
-1	ϵ_1	ϵ_2	$1 - \epsilon_1 - \epsilon_2$

Table 4. The regulation process $e_{i-1,i}$ in G^* is functional inhibition

$(RM_{i-1}\&)M_{i-1}\&R_{i-1}$	$RM_i = +1$	$RM_i = 0$	$RM_i = -1$
$+1$	ϵ_1	ϵ_2	$1 - \epsilon_1 - \epsilon_2$
-1	$1 - \epsilon_1 - \epsilon_2$	ϵ_2	ϵ_1

3.2 Ranking the Route

A Score Based on Conditional Probability. Given (r, m, rm), a set of data observations of the random variables in Bayesian Network G from a specific patient s, we could rank the path G^* with the probability of observing r, m and rm conditioning on the Bayesian network model G, $P(\boldsymbol{R} = \boldsymbol{r}, \boldsymbol{M} = \boldsymbol{m}, \boldsymbol{RM} = \boldsymbol{rm}|G)$. The larger the probability, the more likely the pathway route is perturbed since the observation is highly consistent with the biological logic from G^* encoded in G. One problem of using this probability as a measure, is that the probability will be higher if fewer data are observed. Thus the score displayed in (4) given in [12], will be used instead, where the conditional probability is normalized by $P(\boldsymbol{R}, \boldsymbol{M}, \boldsymbol{RM}$ are consistent$|G)$.

$$Score_s(G^*, r, m, rm) = \frac{P(R = r, M = m, RM = rm \mid G)}{P(R, M, RM \text{ are consistent} \mid G)}$$

$$P(R = r, M = m, RM = rm \mid G) = \sum_{R=r, M=m, RM=rm} P(R, M, RM)$$

$$= \sum_{R=r, M=m, RM=rm} \prod_{Pa^G(R_i)=\emptyset} P(R_i) \prod_{1 \le i < k_{G^*}} P(M_i) \quad (4)$$

$$\prod_{Pa^G(R_i)\neq\emptyset} P(R_i \mid Pa^G(R_i)) \prod_{Pa^G(RM_i)\neq\emptyset} P(RM_i \mid Pa^G(RM_i))$$

where $Pa^G(X)$ is the set containing parent nodes of node X in Bayesian Network G. $P(R, M, RM$ are consistent$|G)$ is the probability that the random variables with observations are fully consistent with the biological logic encoded in the pathway route given $R_1 = +1, M_1 = +1$.

$$P(R, M, RM \text{ are consistent} \mid G)$$
$$= P(R_1 = +1, M_1 = +1, RM_2 = +1 \mid G) \quad (5)$$

(since g_2 is the last node in the route and the interaction e_{12} is phosphorylation, then R_2 and M_2 are not included in the model).

A high score means that the path G^* is highly likely to be perturbed based on the data we observe. A path G^* could only get a high score if the observations, the changes in tumor cells for each gene, are highly consistent with pathway information contained in the Bayesian Network G. Inconsistency between data and the model would lower the score greatly since the conditional probability will be ϵ_1 instead of $1 - \epsilon_1 - \epsilon_2$ during the calculation of the score. Advantages of this measure are

- the analysis could be done across pathways, i.e. after merging pathways in a biologically meaningful way, this measure could recognize a significantly meaningful route across different pathways.
- By decomposing the pathway graph as routes, the conflict that people encounter when treating a pathway as graph is eliminated. And more complicated directional regulations like phosphorylation can be handled.
- even though some observation values are flipped due to random errors from the genomic data (it is observed to be -1 when it is actually $+1$), the whole path would still have a high score if the other genes have consistent observations.

The data here comes from one patient, s, indicating that the score is specifically tailored to the patient s.

Finally we propose the measure for a whole pathway based on the route score. The pathway score for pathway G_B based on data from a group of subjects S, $pScore_S(G_B)$, is displayed in (6):

$$pScore_S(G_B) = \frac{1}{|G_B|} \sum_{G^* \in G_B} I\left(\frac{1}{|S|} \sum_{s \in S} Score_s(G^*) > \beta\right) \quad (6)$$

The above equation is formulated because of following reasons. The pathway could be partitioned to be several routes. We then simply measure the significance of this pathway, G_B, using the proportion of routes that have an average of all the patients' scores, calculated by (4), that is larger than threshold β.

3.3 Data Integration

The observations for each variable in the Bayesian Network G will come from multiple types of data, as shown in Fig. 2(b). The gene expression variable R_i value can be measured by many types of gene expression data, for instance, microarray, mRNA-seq, reverse phase protein array (RPPA), among others. Here mRNA-seq is chosen. R_i's observation r_i is generated with log2 ratio of mRNA-seq RPKM using (9). The threshold is set to be 0.5 to tolerate the random error resulting from sequence processing. If both protein data and mRNA-seq data are available for the same gene of the same patient and these two data have conflicting observations, then we use the protein data observation to overwrite the one from mRNA-seq data. (mRNAs may be degraded while proteins are present for longer half-lives, thus protein data is much more reliable).

The data observed for random variable M_i, m_i, is the congenital functional status for gene g_i. Observation $m_i = -1$ if it can be observed from mutation or CNV data that g_i's DNA causes function loss for the original biological process.

When it comes to the observation of RM_i, rm_i, the data source becomes more complex. The data source will be determined by the specific subtype of **Functional Interaction** $e_{i-1,i}$, namely by the **Evidence Tag**. The general logic is summarized by the following equation in (7).

$$rm_i = Type_{i-1,i} * Tag_{i-1,i} * RawValue_i \tag{7}$$

where $Type_{i-1,i} = +1$ if $e_{i-1,i}$ is activation (arrow) edge and $Type_{i-1,i} = -1$ if $e_{i-1,i}$ is inhibition; $Tag_{i-1,i} = +1$ if $e_{i-1,i}$ has a **Evidence Tag** sign of $(+)$, i.e. $+p, +m, +u$ or $+g$ while $Tag_{i-1,i} = -1$ if $e_{i-1,i}$ is with a **Evidence Tag** of $(-)$ sign, i.e. $-p, -m, -u$ or $-g$. $RawValue_i = +1$ if the database shows that the gene is Phosphorylated, Methylated, Ubiquitinated or Glycosylated and $RawValue_i = -1$ if the database shows that the gene is Dephosphorylated, Demethylated, Deubiquitinated or Deglycosylated. For the Functional Interactions with no Evidence tag, we assume the edge is always working with $r_{i-1} = +1$, $m_{i-1} = +1$ and use the formula in (8) instead.

$$rm_i = Type_{i-1,i} * min(r_{i-1}, m_{i-1}) \tag{8}$$

The formula indicates that given no function loss observation in M_{i-1} and no low expression observation in R_{i-1}, $e_{i-1,i}$ works and determines rm_i.

4   Preliminary Result

The bioinformatics field frequently uses TCGA Breast invasive carcinoma (BRCA) data to test newly developed analysis models. We choose the same

TCGA cancer data set to validate our model. Another cancer data set, Ovarian serous cystadenocarcinoma (OV), is also analyzed with the same methodology for generality. Four types of data sets are used: mRNA-seq, mutation, Copy Number Variation were downloaded from https://gdac.broadinstitute.org/ for both cancer studies and phosphorylation data were extracted from recent work [13] and [14] respectively.

The mRNA-seq data is processed as follows to obtain r_i, the observation for R_i. The cancer vs. normal paired ratios of RPKM are converted to the expression observation with (9). OV mRNA-seq data has no paired normal samples, thus 0 counts observation is encoded as -1. The value for each item is mapped to a node in pathway by official gene symbol.

$$
r_i = \begin{cases} +1 & log_2\left(\frac{Tumor RPKM_i}{Normal RPKM_i}\right) > 0.5 \\ -1 & log_2\left(\frac{Tumor RPKM_i}{Normal RPKM_i}\right) < -0.5 \\ 0 & otherwise \end{cases} \tag{9}
$$

The mutation information is extracted from mutation accessor study [15]. The mutation with a 'medium' or 'high' impact factor is encoded as function loss mutation. Other mutation observations are encoded as no function loss mutation in the data. The value for each item is mapped to a node in pathway by NCBI-protein ID. Copy Number Variation (CNV) data is imported from GISTIC2 study [16], where the copy number variation is quantified by integers varying from -2 to $+2$ and negative values are considered as copy number loss. CNV information determines the observation for M node, m, along with mutation information as we discussed in Sect. 3.3. The value for each item is mapped to a node in pathway by its official gene symbol.

In the end, phosphorylation data is processed. $RawValue_i$(phosphorylation) $= +1$ if the same patient's phosphosite iTRAQ log2 ratio is positive for g_i and otherwise $RawValue_i(phosphorylation) = -1$. Missing values are encoded as 0. The value for each item is mapped to a node in pathway by NCBI-protein ID. However, one challenge one may encounter is that values for different residues within the same protein may be inconsistent, and KEGG pathway fails to provide sufficient information on the specific residue involved for each phosphorylation. As a result, only the consistent signal is considered in the experiment.

All KEGG Homo sapien pathways are used in this study, and the whole pipeline is automated and implemented in R using R package "KEGGgraph" [17] and "gRain" [18]. Next we do a significance analysis similar to that of PARADIGM [3]. We will produce decoy pathways by permuting the genes in the pathway while keeping the interactions. We generate one decoy pathway for each of 308 KEGG pathways. For each pathway, we extract all possible routes in it. Then for each route, we calculate the score for each pathway by (6). We go on to rank the significant real pathways and their corresponding decoy pathway. A threshold is set to do prediction, i.e. the cases with a score higher than the threshold is predicted to be a real pathway.

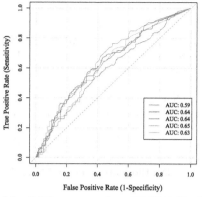 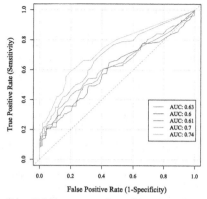

(a) ROC curve for BRCA Significance study. The orange, red, blue, green and purple curve corresponds to $\beta = 0.1, 0.2, 0.3, 0.4, 0.5$ respectively. The threshold is picked from [0,1] with a step of 1/10000.

(b) ROC curve for OV Significance study. The orange, red, blue, green and purple curve corresponds to $\beta = 0.2, 0.4, 0.6, 0.8, 1.0$ seperately. The threshold is picked from [0,1] with a step of 1/10000.

Fig. 3. Preliminary result on BRCA and OV study (Color figure online)

After obtaining False positive rate and True positive rate with various thresholds, the resulting ROC (Receiver operating characteristic) curve can be seen in Fig. 3(a) and (b). The Area Under the Curve (AUC) gets to 0.63(0.74) when taking the threshold $\beta = 0.52(1.00)$ for BRCA (OV). The AUC is reasonable since many real pathways are not differentially regulated for cancer data thus may not be "recognized" from decoy pathways. Similar AUC was reported for SPIA and PARADIGM [3].

5 Discussion

We have further extended the existing deep pathway analysis approach by introducing more detailed information about the pathways. Unlike existing methods, this model has the ability to handle different types of protein functional interactions as well as multiple types of data including expression, mutation, CNV and phosphorylation data. We demonstrated the performance of the model through a significance study with real cancer data. The entire process is automated with R language. Theoretically this framework is capable of handling any omics data as long as the data can be mapped to one of the three steps in the central dogma in Fig. 2(a). When other types of data, namely, ubiquitination, glycosylation, and methylation data are available, the framework will be greatly strengthened. This approach integrates more types of directional regulations and their corresponding data types than existing pathway analysis tools thus it should be useful for cancer community.

References

1. Zhao, Y., Hoang, T.H., Joshi, P., Hong, S.H., Shin, D.G.: Deep pathway analysis incorporating mutation information and gene expression data. In: 2016 IEEE International Conference on Bioinformatics and Biomedicine, BIBM, pp. 260–265. IEEE (2016)
2. Zhao, Y., Hoang, T.H., Joshi, P., Hong, S.H., Giardina, C., Shin, D.G.: A route-based pathway analysis framework integrating mutation information and gene expression data. Methods **124**, 3–12 (2017)
3. Vaske, C.J., Benz, S.C., Sanborn, J.Z., Earl, D., Szeto, C., Zhu, J., Haussler, D., Stuart, J.M.: Inference of patient-specific pathway activities from multidimensional cancer genomics data using PARADIGM. Bioinformatics **26**(12), i237–i245 (2010)
4. Subramanian, A., Tamayo, P., Mootha, V.K., Mukherjee, S., Ebert, B.L., Gillette, M.A., Paulovich, A., Pomeroy, S.L., Golub, T.R., Lander, E.S., et al.: Gene set enrichment analysis: a knowledge-based approach for interpreting genome-wide expression profiles. Proc. Nat. Acad. Sci. **102**(43), 15545–15550 (2005)
5. Li, C., Li, H.: Network-constrained regularization and variable selection for analysis of genomic data. Bioinformatics **24**(9), 1175–1182 (2008)
6. Tarca, A.L., Draghici, S., Khatri, P., Hassan, S.S., Mittal, P., Kim, J.S., Kim, C.J., Kusanovic, J.P., Romero, R.: A novel signaling pathway impact analysis. Bioinformatics **25**(1), 75–82 (2009)
7. Vandin, F., Upfal, E., Raphael, B.J.: Algorithms for detecting significantly mutated pathways in cancer. J. Comput. Biol. **18**(3), 507–522 (2011)
8. Verbeke, L.P., Van den Eynden, J., Fierro, A.C., Demeester, P., Fostier, J., Marchal, K.: Pathway relevance ranking for tumor samples through network-based data integration. PLoS ONE **10**(7), e0133503 (2015)
9. Korucuoglu, M., Isci, S., Ozgur, A., Otu, H.H.: Bayesian pathway analysis of cancer microarray data. PLoS ONE **9**(7), e102803 (2014)
10. Isci, S., Ozturk, C., Jones, J., Otu, H.H.: Pathway analysis of high-throughput biological data within a Bayesian network framework. Bioinformatics **27**(12), 1667–1674 (2011)
11. Kanehisa, M., Goto, S.: KEGG: kyoto encyclopedia of genes and genomes. Nucleic Acids Res. **28**(1), 27–30 (2000)
12. Koller, D., Friedman, N.: Probabilistic Graphical Models: Principles and Techniques. MIT Press, Cambridge (2009)
13. Mertins, P., Mani, D., Ruggles, K.V., Gillette, M.A., Clauser, K.R., Wang, P., Wang, X., Qiao, J.W., Cao, S., Petralia, F., et al.: Proteogenomics connects somatic mutations to signalling in breast cancer. Nature **534**(7605), 55–62 (2016)
14. Zhang, H., Liu, T., Zhang, Z., Payne, S.H., Zhang, B., McDermott, J.E., Zhou, J.Y., Petyuk, V.A., Chen, L., Ray, D., et al.: Integrated proteogenomic characterization of human high-grade serous ovarian cancer. Cell **166**(3), 755–765 (2016)
15. Broad Institute TGDA Center: Mutation assessor (2016)
16. Broad Institute TGDA Center: SNP6 copy number analysis (GISTIC2) (2016)
17. Zhang, J.D., Wiemann, S.: KEGGgraph: a graph approach to KEGG PATHWAY in R and bioconductor. Bioinformatics **25**(11), 1470–1471 (2009)
18. Højsgaard, S.: Graphical independence networks with the gRain package for R. J. Stat. Softw. **46**(10), 1–26 (2012)

Hierarchical Similarity Network Fusion for Discovering Cancer Subtypes

Shuhui Liu and Xuequn Shang$^{(\boxtimes)}$

School of Computer Science and Engineering,
Northwestern Polytechnical University,
Xi'an 710072, People's Republic of China
shang@nwpu.edu.cn

Abstract. Recent breakthroughs in biologic sequencing technologies have cost-effectively yielded diverse types of observations. Integrative analysis of multiple platform cancer data, which is capable of revealing intrinsic characteristics of a biological process, has become an attractive research route on cancer subtypes discovery. Most machine learning based methods need represent each input data in unified space, losing certain important features or resulting in various noises in some data types. Furthermore, many network based data integration methods treat each type data independently, leading to a lot of inconsistent conclusions. Subsequently, similarity network fusion (SNF) was developed to deal with such questions. However, Euclidean distance metrics employed in SNF suffers curse of dimensionality and thus gives rise to poor results.

To this end, we propose a new integrated method, dubbed hierarchical similarity network (HSNF), to learn a fused discriminating patient similarity network. HSNF randomly samples sub-features from different input data to construct multiple input similarity matrixes used as a basic of fusion so that diverse similarity matrixes are generated by multiple random sampling. Then we design a hierarchical fusion framework to make full use of the complementariness of diverse similarity networks from different feature modalities. Finally, based on the final fused similarity matrix, spectral clustering was used to discover cancer subtypes. Experimental results on five public cancer datasets manifest that HSNF can discover significantly different subtypes and can consistently outperform the-state-of-the-art in terms of silhouette, and p-value of survival analysis.

Keywords: Hierarchical similarity network fusion · Multi-platform cancer data
Cancer subtypes discovery · Data integration

1 Introduction

Cancer patients with the same tissue lesions often have the different clinical characteristics and the different response to the same treatment [1]. Many researches show the heterogeneous pathogenesis in the same tissue cancer [1–7]. Specifically, the abnormal gene expression of patients with the same cancer are caused by different mutations of genomic data [3, 7, 8]. Thus, according to precision medicine, we should treat these

© Springer International Publishing AG, part of Springer Nature 2018
F. Zhang et al. (Eds.): ISBRA 2018, LNBI 10847, pp. 125–136, 2018.
https://doi.org/10.1007/978-3-319-94968-0_11

cancer patients in different therapy strategies [9, 10]. Discovering cancer subtypes can provide new insights for new therapeutic treatments and is a useful way to achieve this goal. Recently, breakthroughs in high-through sequencing technologies have generated the diverse types of biological data, such as, genome data (miRNA expression and DNA methylation) and transcriptome data(mRNA expression), used to address biological questions [4, 6, 8, 11]. Particularly, The Cancer Genome Atlas (TCGA) [8, 12, 13] has made large-scale efforts and provided abundant high quality data generated by cross-platforms for discovering complex pathogenesis of human cancers. The multi-platform data integration analysis can explore the pathogenic factors from different data sources to capture the heterogeneity of disease pathogenesis, which has become a critical research direction on the cancer subtypes discovery [12, 14, 15], and is a useful way to achieve precision medicine. Cancer subtypes discovery is a task to cluster patients into meaningful subtypes so that the patients with similar pathogenesis are clustered into the same subtypes while the patients with dissimilar pathogenesis are clustered into the different clusters [1, 4, 5, 12, 14–16]. And subtype information provides useful insights for new therapeutic treatments of cancer patients.

To effectively integrate the biological data from diverse platforms, many data-integrative algorithms have been proposed [6, 8, 12, 16–19]. And data-integration based cancer subtypes discovery methods were presented among these methods. Shen et al. [16] developed a joint latent variable model (iclustering) by incorporating the associations between different data types and the variance-covariance structure within data types. Liang et al. [8] proposed multimodal deep belief network method (DBN) to cluster cancer patients into different subtypes from multi-platform data. Speicher et al. [18] added regularization term on multiple kernel learning (MKL) to avoid overfitting and used several kernels per data type to alleviate choosing the best kernel functions and kernel parameters. Wang et al. [6] proposed a multiplex network-based approach for integrative analysis of heterogeneous omics data. Le Van et al. [19] used rank matrix factorization to identify subtypes based on transformation mutational and gene expression features into ranked data. Xu et al. proposed [17] weighted similarity network fusion method by using the information in the complex miRNA-TF-mRNA regulatory network built by the interactions between features of miRNA, TF and mRNA. Based on the patient similarity networks constructed by each available data type, Wang et al. [12] fused multiple similarity networks into one network that represents the full spectrum of underlying data. However, the above methods either produce inconsistent conclusions because of the independent input data, or suffer from noise due to Euclidean distance. In addition, iclustering needs to preselect genes rather than the full available measurements leading the results sensitive to the step.

Despite powerful of the existing data integration based cancer subtypes discovery algorithms, there are some challenges. The machine learning methods need to represent each data type with a unified form, which will increase information redundancies between data types and dilute the signal-to-noise ratio in some data type [15, 20]. In addition, some network-based methods treated each data type independently, making inconsistent conclusions difficult to integrate. Similarity network fusion [12] (SNF) is a promising method that can ease such problems by integrating patient similarity networks into a fused similarity network. However, SNF calculates patient similarity matrix for each data type using Euclidean distance which leads to poor results in high dimensional biological data [21]. That is, the discrimination of patient similarity networks become blurring. Furthermore, as the sequencing platform differs for different data type available, the different in scales and noise exist in each type of data. Hence, we use random sampling strategy to remove the effect of the noise in each data type, which roots in the ideas of bootstrapping [22] and random forest [23].

In this paper, we propose hierarchical similarity network fusion (HSNF) method, which could benefit from the complementariness of multiple feature modalities [14] by randomly sampling strategy and our proposed hierarchical framework for integrating the numerous similarity networks. Specifically, given the multi-dimensional input data, HSNF can learn a fused similarity network with explicit subtype structure by selecting the appropriate parameters, that is, we should fix the optimal values of sampled features number and random sampling times. Meanwhile, the hierarchical fusion framework is designed by considering the convergence property and parameter settings of SNF method. In particularly, the number of input data types are the number of the similarity matrixes fused by SNF with the default parameters value in [12]. With the selected parameters, the resulting model can effectively learn a fused similarity network. Based on the final fused similarity network, cancer subtypes are identified by spectral clustering. The performance of HSNF on discovering cancer subtypes is examined on five public cancer datasets. Experimental results illustrate that HSNF method discovers significantly different subtypes in survival time and consistently outperforms the state of the art methods in terms of the p-value and silhouette value.

Our contributions are two-fold as follows:

(1) We propose a method, dubbed HSNF, by randomly sampling sub-features from each input data type and then designing hierarchical fusion framework to fuse the multiple similarity matrixes into one discriminative similarity matrix.

(2) HSNF provides a promising route of combing machine learning based technology and network based technology. This route proves to be much effective for discovering cancer subtypes.

The rest of paper is organized as follows. Section 2 describes hierarchical similarity network fusion method and five kinds of cancer datasets. Section 3 presents parameters selection of HSNF method, compares HSNF with the state of the art and examine the stability and expandability of HSNF method. We conclude paper in Sect. 4.

2 Methods

Hierarchical similarity network fusion (HSNF) method is proposed by applying and extending SNF method, which can be regarded as ensemble SNF. In this section, we depict HSNF for discovering cancer subtypes based on multi-dimensional input data. The workflow of HSNF is shown in Fig. 1. First, we introduce how to produce numerous similarity matrixes on r input data types. The produced similarity matrixes in initialization layer are delivered to hierarchical fusion layer. Second, we describe how to fuse the similarity matrixes from initialization layer into one fused similarity matrix. In this step, the hierarchical fusion framework is designed and used to fuse the numerous similarity matrixes into one discriminative similarity matrix and there are

$(n+1)$ layers for randomly selecting r^n times. Finally, cancer subtypes are discovered by spectral clustering based on the final fused similarity matrix.

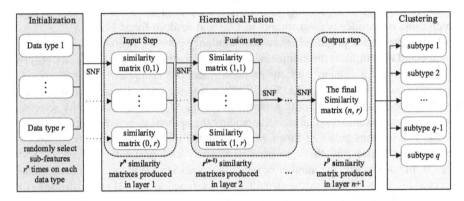

Fig. 1. Workflow of hierarchical similarity network fusion (HSNF).

2.1   Initialization Layer

We use random sampling strategies to ease noise effect from different biological data types. Meanwhile, HSNF can benefit from these numerous and diverse similarity matrixes produced by random sampling strategies. Specifically, given r input data types, HSNF randomly selects sub-features of size $\lceil p_i \times d_i \rceil (i = 1, 2, \cdots, r)$ each time on each data type. d_i denotes the feature number of data type i. p_i is the ratio of the selected sub-features number to all the feature number of data type i. Each time, we respectively select sub-features from all the r type data [12, 14]; calculate similarity matrixes of each selected sub-features by Eq. (1); fuse these r similarity matrixes by SNF.

$$M(i,j) = f(\frac{h(x_i, x_j)^2}{\mu \delta^2}) \tag{1}$$

where $f(x) = exp(-x)$. $h(x_i, x_j)$ calculates distance between patient x_i and x_j. μ is set to 0.5 and δ is learned by the average distance to k-nearest neighborhoods [12, 14]. Function $h(\cdot)$ denotes Euclidean distance. For each sampling, r similarity matrixes from these r data types are fused into one similarity matrix by SNF and we deliver it to hierarchical fusion layer. We repeat r^n times random samplings and r^n similarity matrixes are produced in total. Then we deliver them to hierarchical fusion layer.

2.2   Hierarchical Fusion Layer

For fusing these r^n similarity matrixes from initialization layer, we proposed a hierarchical fusion framework which includes $(n+1)$ layers. In this section, we divide the $(n+1)$ layers into three steps, that is, input step, fusion step and output step respectively. Specifically, input step is corresponding to layer 1 and there are r variables.

When r similarity matrixes are delivered from initialization layer, the input step fuses them into one similarity matrix and delivers it to layer 2. For fusion step, there are $(n-1)$ layer from layer 2 to layer n and each layer in fusion step has the same processing procedure. When r similarity matrixes are delivered from the former layer, the current layer will fuse them into one similarity matrix and deliver it to the next layer. The fusion process will stops if there is no similarity matrix is delivered. That is, r similarity matrixes are exactly inputted to layer n. Then, layer n fuses these r similarity matrix into one final similarity matrix and delivers it to output step. For output step, there is only one layer, that is, layer $(n+1)$ and just one variable which is used to save the final similarity matrix delivered by layer n. The fused similarity is in the layer $(n+1)$ and used to discover cancer subtypes.

2.3   Clustering Layer

For the clustering layer, we cluster patients into different groups on the final fused similarity network by spectral clustering [12, 21] and each patient is only in one group. Patients in the same group are supposed in the same subtype.

Algorithm 1 Hierarchical Similarity Network Fusion (HSNF)

Input: Feature dimensions of r input data: $(d_1, \cdots, d_r)$; the number of random sampling: r^n ;
the number of fused similarity matrixes in hierarchical layer: $(n+1)$; the ratio of selected features to d_i in data type i: $p_i, (i = 1, 2, \cdots, r)$.

Initialization: $\mathbf{IM}_1 = \cdots = \mathbf{IM}_r = []$; $\mathbf{M}_{k,s} = [], 1 \leq k \leq n, 1 \leq s \leq r$; $\mathbf{FM} = []$; $c_1 = \cdots = c_n = 0$; $t = 1$.

1: while iteration $t \leq r^n$ do
2:   randomly sample $\lceil p_i \times d_i \rceil$ features in data type i; update $\mathbf{IM}_i (i = 1, 2, \cdots, r)$ by Eq. (1).
3:   update $\mathbf{M}_{1,\Delta}, (\Delta = \mathrm{mod}(c_1, r) + 1)$ by fusing $[\mathbf{IM}_1, \cdots, \mathbf{IM}_r]$; $c_1 = c_1 + 1$; $\mathbf{IM}_1 = \cdots = \mathbf{IM}_r = []$.
4:   for $j=2$ to n do
5:     if all $\mathbf{M}_{j-1,:}$ are assigned
6:       then update $\mathbf{M}_{j,\Delta}, (\Delta = \mathrm{mod}(c_{j-1}, r) + 1)$ by fusing $[\mathbf{M}_{j-1,1}, \mathbf{M}_{j-1,2}, \cdots, \mathbf{M}_{j-1,r}]$;
7:       and update $c_j = c_j + 1$; $\mathbf{M}_{j-1,1} = \cdots = \mathbf{M}_{j-1,r} = []$.
8:     end if
9:   end for
10:   $t = t + 1$
11: end while
12: update $\mathbf{FM}$ by fusing $[\mathbf{M}_{n,1}, \mathbf{M}_{n,2}, \cdots, \mathbf{M}_{n,r}]$.

Output: Cancer subtypes discovered by spectral clustering based on FSM similarity matrix

---

We give the detailed description of HSNF in Algorithm 1. $\mathbf{IM}_i (i = 1, 2, \cdots, r)$ denotes the variable in the initialization layer and is used to save the similarity matrix produced by data type i. $\mathbf{M}_{k,s}, 1 \leq k \leq n, 1 \leq s \leq r$ is the s-th variable of layer k in hierarchical fusion layer and is used to save the similarity matrix delivered by the former layer. $\mathbf{FM}$ is the variable in the layer $(n+1)$ of output step and is used to save the final fused similarity matrix. $c_i (i = 1, \cdots, n)$ is the counter used to calculate the index of variables so that the similarity matrix can be saved quickly in layer i of hierarchical fusion layer. t counts the iterations. For initialization layer, the similarity matrix with valuable information can be easily produced by fusing these similarity matrixes calculated from different input data types respectively. For hierarchical fusion layer, as the delivered similarity matrixes of the initialization layer are produced from different features modalities by randomly sampling strategy, the fusion process can benefit from the complementariness of these similarity matrixes. The final fused similarity matrix is generated and saved in the output step of hierarchical fusion layer. In this paper, r is simply set based on the number of input data types. For selecting the optimal values of parameters $p_i (i = 1, 2, \cdots, r)$ and n in HSNF, we use grid search. Specifically, we traverse p_i from 0.01 to 1 with the step of 0.01 and traverse n from 1 to 7 with the step of 1. The optimal p_i and n are selected by considering both p-value and silhouette value simultaneously.

2.4 Materials

In this paper, we downloaded five cancer datasets including multi-platform genomic data that are respectively mRNA expression, DNA methylation and miRNA expression collected and processed by Wang et al. [12]. And the corresponding clinical data were also downloaded to evaluate the performance of cancer subtype identification methods. The five cancer types are respectively breast invasive cancer (BIC), glioblastoma multiforme (GBM), kidney renal clear cell carcinoma (KRCCC), colon adenocarcinoma (COAD) and Lung squamous cellcarcinoma (LSCC). Table 1 lists the detailed statistics information of the five cancer datasets on each type input data.

Table 1. Statistics information of five cancer datasets.

Cancer types	The number of mRNA	The number of methylation	The number of miRNA	The number of patients
GBM	12042	1305	534	215
BIC	17814	23094	354	105
KRCCC	17899	24960	329	122
LSCC	12042	23074	352	106
COAD	17814	23088	312	92

3 Results and Discussion

In this section, we firstly discuss the parameters selection of HSNF on GBM dataset. Secondly, based on the selected parameters of HSNF, we compare HSNF with three existing methods in terms of p-value and silhouette. Then, we validate stability of HSNF on all of the five datasets. Finally, we validate expandability of HSNF on GBM dataset.

3.1   Parameters Selection of HSNF

The parameters of HSNF are p_i $(i = 1, 2, \cdots, r)$, r and n respectively and the appropriate parameter values can contribute to the robust HSNF. GBM dataset was used to introduce the parameters selection of HSNF. As GBM dataset includes three data types (mRNA expression, DNA methylation and miRNA expression), r is set to 3. Specifically, each time, three similarity matrixes are fused into one by SNF. Thus, for the parameters in SNF, we follow the experimental parameters setup described in SNF closely (that is, $\mu = 0.5$ and $k = 20$). It is reasonable to select different feature ratio $p_i(i = 1, 2, \ldots r)$ for different data type. However, complexity of selecting optimal parameter values will increase. For simplicity, we don't distinct p_i $(i = 1, 2, \cdots, r)$ instead of setting $p_1 = \cdots = p_r$, which are uniformly denoted by p. Therefore, two parameters (p and n) need to be set, and grid search is used to select the optimal parameter values. We set parameter p ranging from 0.01 to 1 with the step of 0.01 and set parameter n ranging from 1 to 7 with the step of 1. The maximum of n is set to 7 because the sampling time $3^7 = 2187$ is large enough to produce numerous similarity matrixes and lead to the stable results. Moreover, the larger n, the higher complexity of HSNF.

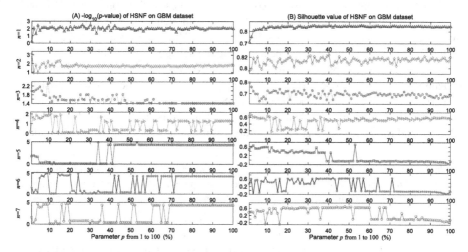

Fig. 2. P-value and silhouette of HSNF respectively vary with parameter p from 0.01 to 1 and with parameter n from 1 to 7 on GBM dataset. (A) $-\log_{10}$(p-value) is reaped by survival analysis of the identified subtypes, (B) silhouette is reaped by analyzing clustering performance.

The parameters p and n are simultaneously selected by considering p-value and silhouette. In particular, p-value is to estimate the significant difference between the subtypes and is calculated by survival analysis of the identified subtypes. The silhouette value [24] is to measure of how similar one patient is to patients in its own subtypes, when compared to patients in other subtypes. Figure 2(A) shows $-\log_{10}$(p-value) varies with the different p and n on GBM dataset. Figure 2(B) shows silhouette varies with the different p and n on GBM dataset. The range spectrums of p-value are smaller with n from 5 to 7 than other n from 1 to 4. And p-value fluctuation is larger of n from 5 to 7 than n from 1 to 4. Generally, the identified subtypes are significant if cox log-rank test p-value of subtypes in survival analysis is less than 0.05. As is well known, the smaller p-value, the more significant the results. And the larger the silhouette, the better the clustering performance. The larger $-\log_{10}$(p-value) is produced with larger p and n from 5 to 7. The smallest p-value is produced by $n = 6$. However, most large $-\log_{10}$(p-value) is corresponding to small silhouette value for n = 6 shown in Fig. 2. Interestingly, the largest $-\log_{10}$(p-value) is produced by $p = 0.09$ and $n = 6$ meanwhile the silhouette value is the largest with the same values of p and n. Thus, we select $p = 0.09$ and $n = 6$ on GBM dataset, and p-value is 9.1e−6 and silhouette value is 0.6620. As shown in Fig. 2, most parameter values in our model deliverer the significant results illustrating the feasibility of HSNF.

3.2   Performance Analysis of HSNF

Comparison with the State of the Art. The number of cancer subtypes of the five public cancer datasets are listed in the parentheses in Table 2 chosen by [12], which is denoted by S. In addition, for HSNF method, the selected values of parameters p and n on five cancer datasets are also listed in the parentheses. We compare HSNF with three existing methods in term of p-value and silhouette on five cancer datasets. Specifically, we focus on comparing HSNF with SNF, ConsensusCluster (CC) and SNF.CC methods which are implemented using R packages CancerSubtypes [25]. The comparison results are presented in Table 2. In general, less p-value and higher silhouette value mean high performance. In terms of silhouette value, HSNF consistently outperforms SNF. Cox log-rank p-value of HSNF is significant on all the five datasets and consistently outperforms than the other three methods. It illustrates that HSNF is effective and robust to discover cancer subtypes based on multiple platform input data.

Table 2. Comparison HSNF with the existing data integrative methods for identifying cancer subtypes on five cancer types in term of silhouette and p-value. The values in the parenthesis respectively denote the number of subtypes of each cancer (S), the values of parameter p and n.

Cancer types	Silhouette values		Cox log-rank p-value			
	HSNF	SNF	HSNF	SNF	CC	SNF.CC
GBM ($S = 3$, $p = 0.09$, $n = 6$)	0.662	0.349	9.10e−6	3.87e−4	7.49e−1	8.72e−4
BIC ($S = 5$, $p = 0.32$, $n = 7$)	0.561	0.227	3.35e−9	1.35e−3	2.81e−5	5.18e−4
KRCCC ($S = 3$, $p = 0.19$, $n = 6$)	0.591	0.390	1.80e−2	3.15e−2	9.30e−1	2.12e−1
LSCC ($S = 4$, $p = 0.04$, $n = 5$)	0.510	0.430	1.20e−2	2.09e−2	1.03e−2	1.64e−2
COAD ($S = 3$, $p = 0.76$, $n = 6$)	0.280	0.086	9.26e−3	3.60e−2	3.72e−2	3.80e−2

Reasonable explanation for such high performance of HSNF method is that HSNF not only benefits from the multiple platform data, but also benefits from the different feature modalities produced by sampling sub-features randomly. Diverse similarity matrixes are generated by sampling sub-features randomly and the proposed hierarchical fusion framework makes full use of complementarity of these similarity matrixes.

The Stability. To validate stability of HSNF, we repeat experiments 50 times with the selected parameters in Table 2 on the five datasets. To observe distribution of the 50 p-value, we plot boxplots. The boxplots of five cancer datasets are shown in Fig. 3. Specifically, each boxplot in Fig. 3 shows values range of the 50 p-value. As shown in Fig. 3, all the boxplots range in the small interval. Especially for Fig. 3(E) COAD, it is interesting that the 50 times repeat experiments output the same p-value. Thus, Fig. 3 shows that the HSNF is stable with the selected parameters. The reason may be that with the numerous sampling times, the probability of the selected features increases and the heterogeneity of the similarity networks becomes small.

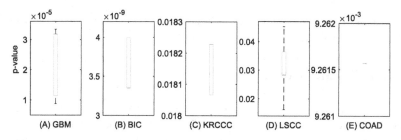

Fig. 3. Boxplot of 50 p-value produced by 50 times repeat experiments of HSNF on (A) GBM dataset. (B) BIC dataset. (C) KRCCC dataset. (D) LSCC dataset. (E) COAD dataset.

The Expandability. For evaluating expandability of HSNF, we add cosine similarity metric to test the expandability of HSNF. Specifically, the extended HSNF distinguishes from original HSNF with two steps. First, we calculate similarity matrixes by both Euclidean distance metric and cosine metric after randomly sampling features and then fuse the similarity matrixes from identical metric. Thus, two fused similarity matrixes are produced from different similarity metrics. Second, we fuse the two produced similarity matrixes into one fused matrix and deliver it to hierarchical fusion layer. The remaining process of the extended HSNF is the same with the original HSNF. We conduct experiment on GBM dataset, which have the largest patient number of the five cancer datasets. The number of subtypes is 3, which is agreement with Wang et al. [12]

We compare the extended HSNF with the original HSNF of p-value by Survival analysis, which shows the significant difference between subtypes in survival profiles. Figure 4 shows survival analysis of the original HSNF and extended HSNF respectively. The survival curves in Fig. 4 manifest that the subtypes identified by the extended HSNF is more significant than the original HSNF, which indicates that HSNF can be well extended. Thus, the fusion process of HSNF can benefit from the complementary information not only generated by the different feature modalities but also produced by the two different similarity metrics.

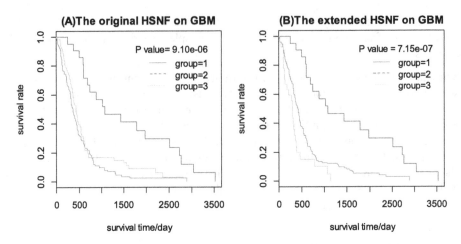

Fig. 4. Kaplan Meier survival curves for analyzing three GBM subtypes identified by (A) The original HSNF method, (B) The extended HSNF method.

4   Discussion and Conclusion

It well known that cancer subtypes discovery can provide useful information for revealing pathogenic mechanism. Although data integration based methods have some advances, some technology issues still need to improve. In this paper, we present a new data integration method, called HSNF, which can be treated as an ensemble SNF. It learns a discriminative similarity network and clusters patients into different subtypes by spectral clustering. We evaluate HSNF on five cancer datasets including GBM, BIC, KRCCC, LUNG, COAD. Each dataset consists of three input data types which are respectively mRNA expression, miRNA expression and DNA methylation. HSNF was validated in the following aspects. First, we discuss the process of selecting the advisable parameters of HSNF based on which desirable p-value and silhouette can be reaped by traversing the preset values. The experimental results show the parameters selection have large effect on HSNF performance. Second, we compare HSNF with the state of the art based on the selected parameters. HSNF consistently outperforms other methods in terms of p-value and silhouette value. In addition, we evaluate the stability of HSNF via observing p-value distribution with 50 times repeated experiments. Finally, HSNF is expanded by adding cosine metric into the original HSNF. Abundant experimental results illustrate HSNF is feasible and reasonable for discovering cancer subtypes.

In fact, we should tune $(r + 2)$ parameters included in HSNF method, that is, r, n and $p_1, \cdots, p_r$. However, the optimal parameters are difficult to choose. In this paper, three data types are used and parameter r is set to 3 by empirical analysis. In addition, for simplicity, parameters p_1, p_2 and p_3 are set to equal value. As the appropriate parameters are critical for HSNF performance, we will explore how to select the optimal parameters in the future work. Furthermore, more similarity metrics can be discussed for extending HSNF and low-rank representation (LRR) [26] is considered to construct patient similarity networks in our future work.

Acknowledgments. The authors would like to thank the anonymous reviewers. This work has been supported by the National Natural Science Foundation of China (Grant No. 61332014 and 61772426).

References

1. Golub, T.R., Slonim, D.K., Tamayo, P., Huard, C., Gaasenbeek, M., Mesirov, J.P., Coller, H., Loh, M.L., Downing, J.R., Caligiuri, M.A.: Molecular classification of cancer: class discovery and class prediction by gene expression monitoring. Science **286**(5439), 531–537 (1999)
2. Maulik, U., Mukhopadhyay, A., Chakraborty, D.: Gene-expression-based cancer subtypes prediction through feature selection and transductive SVM. IEEE Trans. Biomed. Eng. **60**(4), 1111–1117 (2013)
3. Kim, D., Lee, G., Sohn, K.-A., Bang, L., Kim, S.Y.: Identifying subtype-specific associations between gene expression and DNA methylation profiles in breast cancer. BMC Med. Genom. **10**(1), 28 (2017)
4. Cancer Genome Atlas Research Network: Integrated genomic analyses of ovarian carcinoma. Nature **474**(7353), 609–615 (2011)
5. Verhaak, R.G., Hoadley, K.A., Purdom, E., Wang, V., Qi, Y., Wilkerson, M.D., Miller, C. R., Ding, L., Golub, T., Mesirov, J.P.: Integrated genomic analysis identifies clinically relevant subtypes of glioblastoma characterized by abnormalities in PDGFRA, IDH1, EGFR, and NF1. Cancer Cell **17**(1), 98–110 (2010)
6. Wang, H., Zheng, H., Wang, J., Wang, C., Wu, F.-X.: Integrating omics data with a multiplex network-based approach for the identification of cancer subtypes. IEEE Trans. Nanobiosci. **15**(4), 335–342 (2016)
7. Parker, J.S., Mullins, M., Cheang, M.C., Leung, S., Voduc, D., Vickery, T., Davies, S., Fauron, C., He, X., Hu, Z.: Supervised risk predictor of breast cancer based on intrinsic subtypes. J. Clin. Oncol. **27**(8), 1160–1167 (2009)
8. Liang, M., Li, Z., Chen, T., Zeng, J.: Integrative data analysis of multi-platform cancer data with a multimodal deep learning approach. IEEE/ACM Trans. Comput. Biol. Bioinf. (TCBB) **12**(4), 928–937 (2015)
9. Dai, X., Li, T., Bai, Z., Yang, Y., Liu, X., Zhan, J., Shi, B.: Breast cancer intrinsic subtype classification, clinical use and future trends. Am. J. Cancer Res. **5**(10), 2929 (2015)
10. List, M., Hauschild, A.-C., Tan, Q., Kruse, T.A., Baumbach, J., Batra, R.: Classification of breast cancer subtypes by combining gene expression and DNA methylation data. J. Integr. Bioinf. (JIB) **11**(2), 1–14 (2014)
11. Kim, S., Oesterreich, S., Kim, S., Park, Y., Tseng, G.C.: Integrative clustering of multi-level omics data for disease subtype discovery using sequential double regularization. Biostatistics **18**(1), 165–179 (2017)
12. Wang, B., Mezlini, A.M., Demir, F., Fiume, M., Tu, Z., Brudno, M., Haibe-Kains, B., Goldenberg, A.: Similarity network fusion for aggregating data types on a genomic scale. Nat. Methods **11**(3), 333–337 (2014)
13. Cancer Genome Atlas Network: Comprehensive molecular portraits of human breast tumours. Nature **490**(7418), 61–70 (2012)
14. Wang, B., Jiang, J., Wang, W., Zhou, Z.-H., Tu, Z.: Unsupervised metric fusion by cross diffusion. In: 2012 IEEE Conference on Computer Vision and Pattern Recognition (CVPR), pp. 2997–3004. IEEE (2012)

15. Tao, H., Hou, C., Zhu, J., Yi, D.: Multi-view clustering with adaptively learned graph. In: Asian Conference on Machine Learning, pp. 113–128 (2017)
16. Shen, R., Olshen, A.B., Ladanyi, M.: Integrative clustering of multiple genomic data types using a joint latent variable model with application to breast and lung cancer subtype analysis. Bioinformatics **25**(22), 2906–2912 (2009)
17. Xu, T., Le, T.D., Liu, L., Wang, R., Sun, B., Li, J.: Identifying cancer subtypes from miRNA-TF-mRNA regulatory networks and expression data. PLoS One **11**(4), e0152792 (2016)
18. Speicher, N.K., Pfeifer, N.: Integrating different data types by regularized unsupervised multiple kernel learning with application to cancer subtype discovery. Bioinformatics **31** (12), i268–i275 (2015)
19. Le Van, T., van Leeuwen, M., Carolina Fierro, A., De Maeyer, D., Van den Eynden, J., Verbeke, L., De Raedt, L., Marchal, K., Nijssen, S.: Simultaneous discovery of cancer subtypes and subtype features by molecular data integration. Bioinformatics **32**(17), i445–i454 (2016)
20. Zhang, Z., Zhai, Z., Li, L.: Uniform projection for multi-view learning. IEEE Trans. Pattern anal. Mach. Intell. (2016)
21. Law, M.T., Urtasun, R., Zemel, R.S.: Deep spectral clustering learning. In: International Conference on Machine Learning, pp. 1985–1994 (2017)
22. Friedman, J., Hastie, T., Tibshirani, R.: The Elements of Statistical Learning. Data Mining, Inference, and Prediction. Springer, New York (2001)
23. Liaw, A., Wiener, M.: Classification and regression by randomForest. R News **2**(3), 18–22 (2002)
24. Kaufman, L., Rousseeuw, P.J.: Finding groups in Data: An Introduction to Cluster Analysis. Wiley, Hoboken (2009)
25. Xu, T., Le, T.D., Liu, L., Su, N., Wang, R., Sun, B., Colaprico, A., Bontempi, G., Li, J.: CancerSubtypes: an R/bioconductor package for molecular cancer subtype identification, validation, and visualization. Bioinformatics **33**, 3131–3133 (2017)
26. Zhang, Y., Xiang, M., Yang, B.: Low-rank preserving embedding. Pattern Recogn. **70**, 112–125 (2017)

Structure and Interaction

Structure and Interaction

Sprites2: Detection of Deletions Based on an Accurate Alignment Strategy

Zhen Zhang[1], Jianxin Wang[1(✉)], Junwei Luo[1,2], Juan Shang[1],
Min Li[1], Fang-Xiang Wu[3], and Yi Pan[4]

[1] School of Information Science and Engineering, Central South University,
Changsha 410083, China
jxwang@mail.csu.edu.cn
[2] College of Computer Science and Technology, Henan Polytechnic University,
Jiaozuo 454000, China
[3] Division of Biomedical Engineering and Department of Mechanical
Engineering, University of Saskatchewan, Saskatoon SKS7N5A9, Canada
[4] Department of Computer Science, Georgia State University,
Atlanta GA30302, USA

Abstract. Since humans are diploid organisms, homozygous and heterozygous deletions are ubiquitous in the human genome. How to distinguish homozygous and heterozygous deletions is an important issue for current structural variation detection tools. Additionally, due to the problems of sequencing errors, micro-homologies and micro-insertions, breakpoint locations identified with common alignment tools which use greedy strategy may not be the true deletion locations, and usually lead to false structural variation detections. In this paper, we propose a deletion detection method called Sprites2. Comparing with Sprites, Sprites2 adds the following novel function modules: (1) Sprites2 takes advantage of the variance of insert size distribution to determine the type of deletions which can enhance the accuracy of deletion calls; (2) Sprites2 uses a novel alignment strategy based on AGE (one algorithm aligning 5' and 3' ends between two sequences simultaneously) to locate breakpoints which can solve the problems introduced by sequencing errors, micro-homologies and micro-insertions. For testing the performance of Sprites2, simulated and real datasets are used in our experiments, and some popular structural variation detection tools are compared with Sprites2. The experimental results show that Sprites2 can improve deletion detection performance. Sprites2 is publicly available at https://github.com/zhangzhen/sprites2.

Keywords: Structural variation · Deletion detection · Alignment strategy
Sequence analysis

1 Introduction

Structural variations (SVs), refer to the DNA sequence of more than 50 base pairs in the donor genome, which are inserted, deleted, inverted or duplicated, comparing with the reference genome. Deletions are very common in the human genome which are the removal of segments from the donor genome. Deletions can bring about variable

© Springer International Publishing AG, part of Springer Nature 2018
F. Zhang et al. (Eds.): ISBRA 2018, LNBI 10847, pp. 139–150, 2018.
https://doi.org/10.1007/978-3-319-94968-0_12

phenotypes and diseases [1]. For example, Smith-Magenis syndrome (SMS) is linked to deletions at p11.2 region on chromosome 17, leading to mental deficits, facial expression abnormalities, sleep disorders and behavioral problems. Deletions at p11.23 region on chromosome 7 cause Williams-Beuren syndrome (WBS) which is manifested by congenital heart disease, mild mental retardation, hypertension disease [2]. If deletion callers can accurately locate disease-associated deletions and precisely screen candidate genes related to diseases, they have a great value in helping doctors to correctly diagnose and treat human diseases [3].

Currently, several different deletion detection methods have been presented. Pindel [4] deals with paired-end reads that have only one end mapped. By taking the pattern growth approach, Pindel splits the unmapped reads, i.e. one form of split reads, into two segments and maps the two segments separately to determine exact positions of breakpoints with the help of its mapping tool. Delly [5] uses paired-end reads clustering to identify breakpoint-containing reference regions. Within the regions, it searches for putative split reads, i.e. unmapped ends of the paired-end reads. A fast k-mer-based filtering technique is used to choose candidate split reads. From these candidate split reads, a split read consensus is then built. Breakpoints are determined by aligning the consensus sequence to the breakpoint-containing reference region. SVseq2 [6] begins with extracting soft-clipped reads from the input BAM file. For each soft-clipped read, it takes advantage of paired-end reads that span the soft-clipped point of the read to identify the reference region that covers the other breakpoint. The other breakpoint is pinpointed by re-aligning the soft-clipped segment of the read to the breakpoint-containing reference region, i.e. the target region. Sprites [7] takes an approach similar to that of SVseq2 for deletion detection. It starts by dealing with soft-clipped reads, and then uses spanning paired-end reads to identify target sequences, i.e. the reference fragment that covers the other breakpoint of a deletion. Rather than the re-aligning soft-clipped segment strategy, re-aligning the whole read is conducted in determining the location of the breakpoint. This enables Sprites to detect deletions with a micro-homology and deletions with a micro-insertion more accurately than SVseq2.

At present, existing methods have partly solved the problem of locating breakpoints, but for the purposing of accurately locating breakpoints, there are still the following challenges. (1) The sequencing errors cause alignment tools improperly locating breakpoints. It tends to make alignments stop at a sequencing error position. Specifically, for the forward 5'-end soft-clipping reads, the soft-clipped positions given by the alignment tool are shifted to the right relative to true breakpoint locations by a number of base pairs, and for the reverse 5'-end soft-clipping reads, the positions are shifted to the left relative to true locations. (2) Deletions with a micro-homology and deletions with a micro-insertion make it more difficult to exactly determine where soft-clipped points are located, especially for alignment tools based on remapping soft-clipped parts. (3) The features of target sequences have a significant impact on re-alignments. If a target sequence contains tandem duplications, it is very difficult for any kind of alignment algorithm, without the help of additional biological information, to determine breakpoint locations. (4) The greedy strategy used by alignment tools makes locations of soft-clipped points beyond true locations. If the sequence near a true breakpoint is relatively similar to the corresponding soft-clipped read, alignment tools prefer to give a high mapping quality score but a wrong match.

Meanwhile, humans are diploid organisms because all body cells have 23 chromosome pairs. Therefore, there are homozygous and heterozygous structural variations in the human genome.

Sprites uses BWA [8] to re-align soft-clipped reads to target sequences, in order to find the longest prefix or suffix of the read that has a match in the target sequence and further determine the other breakpoint. Although Sprites has partly solved the problem of deletions with a microhomology and deletions with a micro-insertion, it can't eliminate the problem of inaccurate soft-clipped points. AGE [9] is an algorithm for alignment with Gap Excision. It is accurate in general when using AGE to locate deletion breakpoints for two sequences: a target sequence without a deletion and a soft-clipped read containing a deletion. The performance of AGE is affected by the features of target sequences. The target sequence as input can be divided into two cases. In the first case, it is extracted directly from the cluster of overlapping discordant paired-end reads. Regardless of the sizes of deletions, it always contains the whole sequences of deletions. When the deletion size is large, the corresponding target sequence is much longer than reads, making AGE low computational efficiency. In the second case, it is extracted by the cluster of paired-end reads spanning a soft-clipped point. When the deletion size is large, the target sequence does not totally contain the deleted sequence, which makes AGE at a disadvantage, causing that the algorithm can't accurately find the positions of two breakpoints.

In this study, we present a new method, called Sprites2 for detecting deletions, Sprites2 attempts to build two clusters of paired-end reads spanning a soft-clipped point according to their insert sizes, and then uses ANOVA to test whether the clustering is reliable. If the clustering is reliable, it means that the corresponding deletion is heterozygous, else the deletion is homozygous. Then, a new alignment strategy is employed for finding and revising breakpoints. Sprites2 first complements target sequences by adding base pairs to heads or tails, while avoiding introducing problems of micro-homologies. Subsequently it uses AGE for performing precise alignments on soft-clipped reads and complemented target sequences regardless of the sizes of deletions, contributing to reducing run-time and memory consumption effectively. Then, it uses the 1 bp fault-tolerant amplification method to eliminate fake micro-insertions introduced by sequencing errors based on alignment results. Finally, two identical sub-sequences are searched in regions near breakpoints on target sequences based on alignment results to determine whether there is a micro-homology. If there is a micro-homology, the micro-homology sequence is extracted from target sequences.

2 Method

The pipeline of Sprites2 for detecting deletion breakpoints which takes advantage of the alignment tool AGE is outlined as follows.

2.1 Preprocessing

A soft-clipped read has the following characteristics: (1) it has the soft-clipping signature; (2) it is one end of a paired-end read; (3) its partner is uniquely mapped. We search for such reads in the input BAM file. Here, we deal with two types of soft-clipped reads: forward 5'-end soft-clipped reads and reverse 5'-end soft-clipped reads. Forward 5'-end soft-clipped reads mean that the 5'-end of the reads cannot be mapped, and the 3'-end is mapped to the forward strand of the reference. Similarly, reverse 5'-end soft-clipped reads imply that the 5'-end of the reads cannot be mapped, and the 3'-end is mapped to the reverse strand of the reference. Unless otherwise stated, soft-clipped reads refer to 5'-end soft-clipped reads.

The CIGAR string of a soft-clipped read contains an 'S' at the beginning or end. The number before the 'S' represents the length of the read's soft-clipped segment. Sprites2 scans the BAM file only once to choose out reads that have at least 12 (set by default) base pairs soft-clipped, based on their CIGAR string. For each soft-clipped read, the coordinate of the soft-clipped point, its mapping orientation (forward or reverse) and the length of the soft-clipped segment, i.e. the number of soft-clipped base pairs, are saved for the subsequent steps.

The soft-clipped reads whose soft-clipped points have the same coordinate on the reference genome are grouped together. If both forward and reverse soft-clipped reads are placed in the same group, the two deletions supported by forward and reverse soft-clipped reads cannot co-exist. The conflict in the group prevents the group's reads from being further processed and therefore the group needs to be deleted. The previous step outputs groups of forward or reverse soft-clipped reads. These groups are arranged via the soft-clipping position. Given a soft-clipping position of forward soft-clipped reads, paired-end reads that meets the following conditions are selected: (1) one end s_1 is forwardly mapped in the upstream of the soft-clipped position; (2) the other end s_2 is reversely mapped in the downstream of the soft-clipped position, and is required to be mapped within the region $[loc, loc + \mu + 3 * \sigma - len(s_2)]$, where loc stands for the coordinate of the soft-clipped point, and $len(s_2)$ stands for the length of s_2, and μ and σ stand for the mean and standard deviation of insert size, respectively.

Similarly, given a soft-clipping position of reverse 5'-end soft-clipped reads, paired-end reads that meet the following conditions are selected: (1) one end s_2 is reversely mapped in the downstream of the soft-clipped point; (2) the other end s_1 is forwardly mapped in the upstream of the soft-clipped point, and is required to be mapped within the region $[loc - \mu - 3 * \sigma + len(s_1), loc]$. The corresponding target sequence is a fragment of length of $\mu + 3 * \sigma$ ending at $len(s_2) + pos(s_2)$.

2.2 Genotyping Deletions to Be Called

At the end of the previous step, a collection of spanning paired-end reads is obtained for each soft-clipping position. The distance between two mapped ends of each paired-end read is extracted and used to construct a list of insert sizes. Then, the algorithm described below is adopted to split the list into two smaller lists.

First, the list is sorted in ascending order. Second, differences between consecutive elements in the list are calculated. Third, the max difference is found. Last, the elements left and right to the max difference are grouped as the first and second smaller list, respectively.

If the soft-clipping position implies a homozygous deletion, then the insert sizes in the original list are drawn from a single distribution and the mean of the distribution $\mu(i)$ is greater than the library mean μ_L, where i denotes the i-th soft-clipping position. If the soft-clipping position implies a heterozygous deletion, insert sizes in its original list are drawn from two distributions. The mean of one distribution $\mu_1(i)$ is roughly equal to the library mean μ_L, and the mean of the other distribution $\mu_2(i)$ is larger than the library mean μ_L.

Based on the fact described above, the analysis of variance is used to test whether this partition is reasonable. Specifically speaking, it is determined whether the averages $\bar{x}_1, \bar{x}_2$ of the two smaller lists are equal. The test hypothesis is established.

H_0: Two averages $\bar{x}_1, \bar{x}_2$ are equal;

H_1: Two averages $\bar{x}_1, \bar{x}_2$ are not equal.

In addition, we define a significance level α of 0.05 for the hypothesis test. If the p-value is less than α, we reject the null hypothesis H_0. This indicates that the partition is reasonable. If the p-value is larger than α, we fail to reject the null hypothesis H_0. This indicates that the partition is not reasonable.

The reasonable partition is indicative of a heterozygous deletion. The smaller list which has the larger average is selected. The non-reasonable partition implies a homozygous deletion. In such case, the original list is selected. According to the resulting list, the corresponding paired-end reads are extracted for using in the next step.

2.3 Target Sequences Completion

For each soft-clipping position of forward 5'-end soft-clipped reads, spanning paired-end reads along with the coordinates of its far end are saved. For the above spanning pair $(s_1,\ s_2)$ in extracting initial target sequence, a fragment of length $\mu + 3 * \sigma$ starting at $pos(s_1)$ is taken on the reference. In fact, a soft-clipping position always corresponds to more than one target sequence which is needed to be sorted by ending location and be merged until nothing is left to merge. Similarly, given a soft-clipping position of reverse 5'-end soft-clipped reads, the corresponding target sequence is a fragment of length of $\mu + 3 * \sigma$ ending at $len(s_2) + pos(s_2)$.

According to the number of breakpoints contained in target sequences, we divide target sequences into two categories. In the first category, target sequences totally contain deletions, so the number of breakpoints contained is 2. In the second category, target sequences partly contain deletions and one breakpoint because deletions are too large. Target sequences which don't contain any breakpoints will not be considered, since it is assumed that target sequences contain at least one breakpoint.

For target sequences in the first category which contain two breakpoints, the left part of the first breakpoint can be matched with the left segment corresponding to the soft-clipped read, and at the same time. At the right part of the second breakpoint can be matched with the right segment of the soft-clipped read. AGE can be used directly on the first category of target sequences, and it is not necessary to process completion. For target sequences in the second category which contain only one breakpoint of deletions, AGE always can not give accurate alignment results. Hence target sequence completion must be done.

For a forward soft-clipped read, the beginning location and the ending location of its mapped part are c_0 and c_1 respectively, and thus the soft-clipped position is c_0. The length of the unmapped part is l. The ending location of the target sequence is t_1. The sequence that needs to be added to the tail of the target sequence is calculated by the Eq. (1).

$$ExtSeq(S, t_1, c_x, c_1) = \begin{cases} S[c_x, c_1], t_1 < c_1 \\ S[t_1 + 1, c_1], c_x \le t_1 < c_1 \\ \phi, t_1 \ge c_1 \end{cases} \quad (1)$$

Where S is the reference sequence, c_x is the ideal beginning location of the soft-clipped read (assuming the whole soft-clipped read is mapped) which is equal to the difference subtracted l from c_0, ϕ is empty string. As shown in Eq. (1), for a forward soft-clipped read, there are three cases to add sequence to the tail of target sequence. The first case is the end location t_1 of target sequence at the left of c_x, indicating that the deletion is too large. The number of base pairs added at this time is $c_1 - c_x + 1$. When the deletion size is larger than a certain value, the number of added base pairs is no longer associated with the deletion size, but a fixed value which is approximately equal to the read length.

The second case is that t_1 falls between c_x and c_1, the sequence added is from $t_1 + 1$ to c_1 to avoid introducing fake tandem duplications which increases the error probability of AGE algorithm, thus impeding locating breakpoints. The third case is that t_1 is on the right of c_1, that is, the deletion size is small and the target sequence completely contains the deletion without needing to add any base pairs at this time.

It is important to emphasize that a reference segment is added to the end of the target sequence, rather than adding the mapped segment of soft-clipped read. Some base pairs after the soft-clipped point in the initial target sequence are likely to coincide with several base pairs at the left of the mapped part of the soft-clipped read. If the mapped segment is added to the tail of the target sequence, a fake micro-homology is introduced, which increases the complexity of locating breakpoints. The completed target sequence's effects are similar to the reference totally contains deletions simultaneously reducing run time and memory overhead.

Given a reverse soft-clipped read, c_0 and c_1 are the beginning and ending locations of the mapped segment, c_1 is the soft-clipped position, l is the length of the soft-clipped segment, and t_0 is the beginning location of the target sequence. The sequence that needs to be added to the head of the target sequence is calculated by the Eq. (2).

$$ExtSeq(S, t_1, c_x, c_1) = \begin{cases} S[c_x, c_1], t_1 < c_x \\ S[t_1 + 1, c_1], c_x \le t_1 < c_1 \\ \phi, t_1 \ge c_1 \end{cases} \quad (2)$$

Where S is the reference sequence, c_x is the ideal ending location of soft-clipped read, which is equal to a sum of c_1 plus l, ϕ is the empty string. As shown in Eq. (2), for a reverse soft-clipped read, the sequence that needs to be added to the head of the target sequence is also divided into three cases. (1) When t_0 is on the right of c_x, the segment added is the reference sequence from c_0 to c_x, since there are indels, the length is approximately equal to the read length. (2) When t_0 is between c_0 and c_x, the added segment is the reference sequence between c_0 and t_0, (3) When t_0 is on the left of c_0, there is no need to add any base pairs at this time because the target sequence contains two breakpoints.

2.4   Applying AGE for Realignment

We have preliminarily determined one breakpoint of deletions according to soft-clipped point information, employing AGE which inputs complemented target sequences and corresponding soft-clipped reads to amend the breakpoint we've got and locate the other breakpoint. AGE aligns the left and right of two given sequences at the same time, and outputs breakpoint coordinates of aligned regions for two best alignments which are two intervals actually. The endpoint coordinates of intervals are relative, so it is necessary to convert them to absolute coordinates when calculating breakpoint coordinates.

For a forward soft-clipped read, the coordinate of left breakpoint is equal to the sum that end position of the first mapped segment in target sequence (relative coordinate) plus start position of the target sequence obtained by AGE (absolute coordinate). The coordinate of the right breakpoint is equal to the sum of that soft-clipped position and the difference between the length of mapped part of soft-clipped read and the length of the second mapped segment of soft-clipped read obtained by AGE. For a reverse soft-clipped read, the coordinate of left breakpoint is equal to the difference subtracted the difference between the length of the mapped part of soft-clipped read and the length of first mapped segment of soft-clipped read obtained by AGE from soft-clipped position. The coordinate of right breakpoint is equal to the sum of that start position of the second mapped segment in target sequence obtained by AGE (relative coordinate) and start position of the target sequence (absolute coordinate).

2.5   Correct Coordinates of Breakpoints

Given a target sequence v and a soft-clipped read w, the alignment results of AGE are analyzed. It is divided into two cases if the middle part of w has unmatched base pairs. The first case shows unmatched base pairs produced by true micro-insertions, the second case indicates unmatched base pairs caused by sequencing errors, which mistakenly considered to be micro-insertions by AGE. It is necessary to identify and eliminate the second case. Firstly, we extract the unmatched base pairs of w, and the sequence is denoted by w' with length represented by $l(w')$, and then extract a sequence with a length of $l(w')$ from v next to the end of the first mapped segment, which is represented by v'_1. Let n_1 be the number of mismatched base pairs between v'_1 and w'. Subsequently a sequence with a length of $l(w')$ before the start of the second mapped segment of v is extracted, which is represented by v'_2. Let n_2 be the number of mismatched base pairs between v'_2 and w'. (1) If both n_1 and n_2 are greater than 1, it is shown that the unmatched sequence in the middle of w is a micro-insertion, which belongs to the first case where breakpoint coordinates do not need to be corrected. (2) If

n_1 is not greater than 1 and n_2 is greater than 1, indicating that some sequencing errors exist and affect the determination of the coordinate of left breakpoint. (3) If n_1 is greater than 1 and n_2 is not greater than 1, indicating that sequencing error exists and affects the determination of the coordinate of right breakpoint. (4) If neither n_1 nor n_2 is greater than 1, manifesting there are sequencing errors, but it is not possible to determine which breakpoint, left or right, is affected by sequencing errors.

The last three conditions belong to the second case, we develops a slightly different strategy for amending breakpoint coordinates based on two different soft-clipped reads. For a forward soft-clipped read, under the condition of (2), the coordinate of the right breakpoint should be shifted to the left by $l(w')$, and under the condition of (3), the coordinate of the left breakpoint should be shifted by $l(w')$ to the right. Under the condition of (4), the coordinate of the right breakpoint should be shifted to the left by $l(w')$. Likewise for a reverse soft-clipped read, under the condition of (2), the coordinate of the left breakpoint should be shifted to the right by $l(w')$, and under the condition of (3), the coordinate of the right breakpoint should be shifted by $l(w')$ to the left. Under the condition of (4), the coordinate of the left breakpoint should be shifted to the right by $l(w')$.

2.6 Finding Deletion Calls

The target sequence covers two breakpoints and contains the sequences on both sides of breakpoints, and there is no need to re-read the reference sequence from reference genome FASTA file. Given a target sequence t_s with a length of l, and the coordinates of the left breakpoint and the right breakpoint on the target sequence t_s are b_l and b_r (relative coordinates, starting from 0), respectively. l_l indicates the length of the sequence on the left of b_l including b_l, whose value is equal to b_l. l_m represents the length of sequence between b_l and b_r, and its value is equal to $b_r - b_l - 1$. l_r indicates the length of the sequence on the right of b_r including b_r, whose value is equal to $l - b_r$.

Firstly, we get the longest common suffix $LCS(s_l, s_m)$ of the substring s_l with a length of $min(l_m, l_l)$ and the substring s_m with the same length as s_l. The right endpoints of s_m and s_l are b_r and the previous base pair to b_l respectively. Then, we calculate the longest common prefix $LCS(s'_m, s_r)$ of the substring s'_m with a length of $min(l_m, l_r)$ and the substring s_r with the same length as s'_m. The left endpoints of s'_m and s_r are b_r and the next base to b_l, respectively. Finally, we can obtain micro-homology sequence connecting $LCS(s_l, s_m)$ and $LCS(s'_m, s_r)$. The confidence interval of the left breakpoint is set to $[B_l - LCS(s_l, s_m), B_l]$, and the confidence interval of the right breakpoint is set to $[B_r, LCS(s'_m, s_r) + B_r]$. B_l and B_r are initially predicted locations of the left breakpoint and the right breakpoint, respectively.

3 Experiments

We use 17 simulated data sets and 5 real data sets to evaluate the deletions detected by Sprites2, comparing with Sprites, SVseq2, Lumpy, Delly and Pindel. The 17 simulated data sets consist of 5 simulated WGS data sets of a synthetic sample containing homozygous deletions and 12 simulated WGS data sets of a synthetic sample containing heterozygous deletions. The 5 real data sets consist of chromosome 20 data sets of NA19311, NA19312, NA19313, NA19316 and NA19317.

We use double interval overlap to evaluate the breakpoint results detected. One deletion detected is represented by two intervals A and B, and its corresponding real breakpoint position is represented by two intervals A' and B'. When A is overlapped with A', at the same time, B is overlapped with B', the detected deletion refers to a true positive deletion, represented by TP, and otherwise, represented by FP. If a known deletion is failed to report by tools, it is a false negative deletion, represented by FN.

3.1   Results on Simulated Homozygous Deletions

Table 1 shows that Sprites2 has the highest sensitivity (61%, 93.6%, and 98.24%) of simulated homozygous data with 2x, 5x and 10x coverage. It is worth noting that the sensitivity of Sprites2 is higher than Sprites for every sequencing coverage. Sprites2 does not use read depth information when detecting deletions. Thus, the sensitivity of Sprites2 is not the highest at the sequencing depth of 20x and 50x, but only a difference of 0.5% with the highest value (99.4% and 99.52% for Pindel). With 2x and 5x coverage, Lumpy has the lowest FDR (0% and 0.05%).With 10x, 20x and 50x, SVseq2 has the lowest FDR (0.27%, 0.25%, and 0.4%). Although the FDR of Sprites2 increases relative to Sprites and Sprites2, in general, its FDR keeps at a low level. Sprites2 achieves the highest F-scores (0.7572, 0.9636 and 0.9868) at the coverage of 2x, 5x and 10x, while the F-score of Sprites2 at coverage of 20x and 50x is second only to Sprites. The F-score of Sprites2 first raises and then keeps at a high level when the sequencing depth is from low to high, and Delly's F-score first raises and reaches the highest value at 10x coverage, and then significantly falls down.

Table 1. Sensitivity (S) and FDR comparison of simulated homozygous deletions

Tool	2x		5X		10X		20X		50X	
	S	FDR	S	FDR	S	FDR	S	FDR	S	FDR
Sprites2	**61.00**	0.20	**93.6**	0.72	**98.24**	0.88	98.88	1.13	99.00	1.91
Sprites	59.32	0.07	92.08	0.26	97.60	0.28	98.48	0.36	98.52	0.88
SVseq2	44.80	0.09	65.08	0.18	73.00	**0.27**	80.36	**0.25**	89.64	**0.40**
Lumpy	33.56	**0.00**	86.20	**0.05**	96.76	0.33	97.8	1.45	98.08	3.92
Pindel	21.72	0.18	71.88	0.44	96.44	1.07	**99.4**	4.23	**99.52**	18.77
Delly	44.72	0.45	82.96	2.58	91.84	7.34	90.96	21.73	88.72	59.58

3.2   Results on Simulated Heterzygous Deletions

Table 2 shows that Sprites2 has the highest sensitivity for 9 data sets of simulated heterozygous deletion, the sensitivity of Sprites2 is second to the highest value of Delly in the remaining three data sets. In addition, Sprites2 is more sensitive than Sprites in all cases. Table 3 shows that the FDR of Sprites2 is the lowest (0.39%) with 20x coverage and 0.5 SV allele frequency. In most cases, the FDR of Sprites2 is higher than that of Sprites, but the difference between Sprites2 and Sprites is generally below 0.3%. Consistently, SVseq2 has the lowest FDR. The F-score of Sprites2 is the highest in all 12 data sets, indicating that it sacrifices the FDR in exchange for higher sensitivity, and the overall performance of the detection is improved.

Table 2. Sensitivity comparison of simulated heterozygous deletions

Coverage	AF	Sprites2	Sprites	SVseq2	Lumpy	Pindel	Delly
10X	0.05	**9.99**	8.39	7.47	0.47	0.73	6.38
	0.10	**27.59**	24.33	18.91	5.49	3.72	19.62
	0.20	**58.28**	50.73	40.10	29.08	17.26	48.06
	0.50	**90.97**	79.15	62.18	84.63	67.17	89.16
20X	0.05	**28.06**	24.31	19.83	5.20	3.25	18.80
	0.10	**58.45**	50.87	39.2	28.64	17.64	46.63
	0.20	**86.71**	75.45	58.88	74.75	54.33	81.80
	0.50	96.79	84.97	68.24	96.27	93.4	**98.19**
40X	0.05	**58.5**	50.33	39.23	28.35	17.39	47.19
	0.10	**86.44**	76.02	57.87	73.89	53.19	82.20
	0.20	95.78	83.76	67.49	95.52	73.69	**97.44**
	0.50	98.21	86.97	75.76	96.66	97.63	**98.37**

Table 3. FDR comparison of simulated heterozygous deletions

Coverage	AF	Sprites2	Sprites	SVseq2	Lumpy	Pindel	Delly
10X	0.05	1.43	**0.00**	**0.00**	3.70	16.67	26.97
	0.1	0.85	0.37	**0.19**	0.33	6.39	9.76
	0.2	0.40	0.32	**0.18**	0.74	1.04	4.78
	0.5	0.77	0.50	**0.20**	1.39	0.48	2.61
20X	0.05	**0.39**	0.45	0.45	0.69	27.53	32.79
	0.1	0.58	0.39	**0.14**	0.69	5.99	15.51
	0.2	0.60	**0.29**	0.34	1.17	1.90	10.01
	0.5	1.33	0.57	**0.24**	2.50	1.25	8.87
40X	0.05	0.58	0.39	**0.28**	0.70	26.85	44.02
	0.1	0.72	0.35	**0.22**	1.16	10.28	31.37
	0.2	0.87	0.64	**0.32**	1.77	6.12	27.11
	0.5	1.88	1.05	**0.5**	6.26	4.01	27.39

3.3   Deletion Detection Results on Chromosome 20 of 5 Individuals

Table 4 shows that the detection results of Sprites2 are more sensitive than Sprites, in turn, Sprites2 has slightly higher FDR than Sprites. Overall, Sprites2 preforms well from the perspective of F-score. It is worth mentioning that SVseq2 has the highest F-score in NA19311, NA19312 and NA19317.

Table 4. Performance comparison of chromosome 20 of 5 individuals

		Sprites2	Sprites	SVseq2	Lumpy	Pindel	Delly
NA19311	S	**21.00**	19.90	19.90	8.80	5.50	20.40
	FDR	43.50	39.70	27.10	59.00	**23.10**	82.40
	F-score	**0.31**	0.30	**0.31**	0.15	0.10	0.19
NA19312	S	12.20	11.60	11.60	3.30	5.50	**14.90**
	FDR	64.50	58.80	56.20	88.30	**33.30**	84.70
	F-score	**0.18**	**0.18**	**0.18**	0.05	0.10	0.15
NA19313	S	11.10	9.90	8.80	2.20	2.80	**16.60**
	FDR	68.50	**67.30**	77.10	94.00	70.60	87.80
	F-score	**0.17**	0.15	0.13	0.032	0.05	0.14
NA19316	S	6.10	6.10	6.10	0.00	**6.60**	3.90
	FDR	31.30	31.20	35.30	90.00	**25.00**	95.40
	F-score	0.11	0.11	0.11	0.00	**0.12**	0.04
NA19317	S	5.50	5.50	**6.10**	0.60	5.50	3.90
	FDR	60.00	58.30	64.30	84.60	**44.40**	93.50
	F-score	**0.10**	**0.10**	**0.10**	0.01	**0.10**	0.05

4   Conclusion

The factors including diploid organisms, sequencing errors, mapping errors, micro-homology, micro-insertion, tandem repeats, and greedy strategies employed by alignment tools all increase the difficulty in locating breakpoints accurately when detecting deletions. Sprites2 takes advantage of the variance of insert size distribution to determine the type of deletions. When determining breakpoints, Sprites2 firstly completes the target sequence by adding base pairs to the head or tail of the target sequence. Then Sprites2 uses AGE to perform the aligning between the soft-clipped reads and the corresponding target sequences, the running time and the required memory has absolutely no relation to the length of deletions.

According to the comparison results, Sprites2 eliminates fake micro-insertions introduced by sequencing errors through the 1 bp fault-tolerant amplification. In addition, the micro-homology is determined by searching for two identical sequences on the target sequence near breakpoint coordinates.

Based on data sets of simulated homozygous deletions and heterozygous deletions, Sprites2 is found to be better than those based on single-end alignment tools. With the use of AGE to re-align the whole soft-clipped read in simulated data sets and real data sets, experiments illustrate that detection performance has been improved. As good as Sprites, Sprites2 shows excellent performances at data sets with the low coverage, indicating good detection results can be obtained on low sequencing cost condition.

Acknowledgments. This work was supported in part by the National Natural Science Foundation of China under Grant No. 61732009, No. 61622213, No. 61728211, No. 61772552, No. 61772557 and No. 61602156.

References

1. Guan, P., Sung, W.K.: Structural variation detection using next-generation sequencing data: a comparative technical review. Methods **102**, 36–49 (2016)
2. Weischenfeldt, J., Symmons, O., Spitz, F., Korbel, J.O.: Phenotypic impact of genomic structural variation: insights from and for human disease. Nat. Rev. Genet. **14**(2), 125–138 (2013)
3. Alkan, C., Coe, B.P., Eichler, E.E.: Genome structural variation discovery and genotyping. Nat. Rev. Genet. **12**(5), 363–376 (2011)
4. Ye, K., Schulz, M.H., Long, Q., Apweiler, R., Ning, Z.: Pindel: a pattern growth approach to detect break points of large deletions and medium sized insertions from paired-end short reads. Bioinformatics **25**(21), 2865–2871 (2009)
5. Rausch, T., Zichner, T., Schlattl, A., Sttz, A.M., Benes, V., Korbel, J.O.: DELLY: structural variant discovery by integrated paired-end and split-read analysis. Bioinformatics **28**(18), 333–339 (2012)
6. Zhang, J., Wang, J., Wu, Y.: An improved approach for accurate and efficient calling of structural variations with low-coverage sequence data. BMC Bioinf. **13**(S6), S6 (2012)
7. Zhang, Z., Wang, J., Luo, J., Ding, X., Zhong, J., Wang, J., Wu, F., Pan, Y., et al.: Sprites: detection of deletions from sequencing data by re-aligning split reads. Bioinformatics **32**(12), 1788–1796 (2016)
8. Li, H., Handsaker, B., Wysoker, A., Fennell, T., Ruan, J., Homer, N., Marth, G., Abecasis, G., Durbin, R.: The sequence alignment/map format and SAMtools. Bioinformatics **25**(16), 2078–2079 (2009)
9. Abyzov, A., Gerstein, M.: AGE: defining breakpoints of genomic structural variants at single-nucleotide resolution, through optimal alignments with gap excision. Bioinformatics **27**(5), 595–603 (2011)
10. Layer, R.M., Chiang, C., Quinlan, A.R., Hall, I.M.: LUMPY: a probabilistic framework for structural variant discovery. Genome Biol. **15**(6), R84 (2014)
11. Faust, G.G., Hall, I.M.: YAHA: fast and flexible long-read alignment with optimal breakpoint detection. Bioinformatics **28**(19), 2417–2424 (2012)

KSIBW: Predicting Kinase-Substrate Interactions Based on Bi-random Walk

Canshang Deng[1], Qingfeng Chen[1,4], Zhixian Liu[3,4], Ruiqing Zheng[2],
Jin Liu[2], Jianxin Wang[2], and Wei Lan[1(✉)]

[1] School of Computer, Electronics and Information, Guangxi University,
Nanning, People's Republic of China
lanwei@gxu.edu.cn
[2] School of Information Science and Engineering, Central South University,
Changsha, People's Republic of China
[3] School of Electronic and Information Engineering, Qinzhou University,
Qingzhou, People's Republic of China
[4] State Key Laboratory for Conservation and Utilization of Subtropical
Agro-Bioresources, Guangxi University, Nanning, People's Republic of China

Abstract. Protein phosphorylation is an important chemical modification in the organism that regulates many cellular processes. In recent years, many algorithms for predicting kinase-substrate interactions have been proposed. However, most of those methods are mainly focused on utilizing protein sequence information. In this paper, we propose a computational framework, KSIBW, to predict kinase-substrate interactions based on bi-random walk. Unlike traditional methods, the protein-protein interaction (PPI) information are used to measure the similarities of kinase-kinase and substrate-substrate, respectively. Then, the bi-random walk is employed to identify potential kinase-substrate interactions. The experiment results show that our method outperforms other state-of-the-art algorithms in performance.

Keywords: Protein phosphorylation · Kinase-substrate interactions
Bi-random walk · Protein-protein interaction network

1 Introduction

The post-translation modifications (PTMs) of protein, including phosphorylation, glycosylation, ubiquitination, methylation, acetylation and other chemical modifications, etc., are common biological mechanism for protein function regulation [1]. Protein phosphorylation is one of the most basic, universal and important way of regulation in living organisms. It refers to the process of transferring ATP phosphate groups to the amino acid sequence of the substrate by serine (S), tyrosine (Y) and threonine (T) sites under the catalysis of protein kinase. It plays many key roles in cell metabolism, gene expression, and cellular signal transduction, etc. The abnormal of intracellular phosphorylation

F. Zhang et al. (Eds.): ISBRA 2018, LNBI 10847, pp. 151–162, 2018.
https://doi.org/10.1007/978-3-319-94968-0_13

process may cause some serious diseases [2], such as rheumatoid arthritis [3] and diabetes [4].

Considering the important roles of protein phosphorylation in human organisms, several biological methods have been developed to identify phosphorylation sites. It can be classified into two categories: P32 isotope tracer [5] and mass spectrometry [6]. A large number of phosphorylated proteins and their modified sites have been experimentally identified. Several comprehensive databases have been established to store these protein phosphorylation data. Phospho.ELM [7] is a comprehensive knowledgebase that contains eukaryotic cell protein phosphorylation data. It contains 8718 substrate proteins from different species. PhosphoSitePlus [8] is an online resource providing comprehensive information for the study of protein post-translational modifications. It includes 415446 nonredundant PTMs and 20279 proteins. PhosphoNET (http://www.phosphonet. ca/) is a popular kinase-substrate interaction database. It contains over 950000 known and putative phosphorylation sites in over 20000 human proteins that collected from scientific literatures.

A large number of protein phosphorylation data provide a reliable data source for predicting unknown phosphorylation sites by using computational methods [9]. In recent years, many computational methods have been developed for predicting kinase-substrate interactions. Linding et al. [10] proposed an approach (NetworKIN) that augments motif-based predictions with the network context of kinases and phosphoproteins to predict site-specific kinase-substrate interactions. Dang et al. [11] developed a new method to predict phosphorylation sites by using the protein sequence and conditional random fields. Zhou et al. [12] presented a computational method, named GPS, to predict phosphorylation site based on the substitution matrix and Markov Cluster Algorithm. Song et al. [13] extended GPS by integrating PPI information to identify kinase-substrate interactions. Zou et al. [14] proposed a computational framework to identify protein kinase by incorporating support vector machines and the composition of monomer spectrum encoding strategy. Patrick et al. [15] developed a Bayesian network model that integrate cellular context to predict kinase-substrate interactions. Fan et al. [16] proposed a method for kinase-specific phosphorylation sites prediction by using functional information and random forest. Damle et al. [17] proposed a computational method, PhosNetConstruct, to decipher kinase-substrate relationships by analyzing domain-specific phosphorylation network. Li et al. [18] proposed a kernel-based method called SLapRLS, to address the kinase identification problem by using Supervised Laplacian Regularized Least Squares. Song et al. [19] presented a bioinformatics tool to predict kinase-specific substrates and their associated phosphorylation sites by combining protein sequence and functional features. Li et al. [20] proposed a network based method for predicting kinase-substrate interactions based on sequence similarity. Qin et al. [21] presented a computational method for inferring the interactions between kinases and substrates based on protein domains. Existing computational methods for identifying kinase-substrate interactions have achieved great successes. However,

These methods require a large amount of negative samples which are unable to obtain in data [22–24]. Besides, some methods only use local information instead of global information, which may produce more false positives.

In this paper, we propose a computational framework to infer the relationships between kinases and substrates. Firstly, the similarities of kinase-kinase and substrate-substrate are calculated by using short path method based on PPI network, respectively. Then, the bi-random walk algorithm is used to predict potential kinase-substrate interactions. The experimental results demonstrate that our method can effectively predict kinase-substrate interactions.

2   Materials and Methods

2.1   Data Resources

We obtain human phosphorylation data from the Phospho.ELM 9.0 [7]. After removing the redundant data, 216 kinases, 724 substrates and 1256 kinase-substrate interactions are collected in final. The human PPI data are obtained from InWeb_IM [25], it contains 14684 human proteins and 625641 interactions.

2.2   Kinase-Kinase and Substrate-Substrate Similarity Measure

In InWeb_IM, each protein interaction is given a confidence score which is calculated based on the reproducibility of the interaction data between different publications. The PPI network can be described as an undirected graph $G(V, E, W)$, each node $v \in V$ denotes a protein, and each edge $(u, v) \in E$ denotes the interaction between nodes u and v. W denotes the confidence score of the interactions. In order to calculate the similarity between two proteins, we find the shortest path between the two proteins. Then the similarity is calculated as follows:

$$sim(u, v) = \prod_{(i,j) \in SP_{uv}} W(i, j) \tag{1}$$

where SP_{uv} represents the set of edges on the shortest path between u and v, and (i, j) represents the edge between two adjacent nodes on the shortest path. Since both kinases and substrates are proteins, we used the same procedure to calculate kinase similarity and substrate similarity, respectively.

2.3   Construction of Kinase and Substrate Heterogeneous Network

Based on these two similarity matrices, a kinase similarity network and a substrate similarity network are constructed, respectively. For kinase similarity network K, let k_i and k_j represent two different kinases. If the similarity between k_i and k_j is 0, then there is no edge between this two kinases. Otherwise, there is an edge connection between these two kinases, and the weight of the edge is

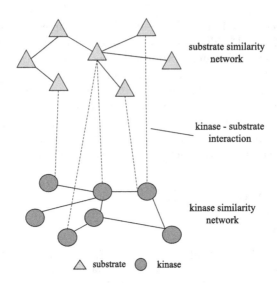

Fig. 1. Illustration of the kinase-substrate heterogeneous network. The triangle and the circle represent the substrate and the kinase, respectively. The solid line shows the similarity between the two proteins, The dotted line shows the kinase-substrate interactions. This two similarity networks are linked by known kinase-substrate interactions.

the similarity value of the two kinases. For substrate similarity network S, If the similarity between substrate s_i and s_j is 0, then there is no edge between this two substrates. Otherwise, there is an edge connection between these two substrates, and the weight of the edge is the similarity value of the two substrates.

Let I denotes the kinase-substrate association network, e_{ij} denotes the edge of I. The initial value of e_{ij} is set to 1 if there is a known interaction between kinase k_i and substrate s_j, otherwise it is set to 0. Based on the association network, we can construct the kinase-substrate heterogeneous network by conjoining kinase similarity network and substrate similarity network. An example of kinase-substrate heterogeneous network is shown in Fig. 1.

2.4 Predicting Kinase-Substrate Interactions Based on Bi-random Walk

Bi-random walk is an extension of the random walk, which is widely used in drug repositioning [26] and miRNA-disease interaction identification [27,28].

We normalize the kinase similarity matrix and the substrate similarity matrix by using Laplace normalization, respectively. For example, we normalize substrate similarity matrix as follows:

$$S_n = D_s^{\frac{-1}{2}} \times S \times D_s^{\frac{-1}{2}} \tag{2}$$

where D_s represents the diagonal matrix of the substrate similarity matrix S and $D_s(i, i)$ is the sum of the i-th row of S. The kinase-normalized similarity matrix K_n can be obtained in the same way.

Different from the previous normalize procedure, the normalization matrix I_n of kinase-substrate interactions matrix I is defined as follows:

$$RW(0) = I_n = I(i, j)/sum(I) \tag{3}$$

where $sum(I)$ denotes the sum of all the elements of I.

After getting normalized matrix S_n, K_n and I_n, we employ bi-random walk to identify kinase-substrate interactions by walking on the kinase similarity network and substrate similarity network simultaneously. Considering that different networks may have different topology structures, the optimal number of steps to walk randomly between these two networks may be inconsistent. Thus, we limit the number of walking steps on two different networks by setting two parameters l and r, where l and r represent the maximum number of random walks on the substrate network and kinase network, respectively. Finally, the bi-random walk procedure is formalized as follows:

Left walk in the substrate similarity network:

$$RW_L(t) = \beta \times S_n \times RW(t-1) + (1-\beta)I_n \tag{4}$$

Right walk in the kinase similarity network:

$$RW_R(t) = \beta \times RW(t-1) \times K_n + (1-\beta)I_n \tag{5}$$

We integrate the left and the right predicted result to acquire the final output:

$$RW(t) = \frac{RW_L(t) + RW_R(t)}{2} \tag{6}$$

Where $RW_L(t)$ and $RW_R(t)$ represent the predicted score of kinase-substrate interactions based on walking on the substrate similarity network and kinase similarity network at the step t, respectively. $RW(t)$ denotes the final predicted result matrix at the step t. The elements of $RW(t)$ represent the probability of the kinase-substrate interactions. The larger the value of $RW(i, j)$ is, the more likely that the substrate s_j is phosphorylated by the kinase k_i. The parameters l, r and β are set to 2, 2 and 0.3 in our experiment, respectively. The flow chart of the bi-random walk algorithm for predicting kinase-substrate interactions is shown in Fig. 2, the pseudocode is outlined in Algorithm 1.

Algorithm 1. Algorithm for predicting kinase-substrate interactions based on bi-random walk.

Input: Kinase similarity matrix K, substrate similarity matrix S, kinase-substrate interaction matix I, parameter β, iteration steps l and r.

Output: Predicted score matrix RW.

1: Normalize K, S, and I to K_n, S_n and I_n
2: **for** $t = 1$ to $max(l, r)$ **do**
3:     **if** $t <= l$ **then**
4:         $RW_L(t) = \beta \times S_n \times RW(t-1) + (1-\beta)I_n$
5:     **else if** $t <= r$ **then**
6:         $RW_R(t) = \beta \times RW(t-1) \times K_n + (1-\beta)I_n$
7:     **end if**
8:     $RW(t) = \frac{RW_L(t) + RW_R(t)}{2}$
9: **end for**
10: **return** RW

---

3   Experiments and Results

3.1   Evaluation Metrics

In this paper, we use ten fold cross-validation and *de novo* test to evaluate the performance of different algorithms. In the ten-fold cross-validation process, known kinase-substrate interactions are randomly divided into ten subsets. In each cross validation trial, nine subsets are used as the training set and the remaining one subset is treated as the test set. After completing the test on the dataset, a predicted scoring matrix is generated. Then we rank the unknown kinase-substrate interactions and test set based on predicted score. For each threshold, the corresponding predicted result of test set is considered true positive (TP) if the predicted score is greater than the threshold. Otherwise, it is considered as false negative (FN). For the unknown kinase-substrate interaction, it is treated as false positive (FP) if its predicted score is greater than the threshold and as true negative (TN) if the predicted value is less than the threshold. By choosing various thresholds, we can calculate different true positive rate (TPR) and false positive rate (FPR). The TPR and the FPR are calculated as follows:

$$TPR = \frac{TP}{TP + FN} \tag{7}$$

$$FPR = \frac{FP}{FP + TN} \tag{8}$$

Finally, the ROC curve is drawed based on the previously calculated TPR and FPR. Then the AUC (Area Under Curve) is calculated to evaluate the performance of different algorithms.

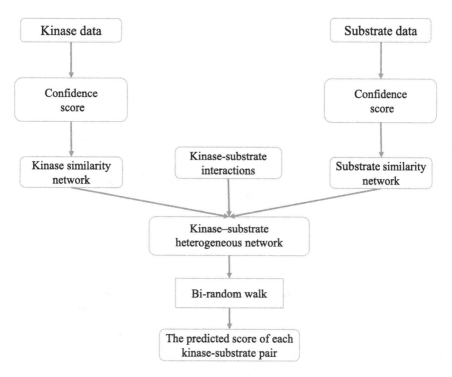

Fig. 2. The flow chart of potential kinase-substrate interactions identification by using bi-random walk.

3.2   Comparison with Network-Based Method

To evaluate the performance of our proposed algorithm, we compare KSIBW with the network-based prediction method (Hetesim-SEQ) [20]. The experimental results of KSIBW and Hetesim-SEQ is showed in Fig. 3. KSIBW achieves the AUC value of 0.836, which is higher than Hetesim-SEQ (AUC = 0.802). It shows that KSIBW performs better than Hetesim-SEQ.

3.3   Comparison with Different Predicted Methods Using De Novo Test

In order to evaluate the power of our method for predicting new kinase-substrate interactions, we perform *de novo* test experiments. In the *de novo* test, we delete all known kinase-substrate interactions of kinase i each time. The rest of kinase-substrate interactions are treated as training set. We compare KSIBW with four popular predicted methods of kinase-substrate interactions, including GPS [12], iGPS [13], NetworKIN [10] and PhosphoPICK [15]. Since these methods only provide web server, we submit the dataset to the corresponding web server for testing. We take six kinase group including Atypical, CAMK, CMGC, Other, STE and TK as examples to illustrate the predictive performance of different methods. The ROC curves for different methods in different kinase groups are

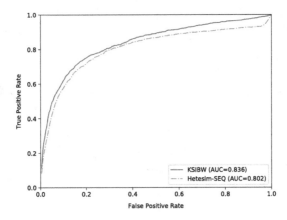

Fig. 3. The ROC curves for predicting kinase-substrate interactions with different methods.

plotted as performance comparisons and displayed in Fig. 4. As shown in Fig. 4, our algorithm performs better than the other four algorithms on different kinase groups.

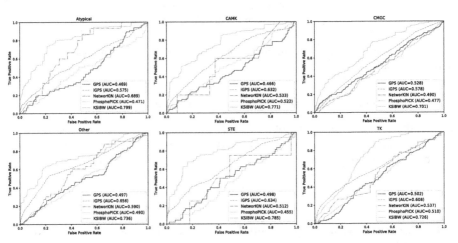

Fig. 4. The ROC curves for kinase group Atypical, CAMK, CMGC, Other, STE and TK with different methods.

3.4   Case Studies

To further validate the ability to predict unknown kinase-substrate interactions, the case study is conducted in here. All known kinase-substrate interactions are used as a training set and the unknown kinase-substrate interactions are used as the test set. We employ KSIBW to predict potential kinase-substrate interactions and acquire the prediction score of all kinase-substrate pair in the test set.

We select the top-20 predicted results which are list in Table 1. By checking the PhosphoNET database, we found that five of the prediction results are recorded by PhosphoNET. For example, the serine site in position 315 of substrate TP53 is catalyzed by kinase CDK2, ATR phosphorylates TP53 at serine site in position 37. The tyrosine site in position 182 of substrate CSNK2A1 is catalyzed by kinase CK2a1. PLCB3 interacts with kinase PRKG1 through serine site in position 26. The tyrosine site in position Y420 of substrate TXK is phosphorylated by the kinases FYN. In addition, although some of the predicted kinase-substrate interactions are not present in PhosphoNET, we validate them based on the recent published literature. For example, TP53 has been found to be regulated by FLT3 [29]. ATR associates with the regulatory protein ATRIP [30]. Polo-like kinase 1 (Plk1) has been verified that physically binds to the tumor suppressor p53 in mammalian cultured cells [31].

In addition, some new interesting kinase-substrate interactions are also discovered from the experimental results. It deserves for biologists to validate by using biological experiments.

Table 1. Top-20 predicted results of unknown kinase-substrate interactions.

Top	Substrate	Prediction kinase	Evidence
1	NEK6	CDK1	PMID:22064517
2	NR4A2	MAPK3	
3	MAP2K1	SRC	
4	ATRIP	ATR	PMID:15743907
5	TP53	CDK2	PhosphoNET
6	TP53	FLT3	PMID:17105820
7	ICAM3	PRKCA	
8	RPS6KB2	PRKCH	
9	TP53	PLK1	PMID:15024021
10	RELA	FGFR4	
11	RRN3	MAPK1	
12	CSNK2A1	CK2a1	PhosphoNET
13	PPP1R14A	GSK3B	
14	MAPK12	ABL1	
15	TEC	LCK	PMID:8636141
16	TP53	NEK2	PMID:24163369
17	PLCB3	PRKG1	PhosphoNET
18	TP53	TGFBR2	
19	TXK	FYN	PhosphoNET
20	TP53	ATR	PhosphoNET

4   Conclusion

Protein phosphorylation plays important regulatory roles in the organism. More and more researchers use computational method to identify kinase substrate interactions. In this article, we proposed a computational framework to predict kinase-substrate interactions based on bi-random walk. The PPI information is used to measure the similarity of pairwise of kinases and substrates, respectively. Then, the bi-random walk method is employed to identify potential kinase-substrate interactions. We evaluate our method in term of ten-fold cross validation and *de novo* prediction. The experimental results demonstrate that our algorithm achieves higher AUC than other state-of-the-art algorithms. Moreover, the case studies has proved the effectiveness of KSIBW in predicting potential kinase-substrate interactions.

Acknowledgement. This work is supported in part by the National Natural Science Foundation of China under Grant No. 61702122, 61751314, 31560317, 61702555, 61662028 and 61762087; Key project of Natural Science Foundation of Guangxi 2017GXNSFDA198033; Key research and development plan of Guangxi AB17195055 and Director Open Fund of Qinzhou City Key Laboratory of Advanced Technology of Internet of Things IOT2017A04.

References

1. Mann, M., Jensen, O.N.: Proteomic analysis of post-translational modifications. Nat. Biotechnol. **21**(3), 255–261 (2003)
2. Lan, W., Wang, J., Li, M., Peng, W., Wu, F.: Computational approaches for prioritizing candidate disease genes based on PPI networks. Tsinghua Sci. Technol. **20**(5), 500–512 (2015)
3. Grabiec, A.M., Korchynskyi, O., Tak, P.P., Reedquist, K.A.: Histone deacetylase inhibitors suppress rheumatoid arthritis fibroblast-like synoviocyte and macrophage IL-6 production by accelerating mRNA decay. Ann. Rheum. Dis. **71**(3), 424–431 (2012)
4. Cohen, P.: The role of protein phosphorylation in human health and disease. FEBS J. **268**(19), 5001–5010 (2001)
5. Aponte, A.M., Phillips, D., Harris, R.A., Blinova, K., French, S., Johnson, D.T., Balaban, R.S.: 32P labeling of protein phosphorylation and metabolite association in the mitochondria matrix. Methods Enzymol. **457**, 63–80 (2009)
6. Lin, J., Xie, Z., Zhu, H., Qian, J.: Understanding protein phosphorylation on a systems level. Brief. Funct. Genomics **9**(1), 32–42 (2010)
7. Dinkel, H., Chica, C., Via, A., Gould, C.M., Jensen, L.J., Gibson, T.J., Diella, F.: Phospho.ELM: a database of phosphorylation sites—update 2011. Nucleic Acids Res. **39**(Suppl_1), D261–D267 (2010)
8. Hornbeck, P.V., Kornhauser, J.M., Tkachev, S., Zhang, B., Skrzypek, E., Murray, B., Latham, V., Sullivan, M.: PhosphoSitePlus: a comprehensive resource for investigating the structure and function of experimentally determined post-translational modifications in man and mouse. Nucleic Acids Res. **40**(D1), D261–D270 (2011)
9. Chen, Q., Wang, Y., Chen, B., Zhang, C., Wang, L., Li, J.: Using propensity scores to predict the kinases of unannotated phosphopeptides. Knowl.-Based Syst. **135**, 60–76 (2017)

10. Linding, R., Jensen, L.J., Ostheimer, G.J., van Vugt, M.A., Jørgensen, C., Miron, I.M., Diella, F., Colwill, K., Taylor, L., Elder, K., et al.: Systematic discovery of in vivo phosphorylation networks. Cell **129**(7), 1415–1426 (2007)
11. Dang, T.H., Van Leemput, K., Verschoren, A., Laukens, K.: Prediction of kinase-specific phosphorylation sites using conditional random fields. Bioinformatics **24**(24), 2857–2864 (2008)
12. Zhou, F.F., Xue, Y., Chen, G.L., Yao, X.: GPS: a novel group-based phosphorylation predicting and scoring method. Biochem. Biophys. Res. Commun. **325**(4), 1443–1448 (2004)
13. Song, C., Ye, M., Liu, Z., Cheng, H., Jiang, X., Han, G., Songyang, Z., Tan, Y., Wang, H., Ren, J., et al.: Systematic analysis of protein phosphorylation networks from phosphoproteomic data. Mol. Cell. Proteomics **11**(10), 1070–1083 (2012)
14. Zou, L., Wang, M., Shen, Y., Liao, J., Li, A., Wang, M.: PKIS: computational identification of protein kinases for experimentally discovered protein phosphorylation sites. BMC Bioinform. **14**(1), 247 (2013)
15. Patrick, R., Lê Cao, K.A., Kobe, B., Bodén, M.: PhosphoPICK: modelling cellular context to map kinase-substrate phosphorylation events. Bioinformatics **31**(3), 382–389 (2014)
16. Fan, W., Xu, X., Shen, Y., Feng, H., Li, A., Wang, M.: Prediction of protein kinase-specific phosphorylation sites in hierarchical structure using functional information and random forest. Amino Acids **46**(4), 1069–1078 (2014)
17. Damle, N.P., Mohanty, D.: Deciphering kinase-substrate relationships by analysis of domain-specific phosphorylation network. Bioinformatics **30**(12), 1730–1738 (2014)
18. Li, A., Xu, X., Zhang, H., Wang, M.: Kinase identification with supervised Laplacian regularized least squares. PLoS ONE **10**(10), e0139676 (2015)
19. Song, J., Wang, H., Wang, J., Leier, A., Marquez-Lago, T., Yang, B., Zhang, Z., Akutsu, T., Webb, G.I., Daly, R.J.: PhosphoPredict: a bioinformatics tool for prediction of human kinase-specific phosphorylation substrates and sites by integrating heterogeneous feature selection. Sci. Rep. **7**(1), 6862 (2017)
20. Li, H., Wang, M., Xu, X.: Prediction of kinase-substrate relations based on heterogeneous networks. J. Bioinform. Comput. Biol. **13**(06), 1542003 (2015)
21. Qin, G.M., Li, R.Y., Zhao, X.M.: PhosD: inferring kinase-substrate interactions based on protein domains. Bioinformatics **33**(8), 1197–1204 (2016)
22. Lan, W., Wang, J., Li, M., Liu, J., Wu, F.X., Pan, Y.: Predicting microRNA-disease associations based on improved microRNA and disease similarities. IEEE/ACM Trans. Comput. Biol. Bioinform. **99**, 1 (2016)
23. Liu, J., Li, M., Lan, W., Wu, F.X., Pan, Y., Wang, J.: Classification of Alzheimer's disease using whole brain hierarchical network. IEEE/ACM Trans. Comput. Biol. Bioinform. **15**(2), 624–632 (2018)
24. Lan, W., Wang, J., Li, M., Liu, J., Pan, Y.: Predicting microRNA-disease associations by integrating multiple biological information. In: 2015 IEEE International Conference on Bioinformatics and Biomedicine, BIBM, pp. 183–188. IEEE (2015)
25. Li, T., Wernersson, R., Hansen, R.B., Horn, H., Mercer, J., Slodkowicz, G., Workman, C.T., Rigina, O., Rapacki, K., Stærfeldt, H.H., et al.: A scored human protein-protein interaction network to catalyze genomic interpretation. Nat. Methods **14**(1), 61–64 (2017)
26. Luo, H., Wang, J., Li, M., Luo, J., Peng, X., Wu, F.X., Pan, Y.: Drug repositioning based on comprehensive similarity measures and Bi-Random walk algorithm. Bioinformatics **32**(17), 2664 (2016)

27. Peng, W., Lan, W., Yu, Z., Wang, J., Pan, Y.: A framework for integrating multiple biological networks to predict microRNA-disease associations. IEEE Trans. Nanobiosci. **16**(2), 100–107 (2017)
28. Peng, W., Lan, W., Zhong, J., Wang, J., Pan, Y.: A novel method of predicting microRNA-disease associations based on microRNA, disease, gene and environment factor networks. Methods **124**, 69–77 (2017)
29. Irish, J.M., Ånensen, N., Hovland, R., Skavland, J., Børresen-Dale, A.L., Bruserud, Ø., Nolan, G.P., Gjertsen, B.T.: Flt3 Y591 duplication and Bcl-2 overexpression are detected in acute myeloid leukemia cells with high levels of phosphorylated wild-type p53. Blood **109**(6), 2589–2596 (2007)
30. Ball, H.L., Myers, J.S., Cortez, D.: ATRIP binding to replication protein A-single-stranded DNA promotes ATR-ATRIP localization but is dispensable for Chk1 phosphorylation. Mol. Biol. Cell **16**(5), 2372–2381 (2005)
31. Ando, K., Ozaki, T., Yamamoto, H., Furuya, K., Hosoda, M., Hayashi, S., Fukuzawa, M., Nakagawara, A.: Polo-like kinase 1 (Plk1) inhibits p53 function by physical interaction and phosphorylation. J. Biol. Chem. **279**(24), 25549–25561 (2004)

XPredRBR: Accurate and Fast Prediction of RNA-Binding Residues in Proteins Using eXtreme Gradient Boosting

Lei Deng[1], Zuojin Dong[1], and Hui Liu[2(✉)]

[1] School of Software, Central South University, Changsha 410075, China
[2] Lab of Information Management, Changzhou University,
Changzhou 213164, Jiangsu, China
hliu@cczu.edu.cn

Abstract. A variety of studies have shown that protein-RNA interactions play a vital role in many fundamental cellular processes, such as protein synthesis, mRNA processing, mRNA assembly, ribosome function and eukaryotic spliceosomes. Identification of RNA-binding residues (RBR) in proteins is an key step to understand the mutual recognition mechanism underlying the protein-RNA interactions. In this paper, we proposed a novel method, XPredRBR, to predict the RNA binding residues in proteins, by exploiting the eXtreme Gradient Boosting (XGBoost) algorithm. Two types of new predictive features derived from residue interaction network and solvent exposures are combined with conventional sequence features and structural neighborhood features to predict RBR. We carried out empirical experiments on two datasets to demonstrate the performance of the proposed method. By 10-fold cross-validations, our method achieved the accuracy of 0.861, sensitivity of 0.872, MCC of 0.584 and AUC of 0.941 on the RBP170 dataset. On another independent test set RBP101, XPredRBR outperformed three traditional classifiers and seven existing RNA-binding residue methods. A case study on the chain E of 3PLA protein illustrated XPredRBR effectively identified most RNA-binding and non RNA-binding sites. Furthermore, XPredRBR is much faster than our previous method PredRBR. These experimental results show that our proposed method achieves state-of-the-art performance in predicting RNA-binding residues in proteins.

Keywords: Protein-RNA interactions · eXtreme gradient boosting
RNA-binding residues

H. Liu—This work was supported by National Natural Science Foundation of China under grants No. 61672541 and No. 61672113, and Natural Science Foundation of Hunan Province under grant No. 2017JJ3287.

F. Zhang et al. (Eds.): ISBRA 2018, LNBI 10847, pp. 163–173, 2018.
https://doi.org/10.1007/978-3-319-94968-0_14

1   Introduction

Protein-RNA interactions play an essential role in a wide variety of fundamental cellular processes, such as transcription, replication, protein synthesis, regulation of gene expression and posttranscriptional modifications [1,2]. Locating the functional residues in proteins interacting with RNA molecules, referred to as RNA-binding residues (RBR), is an key step to understand the detailed recognition mechanism function of various biological activities relevant to protein-RNA interactions. Therefore, there is a pressing need to identify the RNA-binding residues in proteins by an unbiased and systematic manner.

The structure-based methods utilize information derive from the structure, typically based on shape and biophysical characteristics of the protein surface [3]. Chen and Lim [4] proposed a computational method based on structure information including electrostatics, evolution and geometry, which are derived from the protein structure and homologous sequences. Maetschke and Yuan [5] have demonstrated that different geometrical and network topological properties of protein structures, such as retention coefficient, betweenness-centrality, accessible surface area and PSI-BLAST profile, can effectively improve the prediction accuracy of protein-RNA interactions.

Although computational prediction of RNA-binding residues is an established field, the question is far from being well solved. The problem is complicated by the massive diversity in protein recognition folds as well as in RNA conformational states [6]. Besides, the imbalanced problem [7] exists widely in protein-RNA binding sites prediction because the number of interaction residues is usually much smaller than that of non binding sites in proteins. Improvement in addressing imbalance problem should also enhance the performance of RNA-binding sites prediction.

In this paper, we proposed a hybrid approach, referred as to XPredRBR, to predict RNA binding residues in proteins. Our method differs from previous methods by taking advantage of new features and powerful prediction model. Specifically, two types of new features derived from residue interaction network and solvent exposure are exploited to enrich the conventional sequence- and structural features. The eXtreme Gradient Boosting algorithm (XGBoost) [8], the winner of several machine learning competitions [9] in recent years, is used to build the prediction model. We conducted performance evaluation on the training dataset RBP170 and an independent test set RBP101. The empirical experiments show that our proposed method outperformed four classical classifiers.

2   Results

2.1   Performance Evaluation Metrics

To make objective and comprehensive evaluation, we employ different performance measures, including sensitivity (SEN), specificity (SPE), precision (PRE), accuracy (ACC), F1-Score and Matthews Correlation Coefficient

(MCC) score, to assess the prediction results. As SEN and SPE can be used to plot the receiver operating characteristic (ROC) curves and the area under curve (AUC) is also a widely used performance measure, the AUC measure is also used in our evaluation experiments. $F1$-Score is the harmonic mean of SEN and SPE, and the best $F1$-Score point is usually chosen as the cutoff for SEN and SPE in ROC curve. MCC value ranges from -1 (none prediction is correct) to $+1$ (all predictions are correct).

2.2   Comparison with Classical Algorithms

To justify the performance of XPredRBR, we conducted comparison experiments with three traditional classifiers, including Support Vector Machine (SVM) [10], Random Forest (RF) [11], Adaboost [12]. Note that the tested methods utilize all 204 features included in the training dataset (RBP170) and the performance measures are obtained by 10-fold cross-validation. We list in Table 1 the experimental results of five tested methods. It can be found that XPredRBR achieved accuracy 0.861, sensitivity 0.872 and MCC 0.584. The results imply that our proposed method significantly boost the performance in predicting RNA-binding residues in proteins.

Table 1. Performance comparison of various prediction approaches using 10-fold cross-validation on the training set.

Method	ACC	SEN	SPE	PRE	F1-Score	MCC	AUC
Adaboost	0.811	0.730	0.825	0.406	0.514	0.431	0.852
SVM	0.820	0.802	0.838	0.419	0.535	0.476	0.865
Random forest	0.823	0.814	0.845	0.437	0.566	0.517	0.891
XPredRBR	**0.861**	**0.872**	**0.858**	**0.512**	**0.619**	**0.584**	**0.941**

2.3   Evaluation on the Independent Testing Set

We also evaluated the proposed XPredRBR on an independent test dataset. Due to the imbalance between positive samples and negative samples, the receiver operating characteristic (ROC) curve and AUC value are rather appropriate to evaluate the overall performance. Figure 1 shows the ROC curves and AUC scores of tested methods on the RBP101 dataset. XPredRBR, SVM, Adaboost and Random Forest achieve AUC values 0.835, 0.801 and 0.776, 0.765, respectively. The results show that XPredRBR improved AUC values by 2%–6%, compared to three traditional classifiers.

2.4   Time Overhead

We further compared the time overhead of XPredRBR with other methods three classical classifiers and our previous method PredRBR [13].

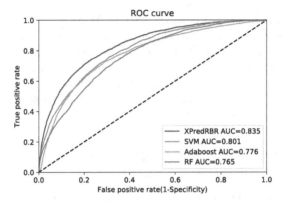

Fig. 1. The ROC curves of XPredRBR and three traditional classifiers on the independent test set sRBP101 dataset.

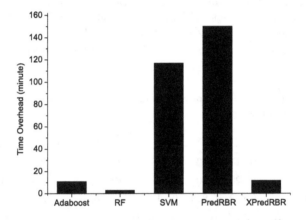

Fig. 2. Comparison of time overhead for training and test of XPredRBR and four counterparts.

The computer configuration is Dell K40 with 2.0 GHZ CPU and 2G memory. We recorded the running time for training on 46,638 samples add test on 32,577 samples of each method, as shown in Fig. 2. It can be seen that XPredRBR is ten times faster than SVM, and 12 times faster than GBT. Although XPredRBR runs at approximately equal speed with Adaboost or a little slower than random forest, its performance is significantly superior than the latter two methods, as shown in Fig. 1.

3   Discussion

Protein-RNA interactions are of great importance in many fundamental cellular processes. As experimental method is time- and cost-consuming for identify RNA-binding residues, quite a number of computational methods have been developed to predict the potential RBRs in proteins. However, the problem is far from being well settled. An effective approach to improve the prediction accuracy is develop new predictive features, in addition to conventional sequence-based and structure-based features [14,15]. In this paper, we proposed two types of new features derived from residue interaction network and solvent exposures. Our empirical experiments on the RBP170 dataset have demonstrated that these new features effectively improve the prediction performance.

Another technical way to promote the performance of computational methods is to make advantage of more powerful predictive models [16]. We adopted the efficient and scalable variant of the Gradient Boosting eXtreme gradient boosting algorithm (XGBoost) [8], to built the prediction model, as XGBoost possess the ability to handle the highly diverse and complex features and produces comparable or even better predictive accuracy. Our performance comparison showed that our method acquire remarkable improvement in both accuracy and efficiency, compared to our previous methods [13] and other counterparts.

Furthermore, a few studies seek to identify Protein-nucleic acid (NA) interactions by combining RNA- and DNA- binding residues into one problem [15], which are always treated as different problems or trained with different datasets within the same framework.

4 Materials and Methods

4.1 Datasets

Two data sets built in our previous studies [13], RBP170 and RBP101, are used to compare the performance of XPredRBR and other methods. PISCES server was used to create RBP170, by extracting protein sequences with $\leq 30\%$ sequence identity and at least resolution of 3.5 Å from all protein-RNA complexes in the PDB [17] (May 2010 release). For inclusion in RBP170, each protein is required to be longer than ≥ 40 amino acids and include more than 3 RNA-binding amino acids. Meanwhile, the RNA molecule in the protein-RNA complex must be longer than 5 nucleotides. Table 2 shows summary statistics of the two data sets.

Table 2. Number of the positive samples and negative samples of the two datasets.

	RBP170	RBP101
Number of PDB files	124	90
Number of protein chains	170	101
Binding site	6 750	2 886
Non-binding site	39 888	29 691

Besides, BPP101 is an independent dataset that consists of 101 RNA interacting chains extracted from 90 RNA-binding proteins. BPP101 was created

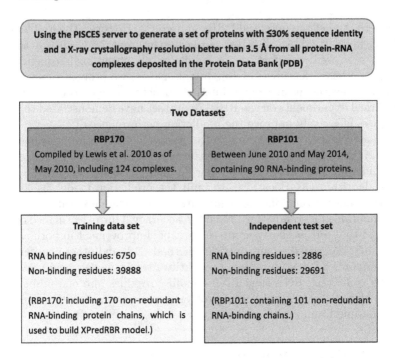

Fig. 3. The flowchart of building training set and testing set. The primary RBPs collection of 214 reviewed RBPs was obtained from Protein Data Bank (PDB).

by selecting the proteins with ≤30% sequence identity and 3.5 Åor higher X-ray crystallography resolution from all protein-RNA complexes deposited in the Protein Data Bank (PDB). Figure 3 shows the flowchart of the construction of the training and testing datasets.

4.2   Features Extraction

In addition to the sequence and structural features used in our previous work [13], two types of novel features are derived from residue interaction network information and solvent exposures. In total, 204 features came from four categories are included. The detailed procedures of the feature extraction are described as below:

Residue Interaction Network Features. Two residues in a structure are defined as physical interaction if the distance between their centers is within 6.5 Å. Residue interaction network (RIN) has been proved to be useful in quite a few bioinformatics applications [16]. In this study, we use NAPS [18] to compute in total 7 topological features that describe the local environment of the target residues in the network, including betweenness, closeness, degree, clustering coefficient, eigenvector centrality, eccentricity and average nearest neighbor degree.

Solvent Exposure Features. Half-sphere exposure (HSE) is predictive in protein stability, conservation among different folds, computational speed and predictability. HSE separates a residue sphere into two half spheres: HSE-up corresponds to the upper sphere in the direction of the chain side of the residue, while HSE-down points to the lower sphere in the direction of the opposite side. HSEpred [19] is used to compute the features of HSEup, HSE-down and CN (coordination number). Based on structure, we use hsexpo to calculate the exposure features such as HSEAU (number of C_α atoms in the upper sphere), HSEAD (number of C_α atoms in the lower sphere), HSEBU (the number of C_β atoms in the upper sphere), HSEBD (the number of C_β atoms in the lower half sphere), CN (coordination number).

4.3   eXtreme Gradient Boosting Algorithm

The eXtreme gradient boosting algorithm (XGBoost) [20] is an ensemble of K Classification and Regression Trees (CART) $T_1(x, y).....T_K(x, y)$, where $x = \{x_1, x_2, \ldots, x_n\}$ $(i = 1..n)$ is the features in the training set associated with a potential RNA-binding residue and its true class label $y = \{y_1, y_2, \ldots, y_n\}$ with $y_i \in \{-1, +1\}$ $(i = 1..n)$ where "-1" represents non-binding site and "$+1$" denotes RNA-binding residue. The purpose of the gradient tree boosting algorithm is to build an effective classifier to predict protein-RNA binding residues. Given that each CART assigns a real score to each leaves (outcome or target), the prediction scores if individual CART are summed up to get the final score and evaluated through K additive functions, as shown in Eq. (1):

$$\widehat{y_i} = \sum_{i=1}^{K} f_k(x_i), f_k \in F \tag{1}$$

where f_k represents an independent tree structure with leaf scores and F is the space of all CART. The regularized objective function to be optimized is as below:

$$Obj(\Theta) = \sum_{i}^{n} l(y_i, \widehat{y_i}) + \sum_{k}^{K} \Omega(f_k) \tag{2}$$

The first term $l(y_i, \widehat{y_i})$ is a differentiable loss function, which measures the difference between the predicted $\widehat{y_i}$ and the true class label y_i. The second is a regularization term $\Omega(f_k)$ which penalizes the complexity of the model so as to avoid overfitting. It is formulated as $\Omega(f) = \lambda T + \frac{1}{2}\lambda \sum_{j=1}^{T} w_j^2$, where T and w_j are the number of leaves and the score of j-th leaf, respectively. γ and λ are two constants used to leverage the degree of penalty. Apart from the use of regularization, shrinkage and descriptor subsampling are two additional techniques used to prevent overfitting [20].

The training procedure in XGBoost is summarized as follows:

1. For each feature sort the numbers and scan the best splitting point (lowest gain).
2. Choose the descriptor with the best splitting point that optimizes the training objective.

3. Continue splitting (as in (I) and (II)) until the specified maximum tree depth is reached.
4. Assign prediction score to the leaves and prune all negative nodes (nodes with negative gains) in a bottom-up order.
5. Repeat the above steps in an additive manner until the specified number of rounds (trees K) is reached.

Since additive training is used, the prediction $\widehat{y}_i^{(t)}$ at step t is expressed as

$$\widehat{y}_i^{(t)} = \sum_{k=1}^{K} f_k(x_i) = \widehat{y}_i^{(t-1)} + f_t(x_i), \tag{3}$$

and thus Eq. (4) can be written as

$$Obj(\Theta)^{(t)} = \sum_{i}^{n} l(y_i, \widehat{y}_i^{(t-1)} + f_t(x_i)) + \Omega(f_t). \tag{4}$$

By taking the Taylors expansion of the loss function to the second order, we have

$$Obj(\Theta)^{(t)} = \sum_{i}^{n} \left[l(y_i, \widehat{y}_i^{(t-1)}) + g_i f_t(x_i) + \frac{1}{2} h_i f_t^2(x_i) \right] + \Omega(f_t) \tag{5}$$

where $g_i = \partial_{\widehat{y}_i^{(t-1)}} l(y_i, \widehat{y}_i^{(t-1)})$ and $h_i = \partial_{\widehat{y}_i^{(t-1)}}^2 l(y_i, \widehat{y}_i^{(t-1)})$ are respectively first and second order terms of the loss function. A simplified objective function without the constant term at step t is as follows:

$$Obj(\Theta)^{(t)} = \sum_{i}^{n} \left[g_i f_t(x_i) + \frac{1}{2} h_i f_t^2(x_i) \right] + \Omega(f_t) \tag{6}$$

The objective function can be written by expanding the regularization term as:

$$\begin{aligned} Obj(\Theta)^{(t)} &= \sum_{i}^{n} \left[g_i f_t(x_i) + \frac{1}{2} h_i f_t^2(x_i) \right] + \gamma T + \frac{1}{2} \lambda \sum_{j=1}^{T} w_j^2 \\ &= \sum_{j=1}^{T} \left[\left(\sum_{i \in I_j} g_i \right) w_j + \frac{1}{2} \left(\sum_{i \in I_j} h_i + \lambda \right) w_j^2 \right] + \gamma T \end{aligned} \tag{7}$$

where $I_j = \{i | q(x_i) = j\}$ is the instance set of leaf j, for a given structure $q(x)$ the optimal leaf weight, w_j^*, and the optimal objective function which measure how good the structure is are given by Eqs. (8) and (9) respectively.

$$w_j^* = -\frac{G_i}{H_i + \lambda} \tag{8}$$

$$Obj^* = -\frac{1}{2} \sum_{j=1}^{T} \frac{G_j^2}{H_j + \lambda} + \lambda T \tag{9}$$

where $G_j = \sum_{i \in I_j} g_j$, $H_j = \sum_{i \in I_j} h_j$.

Equation (10) is used to score a leaf node during splitting. The first, second and third term of the equation stands for the score on the left, right and the original leaf respectively. Moreover, the final term, γ is regularization on the additional leaf.

$$Gain = \frac{1}{2}\left[\frac{G_L^2}{H_L + \lambda} + \frac{G_R^2}{H_R + \lambda} + \frac{(G_L + G_R)^2}{H_L + H_R + \lambda}\right] - \lambda \qquad (10)$$

4.4 The XPredRBR Framework

Figure 4 shows the flowchart of the proposed method experiment. The individual features are extracted from protein-RNA complexes, including sequence features, structural neighborhood features, residue interaction network features and solvent exposure features.

In our experiment, the number of non-binding residues is about six times as much as the number of binding residues on the training dataset RBP170. To deal with the imbalance problem, we use random under-sampling approach

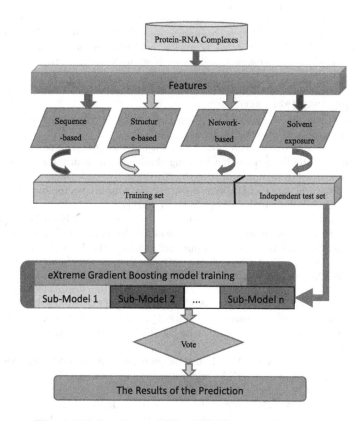

Fig. 4. The flowchart of XPredRBR learning framework.

to create new balanced dataset. Therefore, at each 10-fold cross-validation step, nine subsets are merged as one and then random sub-sampling is run to generate a balanced training dataset, and the remaining subset is used as test set to evaluate the performance of the classifier trained by XGBoost algorithm. Finally, we utilize the independent test dataset (RBP101) to compare the performance of our proposed method with the previous studies.

References

1. Glisovic, T., Bachorik, J.L., Yong, J., Dreyfuss, G.: RNA-binding proteins and post-transcriptional gene regulation. FEBS Lett. **582**(14), 1977–1986 (2008)
2. Re, A., Joshi, T., Kulberkyte, E., Morris, Q., Workman, C.T.: RNA–protein interactions: an overview. In: Gorodkin, J., Ruzzo, W.L. (eds.) RNA Sequence, Structure, and Function: Computational and Bioinformatic Methods. MMB, vol. 1097, pp. 491–521. Humana Press, Totowa, NJ (2014). https://doi.org/10.1007/978-1-62703-709-9_23
3. Miao, Z., Westhof, E.: A large-scale assessment of nucleic acids binding site prediction programs. PLoS Comput. Biol. **11**(12), e1004639 (2015)
4. Chen, Y., Lim, C.: Predicting RNA-binding sites from the protein structure based on electrostatics, evolution and geometry. Nucleic Acids Res. **36**(5), e29 (2008)
5. Maetschke, S., Yuan, Z.: Exploiting structural and topological information to improve prediction of RNA-protein binding sites. BMC Bioinform. **10**, 341 (2009)
6. Miao, Z., Westhof, E.: Prediction of nucleic acid binding probability in proteins: a neighboring residue network based score. Nucleic Acids Res. **43**(11), 5340–5351 (2015)
7. He, H., Garcia, E.A.: Learning from imbalanced data. IEEE Trans. Knowl. Data Eng. **21**(9), 1263–1284 (2009)
8. Babajide Mustapha, I., Saeed, F.: Bioactive molecule prediction using extreme gradient boosting. Molecules **21**(8), 983 (2016)
9. Adam-Bourdarios, C., Cowan, G., Germain-Renaud, C., Guyon, I., Kégl, B., Rousseau, D.: The Higgs machine learning challenge. In: Journal of Physics: Conference Series, vol. 664, no. 7, p. 072015 (2015)
10. Chang, C.-C., Lin, C.-J.: LIBSVM: a library for support vector machines. ACM Trans. Intell. Syst. Technol. (TIST) **2**(3), 27 (2011)
11. Svetnik, V., Liaw, A., Tong, C., Culberson, J.C., Sheridan, R.P., Feuston, B.P.: Random forest: a classification and regression tool for compound classification and QSAR modeling. J. Chem. Inf. Comput. Sci. **43**(6), 1947–1958 (2003)
12. Rätsch, G., Onoda, T., Müller, K.-R.: Soft margins for AdaBoost. Mach. Learn. **42**(3), 287–320 (2001)
13. Tang, Y., Liu, D., Wang, Z., Wen, T., Deng, L.: A boosting approach for prediction of protein-RNA binding residues. BMC Bioinform. **18**(13), 465 (2017)
14. Luo, J., Liu, L., Venkateswaran, S., Song, Q., Zhou, X.: RPI-Bind: a structure-based method for accurate identification of RNA-protein binding sites. Sci. Rep. **7**, 614 (2017)
15. Yan, J., Kurgan, L.: DRNApred, fast sequence-based method that accurately predicts and discriminates DNA- and RNA-binding residues. Nucleic Acids Res. **45**(10), e84 (2017)
16. Pan, X., Shen, H.-B.: RNA-protein binding motifs mining with a new hybrid deep learning based cross-domain knowledge integration approach. BMC Bioinform. **18**(1), 136 (2017)

17. Rose, P., Prlic, A., Altunkaya, A.: The RCSB protein data bank: integrative view of protein, gene and 3D structural information. Nucleic Acids Res. **45**(D1), D271–D281 (2017)
18. Chakrabarty, B., Parekh, N.: NAPS: network analysis of protein structures. Nucleic Acids Res. **44**(W1), W375–W382 (2016)
19. Song, J., Tan, H., Takemoto, K., Akutsu, T.: HSEpred: predict half-sphere exposure from protein sequences. Bioinformatics **24**(13), 1489–1497 (2008)
20. Chen, T., Guestrin, C.: XGBoost: a scalable tree boosting system. In: Proceedings of the 22nd ACM SIGKDD International Conference on Knowledge Discovery and Data Mining, pp. 785–794. ACM (2016)

A Biologically Meaningful Extension of the Efficient Method for Deleterious Mutations Prediction in RNAs: Insertions and Deletions in Addition to Substitution Mutations

Alexander Churkin[1(✉)] and Danny Barash[2]

[1] Sami Shamoon College of Engineering, Beer Sheva, Israel
alexach3@sce.ac.il
[2] Ben-Gurion University of the Negev, Beer Sheva, Israel
dbarash@cs.bgu.ac.il

Abstract. The RNAmute and MultiRNAmute interactive java programs were developed to predict deleterious point mutations in RNA sequences, which intently cause a global conformational rearrangement of the secondary structure of the functional RNA molecules and thereby disrupt their function. RNAmute was designed to deal with only single point mutations in a brute force manner, while the MultiRNAmute tool uses an efficient approach to deal with multiple point mutations. The approach used in MultiRNAmute is based on the stabilization of the suboptimal RNA folding solutions and/or destabilization of the optimal MFE structure of the wild type RNA molecule. Both programs utilize the Vienna RNA package to find the optimal and suboptimal (in case of MultiRNAmute) RNA secondary structures. The main limitation of both programs is their ability to predict only substitution mutations and these programs were not designed to work with deletion or insertion mutations. Herein we develop an efficient approach, based on suboptimal folding solutions, to predict multiple point mutations consisting of deletions, insertions and substitution mutations. All RNAmute algorithms were validated on the TPP-riboswitch and some other functional RNAs.

Keywords: Multiple point mutations · RNA folding predictions
Suboptimal structure · RNA dot plot

1 Introduction

The RNA deleterious mutation prediction problem is a sub-problem of the RNA folding prediction problem, which is fundamental in RNA bioinformatics. Thus, all tools for deleterious mutations analysis utilize methods developed for the RNA folding problem. The most accurate methods for RNA folding prediction are energy minimization methods that use dynamic programming, which are the mfold server [1], RNAstructure [2] and the Vienna RNA package and server [3, 4]. First publicly available methods for the analysis of deleterious mutations in RNAs were the

© Springer International Publishing AG, part of Springer Nature 2018
F. Zhang et al. (Eds.): ISBRA 2018, LNBI 10847, pp. 174–178, 2018.
https://doi.org/10.1007/978-3-319-94968-0_15

RNAmute java tool [5] and a web server called RDMAS [6]. Both these methods utilize the Vienna RNA package for RNA folding prediction and are able to analyze only single point mutations in RNA sequences. To deal with multiple point deleterious mutation, the MultiRNAmute [7] program was developed, which uses an efficient method to find multiple point mutations using suboptimal folding solutions of RNA sequence. A major limitation of the above described methods is that these methods deal only with substitution mutations, but not with insertions or deletions. In this paper, we suggest an extension to our efficient method MultiRNAmute to be able to analyze multiple point deleterious mutations, including deletions, insertions and substitutions.

2   Method

The algorithm consists of four stages. These include the calculations of optimal and suboptimal secondary structure solutions using RNAfold [4], filtering number of suboptimal solutions to reduce computational complexity, finding the stems for optimal and each suboptimal solution and finally gathering all destructive mutations as output. General flowchart of the algorithm is shown in Fig. 1.

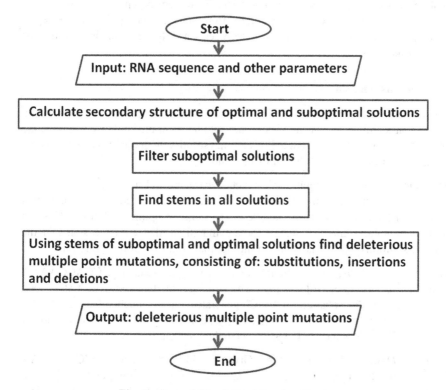

Fig. 1. General flowchart of the algorithm.

2.1   Calculation of Optimal and Suboptimal Solutions

Given the input RNA sequence, the first stage of the algorithm is to find its optimal secondary structure (using RNAfold [4]) and many suboptimal solutions (using RNAsubopt [4]). Both optimal and suboptimal solutions are compactly stored in the dot-bracket notation.

2.2   Filtering

The number of suboptimal solutions collected in previous stage may be huge, thus some filtering is needed. The filtering stage consists of dropping suboptimal solutions that are close to the optimal, unstable solutions that are unlikely to be stabilized by mutations and also solutions that are similar to each other. All distance calculations are based on the base pair distance or tree edit distance (determined by the user) between dot-bracket representations of the secondary structures of two RNA sequences.

2.3   Stems Generation

The next step is to find all stems for the optimal solution and each one of the suboptimal solutions that passed the filtering stage. Each stem is represented by its start and end positions in the secondary structure. Both start and end positions are represented by pair of indexes of nucleotides in RNA sequence, forming the first and the last base-pair in the stem.

2.4   Mutations Analysis

The main idea of our efficient approach is to find deletions, insertions and substitution mutations that stabilize one of the suboptimal folding solutions and/or destabilize the optimal solution in such a way that the suboptimal solution with mutations will become more stable than the optimal solution of the wild type.

Substitutions. Stabilizing mutations in such a case are substitutions that elongate stems of the suboptimal solution, and destabilizing mutations are substitutions that break some of the base-pairs in the middle of the stems of optimal solutions. The best substitution mutations that we search for are mutations that elongate stems in the suboptimal solutions and break stems in the optimal solutions at the same time.

Example of a Stabilizing Substitution: Given a wild type sequence: CCACCAAAA GGCGG and its structure: "((.((....)).))". The structure has two stems with stem loop between the stems. The substitution "A3G" generates a new sequence CCGCCAAAA GGCGG with more stable structure "(((((....)))))", containing one stable stem. Another mutation "C12U" will generate the same secondary structure.

Example of a Destabilizing Substitution: Given a wild type sequence: CGCGCAAAA GCGCG and its structure: "(((((....)))))". The structure has one long and stable stem. The substitution "C3A" generates a new sequence CGAGCAAAAGCGCG with less stable structure "((.((....)).))", containing a stem loop between two short stems.

More examples of stabilizing and destabilizing substitutions are available in [7].

Deletions. Stabilizing deletions are deletions of nucleotides between two stems in the suboptimal solutions such that two stems become connected and form one single stem, or deletions that make a stem-loop between two stems to be more stable. Destabilizing deletions are deletions of nucleotides that shorten some stems in the optimal solutions or generate stem loops inside some stems of the optimal solutions.

Example of a Stabilizing Deletion: Given a wild type sequence: CCCCAAAAGGCGG and its structure: "(((((....)).))". The structure has two stems with a stem loop between the stems. The deletion "C11" generates a new sequence CCCCAAAAGGGG with more stable structure "((((....))))", containing one stable stem.

Example of a Destabilizing Deletion: Given a wild type sequence: CCAC-CAAAAGGUGG and its structure: "(((((....)))))". The structure has one long and stable stem. The deletion "A3" generates a new sequence CCCCAAAAGGUGG with less stable structure "((((....)).))", containing a stem loop between two short stems.

Insertions. Stabilizing insertions are insertions of nucleotides that elongate some of the stems in the suboptimal solutions or connect two stems to one more stable stem. Destabilizing insertions are insertions of nucleotides that shorten or break long stems to two shorter stems in the optimal solutions.

Example of a Stabilizing Insertion: Given a wild type sequence: CCCCAAAA GGCGG and its structure: "((((....)).))". The insertion "3G" generates a new sequence CCGCCAAAAGGCGG with more stable structure "(((((....)))))", containing one long stable stem instead of two shorter stems.

Example of a Destabilizing Insertion: Given a wild type sequence: CCCCAAAA GGGG and its structure: "((((....))))". The insertion "3G" generates a new sequence CCGCCAAAAGGGG with less stable structure "((.((....))))", containing a stem loop between two short stems.

Finally, the algorithm uses different permutations of deletion, insertion and substitution mutations listed above to generate deleterious multiple point mutations. The percentage of deletions, insertions and substitutions may be determined by the user, thus lowering the number of possible multiple point mutations in the output. Also, it is possible to limit the algorithm with maximal desired number of deleterious multiple point mutations, and the algorithm will stop when it reaches this number. Before accepting any of the multiple point mutations to the outputs, the algorithm checks if this mutation indeed deleterious by using RNAfold [4] and calculation of distance between the mutant MFE secondary structure and the wild type. If the limit for maximum number of output mutations is specified, the algorithm will prefer the mutations that at the same time stabilize the suboptimal solution and destabilize the optimal one. The extension is biologically important as it offers many more possibilities for the prediction of deleterious mutations that were not considered by neither RNA-Mute [5] nor MultiRNAMute [7].

References

1. Zuker, M.: Mfold web server for nucleic acid folding and hybridization prediction. Nucl. Acids Res. **31**, 3406–3415 (2003)
2. Reuter, J.S., Mathews, D.H.: RNAstructure: software for RNA secondary structure prediction and analysis. BMC Bioinform. **11**, 129 (2010)
3. Hofacker, I.L., Fontana, W., Stadler, P.F., Bonhoeffer, L.S., Tacker, M., Schuster, P.: Fast folding and comparison of RNA secondary structures. Monatsh. Chem. **125**, 167–188 (1994)
4. Hofacker, I.L.: Vienna RNA secondary structure server. Nucl. Acids Res. **31**, 3429–3431 (2003)
5. Churkin, A., Barash, D.: RNAmute: RNA secondary structure mutation analysis tool. BMC Bioinform. **7**, 221 (2006)
6. Shu, W., Bo, X., Liu, R., Zhao, D., Zheng, Z., Wang, S.: RDMAS: a web server for RNA deleterious mutation analysis. BMC Bioinform. **7**, 404 (2006)
7. Churkin, A., Barash, D.: An efficient method for the prediction of deleterious multiple-point mutations in the secondary structure of RNAs using suboptimal folding solutions. BMC Bioinform. **9**(1), 222 (2008)

Screening of Sonic Hedgehog (Shh) Inhibitors in the Hedgehog Signaling Pathway from Traditional Chinese Medicine (TCM) Database Through Structure-Based Pharmacophore Design

Ilmi Fadhilah Rizki, Mochammad Arfin Fardiansyah Nasution,
Syafrida Siregar, Mega Maulina Ekawati,
and Usman Sumo Friend Tambunan[(✉)]

Faculty of Mathematics and Natural Sciences, Universitas Indonesia, Kampus UI
Depok, Depok, West Java 16424, Indonesia
usman@ui.ac.id

Abstract. Colorectal cancer (CRC) is the second most death-leading cancer in the world. The development of CRC is closely related to the hedgehog signaling pathway. The abnormal activation of the pathway will initiate the binding of Sonic Hedgehog (Shh) to Ptch1 that can trigger the growth of abnormal cells that cause CRC. Traditional Chinese Medicine (TCM) compounds were used to inhibit Shh through structure-based pharmacophore design. This research was started by finding the pharmacophore feature of Shh, continued with docking simulation of Shh with TCM compounds. The best three TCM compounds, which are TCM-8941, TCM-28794 and TCM-32808, give the best ligand interaction and have the lowest Gibbs free binding energy. They also have good pharmacological properties that have been analyzed by using Toxtree v2.6.13, SwissADME and admetSAR. For further research, these TCM compounds may be used as drug candidates in colorectal cancer.

Keywords: Sonic hedgehog · Colorectal cancer
Traditional Chinese Medicine · Structure-Based pharmacophore design

1 Introduction

Colorectal cancer (CRC) is one of the most common cancer and become the second-most cancer that leading cause of death in the world [1]. The development of CRC is closely related to the hedgehog signaling pathway. In mammals, hedgehog ligands include Sonic hedgehog (Shh), Indian hedgehog (Ihh) and Desert hedgehog (Dhh) [2]. Shh is widely overexpressed in colorectal cancer [3]. The binding of Shh ligand to Ptch1 relieves the repression of Smoothened (Smo) protein, leading to the activation of Gli transcription factors and increased transcription of Gli target genes, including Gli1 and Ptch1 [4]. Traditional Chinese Medicine (TCM) as one of the oldest traditional medication from China has been adopted worldwide. Nowadays, TCM has increasingly

© Springer International Publishing AG, part of Springer Nature 2018
F. Zhang et al. (Eds.): ISBRA 2018, LNBI 10847, pp. 179–184, 2018.
https://doi.org/10.1007/978-3-319-94968-0_16

become popular over the world including in cancer patients, because it has important function in minimizing disability, protecting cancer patient against suffering from complications, and helping patients to live well [5].

2 Research Methodology

2.1 Preparation of the Traditional Chinese Medicine and Shh Protein

There were 30,620 Traditional Chinese Medicine compounds which were collected from TCM database (http://tcm.cmu.edu.tw/) [6]. Then, all TCM compounds were further analyzed based on druglikeness and toxicity properties by using the DataWarrior v4.6.1 software [7]. All ligands were optimized using MMFF94x (modified) force field with root mean square (RMS) gradient of 0.001. The optimized TCM ligands were later stored in .mdb database format. This research used the 3D structure of the Shh-Hhip complex taken from RCSB PDB (PDB ID: 3HO5) [8]. The protein optimization process used Amber10:EHT forcefield in 'Gas Phase' solvation in the potential setup. Protein optimization used LigX parameters, such as protonation of the protein, energy minimization for geometry optimization, and using the RMS gradient of 0.05 kcal/mol Å. Then, the optimized Shh protein was saved in .moe format.

2.2 Pharmacophore Generation

The starting point for generating pharmacophore was done by docking simulation of Shh to robotnikinin as standard ligands. This simulation was begun with determining of potential sites for ligand binding by using 'SiteFinder' feature in MOE 2014.09 software [9]. The pharmacophore features can be determined by using Pharmacophore Query Editor based on the binding of robotnikinin with Shh protein. Later, this pharmacophore feature was saved in .ph4 file, which was used for docking simulation.

2.3 Pharmacophore-Based Molecular Docking Simulation

Pharmacophore-based molecular docking simulations were carried out using MOE 2014.09 with Amber10:EHT forcefield in 'Gas Phase' solvation. The molecular docking simulation was conducted twice by using 'Rigid Receptor' and continued with 'Induced Fit' protocol. All docking simulations must be in pharmacophore editor and placement.

2.4 Pharmacological Properties, ADME and Toxicity Prediction

The three best ligands that obtained through this simulation underwent computational pharmacological properties screening by using Toxtree v2.6.13 to predict their carcinogenicity/mutagenicity properties. Furthermore, SwissADME [10] and admet-SAR [11] software were used as well to predict the bioavailability and pharmacokinetic properties of the three best ligands obtained.

3   Results and Discussions

3.1   Preparation of the Standard Ligand and the Traditional Chinese Medicine

There were 30,620 compounds that obtained from TCM database. First, all TCM compounds must be screened for druglikeness and toxicity properties by using DataWarrior v4.6.1. Throughout this screening process, about 6,341 TCM compounds were retrieved that will be further optimized. Three-dimensional structure of Shh protein was obtained from Research Collaboratory for Structural Bioinformatics in the Protein Data Bank (RCSB-PDB) with PDB ID: 3HO5. The structure from PDB data were imported into MOE 2014.09 software. The protein optimization process was used Amber10:EHT forcefield in gas phase solvation as potential setup. These parameters were chosen because it is suitable for proteins and nucleic acids. Solvation model which used for protein optimization was 'Gas Phase'. Both 'Fix Hydrogens' and 'Fix Charges' must be done in this optimization in order to add the missing hydrogen atoms and to rectify the charge in the protein system, respectively. The last step of protein optimization used LigX parameters. In this research, the strength value choosen was 100,000. The greater the strength the less atom will deviate from its initial coordinates. Root mean square (RMS) gradient was using 0.05 kcal/mol.Å. The RMS value is a standard choice which appropriates for the protein and related to the energy minimization of the protein optimization.

3.2   Structure-Based Pharmacophore Generation

The pharmacophore generated by MOE 2014.09 using the protein structure obtained from molecular docking simulation of Shh protein with robotnikinin as a standard. The docking simulation was performed with Amber10:EHT force field in gas phase solvation. The determination of Shh binding site was done by 'Site Finder' feature. By using this feature, it was determined that the binding site of this protein consists of 13 amino acid residues and one cofactor (Glu53, His133, His134, Ser135, Ser138, Leu139, His140, Asp147, Trp172, Tyr174, Glu176, His180, His182, Zn400). The selected residues were used for molecular docking simulation with induced fit docking protocols. Further, the structure-based pharmacophore models were generated using the interaction of robotnikinin to Shh protein from previous molecular docking simulation. Two acceptor (Acc) features were generated based on the interaction points available from the active site. The first Acc feature was interacted to Zn400 and the second Acc feature was interacted to Ser138. The pharmacophore generation and validation processes were done by using robotnikinin as the standard. After two features generated from 'Pharmacophore Editor', those were two Acc features, then continued by creating excluded volumes around pocket atoms. The pharmacophore features in this research were validated because robotnikinin molecules fit in all features generated. Then, database searching was done with 6,341 TCM compounds and obtained 4,474 compounds that hits to the pharmacophore features.

3.3 Pharmacophore-Based Molecular Docking Simulation

There are 4,474 TCM compounds obtained from pharmacophore search as ligand database. In the beginning of molecular docking simulation, these ligands are used for virtual screening simulation. From this simulation, 2,722 ligands were picked out based on the Gibbs free binding energy ($\Delta G_{binding}$) value lower than robotnikinin as a standard. The most negative $\Delta G_{binding}$ result indicates that the ligand conformation is the most stable. Furthermore, 2,722 ligands were used in molecular docking simulation by using 'Rigid Receptor' and continued with 'Induced Fit' protocol with. This led three best ligands that have the lowest $\Delta G_{binding}$ among others and have Root Mean Square Deviation (RMSD) value that lower than 2. The docking simulation result of three best ligands is shown in Table 1.

Table 1. $\Delta G_{binding}$ and RMSD value from the docking simulation

Ligand	$\Delta G_{binding}$ (kcal/mol)	RMSD
TCM-8941	−14.920	1.874
TCM-28794	−11.520	0.755
TCM-32808	−8.952	1.222
Robotnikinin[a]	−7.344	1.603

[a]Standard ligand

Besides Gibbs free binding energy and RMSD value, the molecular interaction of Shh protein and ligand must be observed. This molecular interaction includes hydrogen bond and Van der Waals interaction between amino acid residu in Shh protein with each ligand. The molecular interaction between Shh protein and the three best ligands of TCM, as well as the standard ligands, is shown in Table 2.

Table 2. Ligand interaction of Shh protein binding site with TCM compounds

Ligand	Interaction Site
TCM-8941	Glu89, Lys87, Glu126, Thr125, His180, Ala179, His133, His134, Ser177, Asp147, Ser138, His140, Ser135, His182, Glu176, Zn400, Ca401
TCM-28794	Arg155, Arg153, His180, Glu176, Trp172. Arg123, Tyr158, Thr149, Ala179, Lys87, Glu89, His133, His134, Thr125, Asp147, His140, Tyr174, His182, Ser135, Glu126, Ser138, Glu53, Leu139, Glu137, Zn400, Ca401
TCM-32808	Glu176, His180, Lys87, Glu89, His133, Ser138, His182, His140, Thr125, Asp147, Ser135, His134, Glu126, Zn400, Ca401

Red color indicates the active site of Shh protein contact.

3.4   Pharmacological Properties, ADME and Toxicity Prediction

In this study, all three ligands were going through some pharmacological and toxicity prediction by using Toxtree for carcinogenicity and mutagenicity prediction; SwissADME for pharmacokinetics prediction and admetSAR for toxicity prediction. All these pharmacological prediction are shown in Tables 3, 4, and 5.

Table 3. Carcinogenicity and mutagenicity prediction by Toxtree

Parameters	TCM-8941	TCM-28794	TCM-32808
Negative for genotoxic carcinogenicity	No	No	No
Negative for nongenotoxic carcinogenicity	No	No	No
Potential salmonella typhimurium TA100 mutagen based on QSAR	No	No	No
Potential carcinogen based on QSAR	No	No	No

Table 4. Pharmacokinetics prediction by SwissADME

Ligand	Pharmacokinetics						
	1	2	3	4	5	6	7
TCM-8941	Low	No	No	No	No	No	No
TCM-28794	Low	Yes	No	No	No	No	No
TCM-32808	Low	No	No	No	No	No	No

1 = GI Absorption; 2 = P-gp substrate; 3 = CYP1A2 inhibitor; 4 = CYP2C19 inhibitor;
5 = CYP2C9 inhibitor; 6 = CYP2D6 inhibitor; 7 = CYP3A4

Table 5. Toxicity prediction by admetSAR

Ligands	Toxicity			
	AMES Toxicity	Carcinogens	Biodegradation	Acute Oral Toxicity
TCM-8941	Non toxic	Non Carc.	Ready	III
TCM-28794	Non toxic	Non Carc.	Not ready	III
TCM-32808	Non toxic	Non Carc.	Ready	III

4   Conclusion

In this research, pharmacophore generated based on the structure-based approaches. There are two pharmacophore features generated from the binding site of Shh protein with robotnikinin by molecular docking simulation. All pharmacophore features were fit into robotnikinin and used for database screening using TCM compounds. There are 4,474 TCM compounds as a result of hits from the pharmacophore database screening.

These compounds were further analyzed to select the best three ligands that can be used to inhibit Shh protein through molecular docking simulation series. The docking simulation were consist of rigid receptor docking protocols and induced fit docking protocol. Finally, three best ligand obtained, there are TCM-8941, TCM-28794 and TCM-32808, with the lowest Gibbs free binding energy and also have the best molecular interaction with Shh protein. The three ligands, analyzed with Toxtree, SwissADME, and admetSAR, have non toxic, non carcinogenic, non mutagenic, and also have a good drug metabolism to human body.

Acknowledgement. This research is financially supported by the Directorate of Research and Community Engagement of Universitas Indonesia (DRPM UI) through Hibah Publikasi Internasional Terindeks Untuk Tugas Akhir Mahasiswa (PITTA) UI 2327/UN2.R3.1/HKP.05.00/2018. Herewith, the authors declare that there is no conflict of interest regarding of the manuscript.

References

1. Seow, H.F., Yip, W.K., Fifis, T.: Advances in targeted and immunobased therapies for colorectal cancer in the genomic era. Onco Targets Ther. **9**, 1899–1920 (2016)
2. Kolterud, Å., Grosse, A.S., Zacharias, W.J., Walton, K.D., Katherine, E., Madison, B., Waghray, M., Ferris, J.E., Hu, C., Merchant, J.L., Dlugosz, A., Kottmann, A.H., Gumucio, D.L.: Paracrine Hedgehog signaling in stomach and intestine: new roles for hedgehog in gastrointestinal patterning. Gastroenterology **137**, 618–628 (2009)
3. Berman, D.M., Karhadkar, S.S., Maitra, A., De Oca, R.M., Gerstenblith, M.R., Briggs, K., Parker, A.R., Shimada, Y., Eshleman, J.R., Watkins, D.N., Beachy, P.A.: Widespread requirement for Hedgehog ligand stimulation in growth of digestive tract tumours. Nature **425**, 846–851 (2003)
4. Song, J., Zhang, J., Wang, J., Wang, J., Guo, X., Dong, W.: β1 integrin mediates colorectal cancer cell proliferation and migration through regulation of the Hedgehog pathway. Tumour Biol. **36**, 2013–2021 (2015)
5. Li, X., Yang, G., Li, X., Zhang, Y., Yang, J., Chang, J., Sun, X., Zhou, X., Guo, Y., Xu, Y., Liu, J., Bensoussan, A.: Traditional Chinese medicine in cancer care: a review of controlled clinical studies published in Chinese. PLoS ONE **8**, e60338 (2013)
6. Chen, C.Y.C.: TCM Database@ Taiwan: the world's largest traditional Chinese medicine database for drug screening in silico. PLoS ONE **6**, e15939 (2011)
7. Sander, T., Freyss, J., Von Korff, M., Rufener, C.: DataWarrior: An open-source program for chemistry aware data visualization and analysis. J. Chem. Inf. Model. **55**, 460–473 (2015)
8. Bishop, B., Aricescu, A.R., Harlos, K., O'Callaghan, C.A., Jones, E.Y., Siebold, C.: Structural insights into hedgehog ligand sequestration by the human hedgehog-interacting protein HHIP. Nat. Struct. Mol. Biol. **16**, 698–703 (2009)
9. Molecular Operating Environment (MOE), 09 2014 (2015)
10. Daina, A., Michielin, O., Zoete, V.: SwissADME: a free web tool to evaluate pharmacokinetics, drug-likeness and medicinal chemistry friendliness of small molecules. Sci. Rep. **7**, 1–13 (2017)
11. Cheng, F., Li, W., Zhou, Y., Shen, J., Wu, Z., Liu, G., Lee, P.W., Tang, Y.: AdmetSAR: A comprehensive source and free tool for assessment of chemical ADMET properties. J. Chem. Inf. Model. **52**, 3099–3105 (2012)

Novel Inhibitors of T315I Mutant BCR-ABL1 Tyrosine Kinase for Chronic Myeloid Leukemia Disease Through Fragment-Based Drug Design

Satya Anindita, Atika Marnolia, Hersal Hermana Putra,
Muhammad Chandra Haikal, and Usman Sumo Friend Tambunan[✉]

Faculty of Mathematics and Natural Sciences, Universitas Indonesia,
Kampus UI Depok, Depok, West Java 16424, Indonesia
usman@ui.ac.id

Abstract. The hallmark genetic abnormality of CML is named Philadelphia chromosome. Philadelphia chromosome occurs as a result of recombination of two genes, namely the cellular ABL gene on chromosome 9 and BCR gene located on chromosome 22. The Philadelphia chromosomal translocation is responsible for the ABL and BCR fusion. The ABL and BCR proteins play a central role in the pathogenesis of CML. The malignant transformation by BCR-ABL is critically dependent on its protein tyrosine kinase activity. It makes ABL kinase is an attractive target for therapeutic intervention. In this research, about 653,214 leadlike compounds were obtained from MOE database. The compounds were screened using Data Warrior v.4.6.1 and also docked to predict their binding affinity to BCR-ABL1 tyrosine kinase protein using MOE 2014.09 software. Fragment-based drug design was applied to find a new drug candidate. Finally, five new compounds were generated from this method. The compound LUT-1 has the highest potential due to the low ΔG binding score, acceptable RMSD score, and ADME-Tox result.

Keyword: CML · BCR-ABL1 · Docking · Fragment-based

1 Introduction

Chronic myeloid leukemia (CML) is associated with a cytogenetic abnormality which is known as Philadelphia (Ph) chromosome [1]. Philadelphia chromosome firstly described by Nowell and Hungerford in 1960. In 1973, Rowley reported that Ph chromosome is generated from a balanced reciprocal translocation involving chromosomes 9 and 22 [2]. The study of the t(9;22) translocation associated with CML led to the discovery of BCR-ABL, the first gene of protein tyrosine kinase fusion more than 25 years ago [3].

BCR-ABL tyrosine kinase is a proven target for drug development of CML disease [4]. Targeting the tyrosine kinase activity of BCR-ABL exhibits a very promising therapeutic strategy in CML [5]. The BCR-ABL kinase domain mutations may cause resistance to tyrosine kinase inhibitors (TKIs) patients. Imatinib, nilotinib, and dasatinib as TKIs are approved for the treatment of diagnosed CML patients. Nilotinib and dasatinib as the second generation of TKIs were developed to overcome imatinib resistance [6]. The third generation ABL1 TKI, ponatinib, is still vulnerable to certain

© Springer International Publishing AG, part of Springer Nature 2018
F. Zhang et al. (Eds.): ISBRA 2018, LNBI 10847, pp. 185–190, 2018.
https://doi.org/10.1007/978-3-319-94968-0_17

BCR-ABL1 compound mutants. An experiment done by Pemovska et al. found that axitinib effectively inhibited BCR-ABL1 (T315I) with a distinct binding conformation [7]. Fragment-based drug discovery (FBDD) has become a tool for discovering drug leads. The approach first identifies fragments, tiny molecules, which are about half size of common drugs. The fragments are then linked together to generate drug leads [8].

2   Materials and Method

The 3D structure of the protein was downloaded from RCSB protein data bank with PDB ID 3QRJ and only chain A was used for the next process [9]. The best protein model result was optimized through protonation and minimization process in Molecular Operating Environment (MOE) 2014.09 software [10]. The binding site of the protein had been known through the original ligand binding site in the protein.

The leadlike compounds were collected from MOE database. All the compounds were screened using OSIRIS Data Warrior v4.6.1 with the rule of three parameters [11]. The parameters comprise several requirements such as (1) Molecular weight of the compounds should be ≤ 300 daltons. (2) The clogP score ranges from -0.5 until 3. (3) Hydrogen bond acceptors should be ≤ 3. (4) Hydrogen bond donors should be ≤ 3. (5) Topological polar surface area should be ≤ 60 ^{2}. (6) Rotatable bond should be ≤ 3. (7) Druglikeness of the compounds should be in a positive score.

All the selected fragments were optimized using an MMFF94x force field. Standard molecules such as imatinib, dasatinib, nilotinib, axitinib and DCC-2036 were also prepared in the same steps with the fragments [9].

Molecular docking simulations of leadlike fragments were conducted using virtual screening docking, rigid receptor docking with retaining pose of 30 and 100. These simulations were performed using Triangle Matcher placement and London dG rescoring methods with the Generalized Born Solvation Model/Weighted Surface Area (GBVI/WSA) force field parameter.

In the linking fragment process, two fragments were linked using MOE 2014.09 software to generate the new ligands. The fragments that were used to link did not overlap each other. The linkers were obtained from MOE linker database. After that, molecular docking simulations were conducted. The simulations were conducted using rigid receptor docking with retaining pose of 30 and 100 and induced fit docking with retaining pose of 100. The linked compounds which had been docked from this step were screened once again with Lipinski rule of five parameters [12].

All the linked compounds were tested with the rule of five parameters. These parameters set requirements such as (1) Molecular weight of the compounds should be ≤ 500 daltons. (2) The clogP score ranges from -0.5 to 5.6. (3) Hydrogen bond acceptors should be ≤ 10. (4) Hydrogen bond donors should be ≤ 5. The Veber rule was also applied, such as (1) Topological polar surface area of the compounds should be ≤ 140 ^{2}. (2) Rotatable bond should be ≤ 10. The druglikeness of the compounds should be in a positive score [13]. Linked compounds which had nasty functions were also eliminated.

At last, five best-linked compounds were selected to be analyzed for toxicity. Toxtree v2.6.13 software was used to predict structural alert for genotoxic carcinogenicity, structural alert for nongenotoxic carcinogenicity, potential S. *typhimurium*

TA100 mutagen based on QSAR and potential carcinogen based on QSAR. VEGA v1.1.4 software was used to predict the developmental/reproductive toxicity of the compounds. The ADME analysis was done by using the SwissADME software. The default parameters such as mutagenic, tumorigenic, reproductive effective and irritant of the compounds were predicted using Data Warrior v4.6.1.

3 Results and Discussion

According to Chan et al., the key binding interaction of BCR-ABL1 tyrosine kinase involved Lys271, Glu282, Glu286, Met290, Met318, His361, Arg362, Phe382, Arg386, Tyr393 [9].

About 653,214 leadlike compounds were collected from MOE database. From the rule of three screening step, 647,354 compounds were eliminated. Some 5,860 leadlike fragments were retrieved to be used in molecular docking simulations. From virtual screening docking, 5,075 fragments were eliminated based on the ΔG binding score of standard molecules. After that, the 785 leadlike fragments underwent rigid receptor docking simulations. Only 765 leadlike fragments were retrieved that could be used in the linking fragments process. Linked compound molecular weights should be lower than 500 daltons. A total of 57 linked compounds were developed.

There were 57 linked compounds which underwent molecular docking simulations with retaining pose of 30 and 100. A total of 53 compounds were obtained from these simulations. The further simulation was induced fit docking with retaining pose of 100. Only 19 compounds were retrieved from this simulation. Finally, five best-linked compounds were selected to be lead inhibitors for a BCR-ABL1 tyrosine kinase. All the five compounds met the rule of five requirements. The linked compounds are shown below.

Note: Blue colors show the fragments while red colors show the linkers.

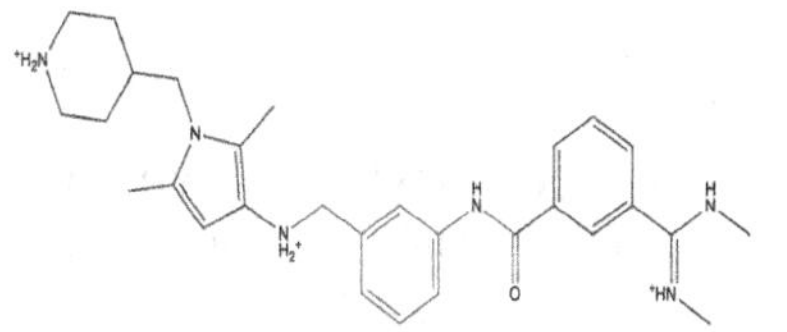

LUT-1

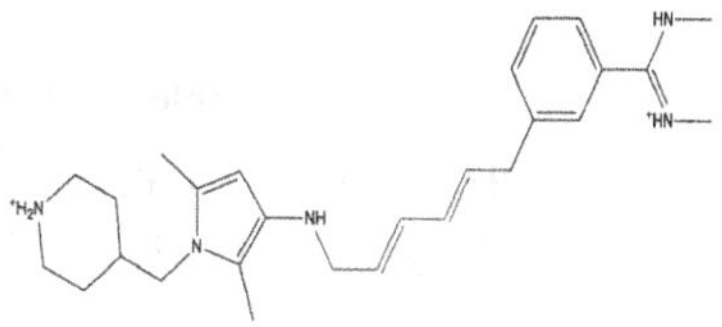

LAM-2

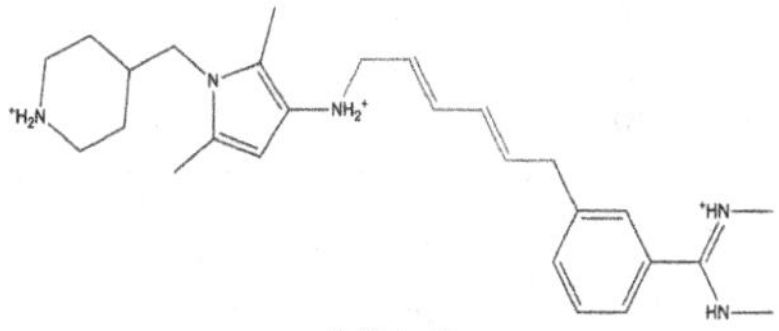

LSA-3

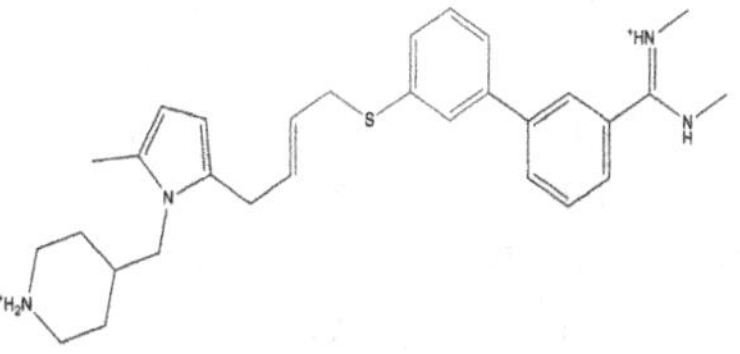

LHP-4

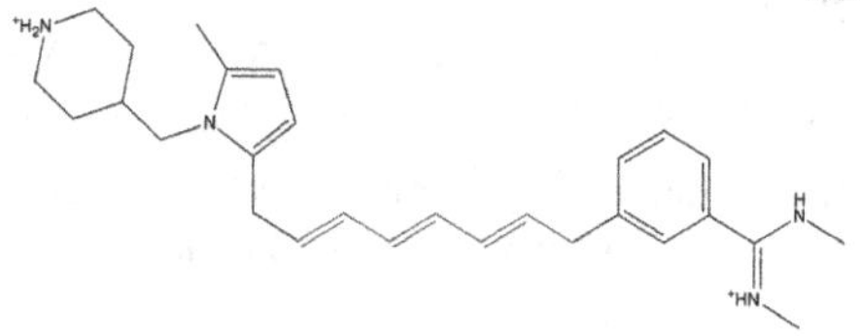

LCH-5

The LUT-1, LAM-2, LSA-3, LHP-4, and LCH-5 compounds were analyzed for the toxicities. The results are shown in the tables below (Table 1).

Table 1. Toxicity of linked compounds and standard molecules.

Name	M	T	RE	I	DRT	SAGC	SANC	PSTA	PC
LUT-1	N	N	N	N	NT	No	No	No	No
LAM-2	N	N	N	N	NT	Yes	No	No	No
LSA-3	N	N	N	N	NT	Yes	No	No	No
LHP-4	N	N	N	N	NT	No	No	No	No
LCH-5	N	N	N	N	NT	Yes	No	No	No
DCC-2036	H	N	N	N	NT	No	No	No	No
Imatinib	N	N	N	N	NT	No	No	No	No
Nilotinib	N	N	N	N	NT	No	No	No	No
Axitinib	N	N	N	N	NT	No	No	No	No
Dasatinib	N	N	H	H	NT	No	No	No	No

Note: M = mutagenic, T = tumorigenic, RE = reproductive effective, I = irritant, DRT = developmental/reproductive toxicity, SA GC = structural alert for genotoxic carcinogenicity, SANC = structural alert for nongenotoxic carcinogenicity, PSTA = potential S typhimurium TA100 mutagen based on QSAR, PC = potential carcinogen based on QSAR, N = none, H = high, NT = non toxicant.

From the toxicity test, three compounds had a structural alert for genotoxic carcinogenicity. Those compounds were LAM-2, LSA-3, and LCH-5. After the toxicity test, ADME test was conducted using the online SwissADME software (Table 2).

Table 2. ADME test result.

Name	GI abs.	BBB perm.	P-gp subs.	CYP1A2 inb.	CYP2C19 inb.	CYP2C9 inb.	CYP2O6 inb.	CYP3A4 inb.
LUT-1	High	No	Yes	Yes	No	No	No	No
LAM-2	High	Yes	Yes	No	Yes	No	No	No
LSA-3	High	No	Yes	No	No	No	No	No
LHP-4	High	Yes	Yes	Yes	Yes	No	No	No
LCH-5	High	Yes	Yes	No	Yes	No	No	Yes
DCC-2036	Low	No	No	No	Yes	Yes	No	Yes
Imatinib	High	No	Yes	Yes	Yes	No	No	No
Nilotinib	Low	No	No	No	Yes	Yes	Yes	Yes
Axitinib	High	No	No	Yes	Yes	Yes	Yes	Yes
Dasatinib	High	No	No	No	No	Yes	No	Yes

Note: GI abs. = gastrointestinal absorption, BBB perm. = blood-brain barrier permeation, P-gp = permeability glycoprotein., CYP1A2 inb. = CYP1A2 inhibitor.

Predictions of passive human gastrointestinal absorption (GI abs.), blood-brain barrier permeation (BBB perm.) and skin permeation (skin perm.) indicate the human body ability to absorb the compounds. The result showed that all of our novel compounds predicted would be absorbed by human gastrointestinal highly. Compound LAM-2, LHP-4, and LCH-5 had blood-brain barrier permeability. The less skin permeant was compound LUT-1 with log Kp −6.82 cm/s while the highest skin permeant was compound LCH-5.

The knowledge about compounds being substrate or non-substrate of the permeability glycoprotein (P-gp) is vital to appraise active efflux through biological membranes [14]. Compounds which are substrates of permeability glycoprotein will be transported to another side of the body. All of our novel compounds were predicted as substrates of the permeability glycoprotein.

Also essential is the knowledge about molecules interaction with cytochromes P450 (CYP). There are five major isoforms of the enzyme namely CYP1A2, CYP2C19, CYP2C9, CYP2D6, CYP3A4 [14]. Inhibition of these enzymes is leading to toxic or other unwanted adverse effects. Compound LSA-3 has no propensity to inhibit cytochromes P450 (CYP) while compound LUT-1 and LAM-2 tend to inhibit an isoform of the enzyme. Furthermore, compound LHP-4 and LCH-5 inhibit two isoforms of the enzyme.

The best compound was selected to become drug candidates. The compound was LUT-1. There were 25 amino acid residues involved in the interaction between LUT-1 and protein 3QRJ.A. They were Leu 248, Gly 249,Tyr 253, Val 256, Ala 269, **Lys 271**, **Glu 286**, Val 289, **Met 290**, Leu 298, Val 299, Ile 315, Glu 316, Phe 317, **Met 318**, Gly 321, Asn 322, Leu 354, Phe 359, Ile 360, **His 361**, Leu 370, Ala 380, Asp 381, and **Phe 382**. Four amino acid residues, which were Tyr253, Glu286, Met290, and Asp381, interacted with LUT-1 through hydrogen bonds.

4 Conclusion

In this study, 653,214 leadlike compounds were collected from MOE database. They were subjected to in silico method to find a potential inhibitor of BCR-ABL1 tyrosine kinase. After the rule of three screening, first molecular docking simulations, linking process, second molecular docking simulations, the rule of five screening and ADME-Tox test, five compounds which were potential to become drug candidates for Leukemia targeting BCR-ABL1 tyrosine kinase were obtained. Compound LUT-1 had the highest potential due to the low ΔG binding score, acceptable RMSD score, and ADME-Tox result. A further in silico experiment is urgently needed to validate their stability in BCR-ABL1 tyrosine kinase protein through molecular dynamics simulations.

Acknowledgment. This research is financially supported by the Direktorat of Research and Community Engagement of Universitas Indonesia (DRPM UI) by Hibah Publikasi Internasional Terindeks untuk Tugas Akhir Mahasiswa (PITTA UI) no. 2327/UN2.R3.1/HKP.05.00/2018.

References

1. Chasseriau, J., Rivet, J., Bilan, F., Chomel, J.C., Guilhot, F., Bourmeyster, N., Kitzis, A.: Characterization of the different BCR-ABL transcripts with a single multiplex RT-PCR. J. Mol. Diagn. **6**, 343–347 (2004)
2. Crisan, D., Carr, E.R.: BCR/*abl* gene rearrangement in chronic myelogenous leukemia and acute leukemias. Lab. Med. **23**, 730–736 (1992)
3. Medves, S., Demoulin, J.B.: Tyrosine kinase gene fusions in cancer: translating mechanisms into targeted therapies. J. Cell Mol. Med. **16**, 237–248 (2012)
4. Banavath, H.N., Sharma, O.P., Kumar, M.S., Baskaran, R.: Identification of novel tyrosine kinase inhibitors for drug-resistant T315I mutant BCR-ABL: a virtual screening and molecular dynamics simulations study. Sci. Rep. **4**, 1–11 (2014)
5. Roche-Lestienne, C., Soenen-Cornu, V., Grardel-Duflos, N., Laï, J.L., Philippe, N., Facon, T., Fenaux, P., Preudhomme, C.: Several types of mutations of the Abl gene can be found in chronic myeloid leukemia patients resistant to STI571, and they can pre-exist to the onset of treatment. Blood **100**, 1014–1018 (2002)
6. Soverini, S., Hochhaus, A., Nicolini, F.E., Gruber, F., Lange, T., Saglio, G., Pane, F., Mu, M.C., Ernst, T., Rosti, G., Porkka, K., Baccarani, M., Cross, N.C.P., Martinelli, G.: BCR-ABL kinase domain mutation analysis in chronic myeloid leukemia patients treated with tyrosine kinase inhibitors: recommendations from an expert panel on behalf of European LeukemiaNet. Blood **118**, 1208–1215 (2011)
7. Pemovska, T., Johnson, E., Kontro, M., Repasky, G.A., Chen, J., Wells, P., Cronin, C.N., Mctigue, M., Kallioniemi, O., Porkka, K., Murray, B.W., Wennerberg, K.: Axitinib effectively inhibits BCR-ABL1(T315I) with a distinct binding conformation. Nature **519**, 102–105 (2015)
8. Erlanson, D.A.: Introduction to fragment-based drug discovery. Top. Curr. Chem. **317**, 1–32 (2012)
9. Chan, W.W., Wise, S.C., Kaufman, M.D., Ahn, Y.M., Ensinger, C.L., Haack, T., Hood, M. M., Jones, J., Lord, J.W., Lu, W.P., Miller, D., Patt, W.C., Smith, B.D., Petillo, P.A., Rutkoski, T.J., Telikepalli, H., Vogeti, L., Yao, T., Chun, L., Clark, R., Evangelista, P., Gavrilescu, L.C., Lazarides, K., Zaleskas, V.M., Stewart, L.J., Van Etten, R.A., Flynn, D.L.: Conformational control inhibition of the BCR-ABL1 tyrosine kinase, including the gatekeeper T315I mutant, by the switch-control inhibitor DCC-2036. Cancer Cell **19**, 556–568 (2011)
10. Tambunan, U.S.F., Toepak, E.: In silico design of fragment-based drug targeting host processing α-glucosidase I for dengue fever, vol. 172, pp. 1–10. IOP Publishing Ltd (2017)
11. Congreve, M., Carr, R., Murray, C., Jhoti, H.: A "rule of three" for fragment-based lead discovery? Drug Discov. Ther. **8**, 876–877 (2003)
12. Lipinski, C.A., Lombardo, F., Dominy, B.W., Feeney, P.J.: Experimental and computational approaches to estimate solubility and permeability in drug discovery and development setting. Adv. Drug Deliv. Rev. **46**, 3–26 (2001)
13. Veber, D.F., Johnson, S.R., Cheng, H., Smith, B.R., Ward, K.W., Kopple, K.D.: Molecular properties that influence the oral bioavailability of drug candidates. J. Med. Chem. **45**, 2615–2623 (2002)
14. Daina, A., Michielin, O., Zoete, V.: SwissADME: a free web tool to evaluate pharmacokinetics, drug-likeness and medicinal chemistry friendliness of small molecules. Sci. Rep. **7**, 1–13 (2017)

HPC and CryoEM

On k-Mismatch Shortest Unique Substring Queries Using GPU

Daniel W. Schultz and Bojian Xu[(✉)]

Department of Computer Science, Eastern Washington University,
Cheney, WA 99004, USA
{dschultz1,bojianxu}@ewu.edu

Abstract. k-mismatch shortest unique substring (SUS) queries have been proposed and studied very recently due to its useful applications in the subfield of computational biology. The k-mismatch SUS query over one given position of a string asks for a shortest substring that covers the given position and does not have a duplicate (within a Hamming distance of k) elsewhere in the string. The challenge in SUS query is to collectively find the SUS for every position of a massively long string in a both time- and space-efficient manner. All known efforts and results have been focused on improving and optimizing the time and space efficiency of SUS computation in the sequential CPU model. In this work, we propose the first parallel approach for k-mismatch SUS queries, particularly leveraging on the massive multi-threading architecture of the graphic processing unit (GPU) technology. Experimental study performed on a mid-end GPU using real-world biological data shows that our proposal is consistently faster than the fastest CPU solution by a factor of at least 6 for exact SUS queries ($k = 0$) and at least 23 for approximate SUS queries over DNA sequences ($k > 0$), while maintaining nearly the same peak memory usage as the most memory-efficient sequential CPU proposal. Our work provides practitioners a faster tool for SUS finding on massively long strings, and indeed provides the first practical tool for approximate SUS computation, because the any-case quadratical time cost of the state-of-the-art sequential CPU method for approximate SUS queries does not scale well even to modestly long strings.

Keywords: String · Shortest unique substring · Parallel computing
GPU · CUDA

1 Introduction

k-mismatch shortest unique substring queries search for the shortest substring(s) that covers a particular string position and does not have a duplicate (within a Hamming distance of k) elsewhere in the string. It has a breadth of practical applications in subfield of computational biology. As discussed in [4], shortest unique substrings are used to facilitate gene sequence comparison without

---

Author names are in alphabetical order. Due to page limit, the pseudocode-level details of all algorithms of this work are presented in the full version of this paper.

© Springer International Publishing AG, part of Springer Nature 2018
F. Zhang et al. (Eds.): ISBRA 2018, LNBI 10847, pp. 193–204, 2018.
https://doi.org/10.1007/978-3-319-94968-0_18

having to align the sequences. In molecular biology, shortest unique substrings found in DNA sequences can be used to aid the design of polymer chain reaction (PCR) [9]. Shortest unique substrings can also be used to compare closely related organisms and identify unique patterns. One of the primary ways that researchers work to identify a specific gene's function is to study the genomes of closely related organisms. The SUS for specific segments and regions are distinct from organism to organism. Similar organisms are compared for the subtle differences that define the distinct features that make the organisms unique. When comparing sequences, the shortest unique segments are sought. This comparative study allows scientists to theorize the functions of specific genome regions, potentially aiding research to find a specific gene that causes a disease.

When comparing distinct organisms, the usefulness of exact comparisons is limited. If researchers compared organisms only using exact shortest unique substrings, possible patterns could be omitted. Expanding the search area and using the approximate SUS could lead to more interesting pattern discovery. The notation of k-mismatch SUS provides a more general sense of unique substring for biological sequences that allow mutations to be considered in the SUS search.

1.1   Prior Work

The work of Haubold et al. [4] uses suffix tree and a hashing technique to find the exact SUS for any string in linear time, where their focus is to find the global SUS of the entire string without considering the coverage of any particular string positions. The SUS finding in their work is mainly used for research of comparing genome segments without alignment. This research is written for biologists and thus lacks rigorous algorithmic analysis.

Pei et al. [9] proposed and studied exact shortest unique substring queries for specific string positions, where mismatches are not allowed. They proposed a series of algorithms for finding exact shortest unique substrings using suffix tree. For a string of size n, their work can answer any shortest unique substring query covering a particular string position using $O(n)$ time. Their work can also be further extended to finding the SUS query answer for every of the n string positions using a total of $O(h \cdot n)$ time, where $h = O(n)$. Although they show that in practice h is much smaller than n and can be treated as a constant, the theoretical time complexity of their work is $O(n^2)$. The work by Pei et al. is later improved by [7,11], two independent studies that reduced the worst-case time complexity to be $O(n)$, using suffix array rather than the more space-consuming suffix tree data structure. Hu et al. [6] studied a generalized version of SUS query that asks for a SUS covering a string position interval rather than a single position. They presented algorithms that preprocess any string of size n using $O(n)$ time and space, and then can answer any position interval-based SUS query in $O(1)$ time. The idea of their work is to reduce the position interval-based SUS query problem to the well-known range minimum queries from computational geometry. Hon et al. [5] presented a new approach that allows in-place computation of k-mismatch SUS—a new type of SUS queries they proposed. Their in-place algorithms still keep the $O(n)$ time cost for exact SUS queries ($k = 0$), while having to spend $O(n^2)$ time cost for approximate SUS queries where mismatches are considered.

1.2 Our Contribution

All prior efforts on SUS computation have been focused on improving and optimizing the computation in the sequential CPU model. In this work, we proposed the first parallel method for SUS queries in the shared-memory model, particularly leveraging on the massive multi-threading GPU technology. Our work offers practitioners with a faster tool for SUS finding, which becomes essential when massively long strings such as genomic sequences are involved.

- Experimental study performed on a mid-end GPU using real-world biology data shows that our proposal is consistently faster than the fastest CPU solution by a factor of at least 6 for exact SUS queries ($k = 0$), and at least 23 for approximate SUS queries ($k > 0$) over DNA sequences, while maintaining nearly the same peak memory usage as is needed by the most memory-efficient sequential CPU method.
- The speedup of our proposal for approximate SUS queries even becomes increasingly significant as the string size increases, due to a smaller time cost growth rate of our proposal than the existing CPU method.
- Our proposal indeed provides the first useful tool in practice for approximate SUS finding over long strings such as those from biology study. The state-of-the-art sequential CPU solution for approximate SUS query has a quadratic time complexity in any case and for any value $k > 0$, and can take months to process a string whose size is only around 100 MB.

2 Problem Formulation

We consider a **string** $S[1..n]$, where each character $S[i]$ is drawn from an alphabet $\Sigma = \{1, 2, \ldots, \sigma\}$. We say the character $S[i]$ **occupies** the string position i. A **substring** $S[i..j]$ of S represents $S[i]S[i+1]\ldots S[j]$ if $1 \le i \le j \le n$, and is an empty string if $i > j$. We call i the **start position** and j the **ending position** of $S[i..j]$. We say the substring $S[i..j]$ **covers** the kth position of S, if $i \le k \le j$. String $S[i'..j']$ is a **proper substring** of another string $S[i..j]$ if $i \le i' \le j' \le j$ and $j' - i' < j - i$. The **length** of a non-empty substring $S[i..j]$, denoted as $|S[i..j]|$, is $j - i + 1$. We define the length of an empty string as zero.

The **Hamming distance** of two non-empty strings A and B of equal length, denoted as $H(A, B)$, is defined as the number of string positions where the characters differ. A substring $S[i..j]$ is k-**mismatch unique**, for some $k \ge 0$, if there does not exist another substring $S[i'..j']$, such that $i' \ne i$, $j - i = j' - i'$, and $H(S[i..j], S[i'..j']) \le k$. A substring is a k-**mismatch repeat** if it is not k-mismatch unique.

Definition 1 (k-Mismatch SUS). *For a particular string position p in S and an integer k, $0 \leq k \leq n - 1$, the k-mismatch shortest unique substring (SUS) covering position p, denoted as SUS_p^k, is a k-mismatch unique substring $S[i..j]$, such that (1) $i \leq p \leq j$, and (2) there does not exist another k-mismatch unique substring $S[i'..j']$, such that $i' \leq p \leq j'$ and $j' - i' < j - i$.*

We call the case of $k = 0$ **exact** and the case of $k > 0$ **approximate**.

Problem (k-Mismatch SUS Finding on GPU): Given a string $S[1..n]$ and the value of $k \geq 0$, design a time- and space-efficient algorithm for shared-memory parallel processing on GPU to find SUS_p^k, for every $p = 1, 2, \ldots, n$. If there are multiple choices for any SUS_p^k, we pick the leftmost one[1].

3   Preparation

The **suffix array** $SA[1..n]$ of the string S is a permutation of $\{1, 2, \ldots, n\}$, such that for any i and j, $1 \leq i < j \leq n$, we have $S[SA[i]..n] < S[SA[j]..n]$. That is, $SA[i]$ is the start position of the ith smallest suffix in the lexicographic order. The **rank array** $RA[1..n]$ is the inverse of the suffix array, i.e., $RA[i] = j$ iff $SA[j] = i$.

Definition 2. *The k-mismatch longest common prefix (LCP) between two strings A and B, $k \geq 0$, denoted as $LCP^k(A, B)$, is the longest prefix of A and B within Hamming distance k.*

Definition 3 (k-Mismatch LSUS). *For a particular string position p in S and an integer k, $0 \leq k \leq n - 1$, the k-mismatch left-bounded shortest unique substring (LSUS) starting at position p, denoted as $LSUS_p^k$, is a k-mismatch unique substring $S[p..j]$, such that either $p = j$ or any proper prefix of $S[p..j]$ is not k-mismatch unique.*

Lemma 1 ([5]). *For any k and p: (1) $LSUS_1^k$ always exists. (2) If $LSUS_p^k$ exists, then $LSUS_i^k$ exists, for all $i \leq p$. (3) If $LSUS_p^k$ does not exist, then none of $LSUS_i^k$ exists, for all $i \geq p$.*

4   Parallel SUS Computation

In this section, we present our GPU solution for the k-mismatch SUS queries. Figure 1 shows the high-level picture of the strategy in the computation for both exact and approximate SUS. Our strategy makes full use of the GPU resource in the sense that all computation involved happens on the GPU. Given the string S, the only work for the CPU host is to ship the string S to the GPU memory as the input for the rest of the computation.

---

[1] Since any SUS may have multiple choices, it is our arbitrary decision to resolve the ties by picking the leftmose choice. However, our solution can also be easily modified to find any other choice.

Our GPU solution for k-mismatch SUS queries takes two different routes for exact SUS ($k = 0$) and approximate SUS ($k > 0$) computation. Each route follows the same strategy of computing LSUS (exact or approximate) first, then use the result from the LSUS computation to find SUS query answers. The two routes differ at the part where LSUS is computed. For the exact LSUS part, the computation takes advantage of efficient string data structures including suffix array, rank array, and implicit LCP array. The use of implicit LCP array is not picturized in Fig. 1

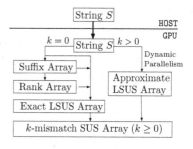

Fig. 1. Diagram of k-mismatch SUS computation on GPU

but will be explained with more details later in this section. For the approximate LSUS part, data structures such as suffix array and rank arrays are no longer useful as their design does not allow mismatches. Thus, the computation of approximate LSUS needs different approaches that are parallelizable in the GPU architecture. The two routes share the same algorithm for the step of computing SUS using LSUS.

In the rest of this section, we will explain the algorithmic and design details of each route, from the input string S to the answers for SUS queries. We start with the following notation that is used frequently in the rest of this section.

Definition 4 (Subrange of String Positions). *For some integers m, t, and j, $1 \le m \le n$, $1 \le t \le m$, and $1 \le j \le t$, let $\mathcal{P}(m,t,j) = \{(j-1)m/t + i \mid 1 \le i \le m/t\}$. Namely, if we cut the range of the m string positions $\{1, 2, \ldots, m\}$ into t equally sized subranges, where each subrange has size m/t, $\mathcal{P}(m,t,j)$ is the collection of string positions that belong to the jth subrange[2].*

4.1 Exact LSUS

The first step in calculating exact SUS is the construction of the suffix array $SA[1..n]$ of the input string S. We use the Nvidia-endorsed recent breakthrough [12] for GPU suffix array construction. In addition to a faster construction, the GPU suffix array construction also helps avoid the extra latency caused by the transmission of the massive suffix array from the CPU host to the GPU memory.

After the suffix array is constructed and resides in the GPU global memory, the rank array, which is the inverse of the suffix array, can be easily calculated: $RA[SA[i]] = i$, $1 \le i \le n$, in a parallel manner. In particular, if we hire t worker threads on GPU, each worker thread j only needs to do these assignments:

$$\text{For every } i \in \mathcal{P}(n,t,j): \ RA[SA[i]] = i.$$

---

[2] For simplicity and clearness of exposition, we neglect the details on the rounding of floating point numbers that happens in division operations of integers. It is trivial to handle such engineering details in implementation.

Because the access to the suffix array within each worker thread is a continuous subarray of the suffix array and the access is in a sequential manner, this parallel construction of the rank array is cache efficient. After the suffix array and rank array are constructed, we are ready to construct the exact LSUS array.

Definition 5

$$x_i = \begin{cases} \left| \, LCP^0 \left(S[i..n], S\left[SA\left[RA[i] \right] - 1 \right]..n \right) \right|, & \text{if } RA[i] > 1 \\ 0, & \text{otherwise} \end{cases}$$

$$y_i = \begin{cases} \left| \, LCP^0 \left(S[i..n], S\left[SA\left[RA[i] \right] + 1 \right]..n \right) \right|, & \text{if } RA[i] < n \\ 0, & \text{otherwise} \end{cases}$$

That is, x_i (y_i, resp.) is the length of the exact LCP of $S[i..n]$ and its lexicographically preceding (succeeding, resp.) suffix, if the preceding (succeeding, resp.) suffix exists. The next fact shows that $LSUS_i^0$ can be obtained directly provided with x_i and y_i, for any i.

Fact 1 ([5]). *For every string position i, $1 \leq i \leq n$:*

$$LSUS_i^0 = \begin{cases} S\left[i..i + \max\{x_i, y_i\}\right], & \text{if } i + \max\{x_i, y_i\} \leq n \\ \text{not existing}, & \text{otherwise}. \end{cases}$$

Using the idea behind the linear-time LCP array construction [8], both sequences of x_i and y_i, $i = 1, 2, \ldots, n$, can be computed sequentially in linear time [5], and thus so is the exact LSUS array construction in $O(n)$ time. The GPU method for exact LSUS computation is to hire t worker threads, and each thread j works on the computation for x_i, y_i, and $LSUS_i^0$ for every $i \in \mathcal{P}(n,t,j)$. Within each thread j, only the computation of $x_{\min(\mathcal{P}(n,t,j))}$, $y_{\min(\mathcal{P}(n,t,j))}$, and $LSUS_{\min(\mathcal{P}(n,t,j))}^0$ may take $O(n)$ time, which however is usually a small number for real-world data [10]. The computation of each of other x, y, and $LSUS^0$ only takes $O(1)$ time. Therefore, the total workload of this parallelized method for exact LSUS computation on GPU remains to be $O(n)$ time, practically.

4.2 Approximate LSUS

Definition 6 (k-Mismatch LLR). *For a particular string position p in S and an integer k, $0 \leq k \leq n - 1$, the k-mismatch left-bounded longest repeat (LLR) starting at position p, denoted as LLR_p^k, is a k-mismatch repeat $S[p..j]$, such that either $j = n$ or any $S[p..q]$ is a k-mismatch unique, $q > j$.*

Fact 2. *If $LLR_p^k = S[p..j]$ and $j < n$, then $LSUS_p^k = S[p..j+1]$. In other words, any $LSUS_p^k$ is a one-character right extension of LLR_p^k if that extension exists.*

Our approach for approximate LSUS finding is comprised of $n - 1$ *rounds* of computation, indexed as $1, 2, \ldots, n - 1$. We work through these n rounds one after another, sequentially, and parallelism happens within each round. Each round i finds $LSUS_{n-i+1}^k$, via the finding of LLR_{n-i+1}^k. By Fact 2, we know once LLR^k_{n-i+1} is found, $LSUS^k_{n-i+1}$ is just the one-character right extension of LLR^k_{n-i+1} if the right boundary of LLR^k_{n-i+1} is not n; otherwise, $LSUS^k_{n-i+1}$ simply does not exist. The search of LLR^k_{n-i+1} is conducted by finding the k-mismatch LCP between the suffix S_{n-i+1} with each of all other suffixes of the string S. And, LLR^k_{n-i+1} is the longest one among all these k-mismatch LCPs. Next, we describe the details of one particular round i that finds LLR^k_{n-i+1} (and thus $LSUS^k_{n-i+1}$).

Round i: We first describe the sequential behavior of round i for clearness in concept. Recall that the goal of round i is to find LLR_{n-i+1} (and thus $LSUS^k_{n-i+1}$) via the comparison between suffix S_{n-i+1} and all other suffixes of string S. Note that, when we work at round i, all the previous rounds $1, 2, \ldots, i-1$ have been finished. That means, the comparisons between suffix S_{n-i+1} with all the suffixes $S_j, j > n-i+1$, have been conducted and thus we do not have to redo them if the results have been saved. Using this idea for the purpose of better performance, we allocate an array $R[1..n]$, where each element $R[x]$ always saves the longest one among all the k-mismatch LCPs between suffix S_x and any other suffix that S_x has been compared with over the rounds of computation. In particular, we want to maintain the following invariant by the time round i is finished:

1. Each array element $R[x]$, $1 \leq x \leq n-i$, has stored the longest one among all the k-mismatch LCPs between between S_x and S_j, $j \geq n-i+1$.
2. Each array element $R[x]$, $x \geq n-i+1$, has stored LLR^k_x.

By using array R, each round i only needs to compare suffix S_{n-i+1} with all suffix $S_j, j \leq n-i$.

Next, we describe the parallel implementation of round i on GPU. We equally divide the work of comparing S_{n-i+1} and all other suffixes $S_j, j \leq n-i$, to t worker threads. Each thread handles the work of $(n-i)/t$ suffix comparisons. Namely, each thread α, $1 \leq \alpha \leq t$, compares suffix S_{n-i+1} with suffix S_x, for every $x \in \mathcal{P}(n-i, t, \alpha)$. Within a particular thread α, for every $x \in \mathcal{P}(n-i, t, \alpha)$, it always picks the longer one between $R[x]$ and $LCP^k(S_{n-i+1}, S_x)$, and saves the result back into $R[x]$. We also allocate an array L of size t. Each thread α also keeps track of the longest one among all LCPs between S_{n-i+1} and S_x, over all $x \in \mathcal{P}(n-i, t, \alpha)$, and save the result in $L[\alpha]$. After all the t threads are finished and synchronized, we use *parallel reduction* over the array L to find $LLR_{n-i+1} = \max(R[n-i+1], \max\{L[\alpha] \mid 1 \leq \alpha \leq t\})$ and save the result back into $R[n-i+1]$.

Dynamic Resource Management in Different Rounds. The number of suffix comparisons changes from round to round. Each round i has $n-i$ suffix comparisons. Since the workload is not static, the algorithm dynamically decides how many threads to launch each round. At the beginning of each round, there is a check to determine if the kernel launch parameters will ensure optimal resource use, and if not, t is adjusted. Due to page limit, we present the details of this engineering effort in the full version of this paper.

Launching Rounds via Dynamic Parallelism. For the approximate LSUS calculation, there are $n - 1$ rounds of work for finding the approximate LLR. If dynamic parallelism is not used, each round of work incurs one round of CPU-GPU kernel launch and communication. For better efficiency, we thus hired the dynamic parallelism in the approximate LSUS calculation as follows. The CPU uses one thread to launch a single kernel on the GPU. The GPU thread handles all kernel launches for all the rounds in the LLR computation. Once the parent kernel is launched on the GPU, there is zero communication needed until all rounds are finished. This significantly reduces the latency experienced over the course of obtaining the approximate LSUS.

4.3   Parallel SUS Computation from LSUS

Now we discuss the GPU procedure for finding SUS from LSUS. The algorithm accomplishing this step works for both exact and approximate SUS finding.

Definition 7 (k-Mismatch SLS). *For a particular string position p in S and an integer k, $0 \leq k \leq n - 1$, we use SLS_p^k to denote the* shortest k-mismatch LSUS covering position p.

Lemma 2 ([5]). *For any k and p, SUS_p^k is either SLS_p^k or $S[x..p]$, for some x, such that $x + |LSUS_x^k| - 1 < p$.*

Lemma 2 says every SUS_p^k is either the shortest k-mismatch LSUS that covers position p or a right-extension (through position p) of a k-mismatch LSUS. In the GPU method for finding k-mismatch SUS from k-mismatch LSUS, we hire t worker threads, where each thread i is assigned with the task of computing SUS_p^k for every $p \in \mathcal{P}(n, t, i)$. When thread i computes a particular SUS_p^k for some $p \in \mathcal{P}(n, t, i)$, it walks from position p toward left, checking $LSUS_p^k$ toward $LSUS_1^k$. The walk stops at a position, say z, such that $LSUS_z^k$ does not cover position p or $z = 1$. By Lemma 2, we can assert that SUS_p^k is the shortest one among all LSUSes we have seen during the walk and the right-extension (through position p) of $LSUS_z^k$ if $LSUS_z^k$ does not cover position p.

$$SUS_p^k = \begin{cases} SLS_p^k, \ if \ LSUS_z^k \ \text{covers position } p. \\ \text{The shorter one of } SLS_p^k \text{ and } S[z..p], \ \text{otherwise.} \end{cases}$$

Because we simply ignore any non-existing LSUS during the walk, a practical improvement to the performance of the walk can be as follows. Within each thread i, before doing any walk, we first find: $q = \max\{p \mid LSUS_p^k \ exists \ and \ p \leq \max(\mathcal{P}(n, t, i))\}$, i.e., q is the largest string position such that: (1) There is an LSUS starting from that position. (2) It is on or before the right boundary of the work area of thread i. We can easily find q by checking $LSUS_{\max(\mathcal{P}(n,t,i))}$ towards the left until we meet q at which we observe the first existing LSUS. After that, the walk within each thread i searching for SUS_p^k, for each $p \in \mathcal{P}(n, t, i)$, only needs to start from the string position $\min(p, q)$ toward the left.

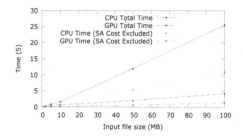

Fig. 2. Time cost of exact SUS computation on protein

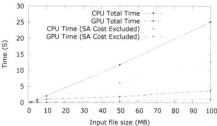

Fig. 3. Time cost of exact SUS computation on DNA

5 Experimental Study

Our experiments were conducted on a MSI GT Series GT62VR Dominator Pro laptop that has a Quad Core Intel i7 6700HQ CPU with 8 MB L3 Cache, 16 GB DDR4 Main System Memory Running @ 2400 MHz, and a Nvidia GTX 1070 Mobile GPU with 8 GB Video Memory. The machine runs Microsoft Windows 10 64-bit Operating System. The CUDA C API was used for the implementation of our GPU proposal[3]. The C implementation of IPSUS, which is currently the fastest and most space-saving sequential solution for exact SUS computation and is the only prior work on approximate SUS computation, was given by the authors of IPSUS. The `libdivsufsort8` library [1] was used for the sequential suffix array construction on CPU, whereas `CUDPP` [3,12] was used for parallel suffix array construction on GPU. The real-world protein and DNA sequences from Pizza&Chili Corpus [2] were used as input strings in all experiments. Each experiment was repeated ten times and the average of the results from the ten repeats were taken as the performance of the relevant algorithm.

5.1 Experimental Results on Exact SUS

Time. (1) We observe that both IPSUS and our GPU proposal have a linear time cost over string size, but our GPU proposal is consistently faster than IPSUS by a factor of at least 6. More precisely, our GPU proposal has a speedup of 6.09x and 6.72x for protein and DNA sequences, respectively. Figures 2 and 3 show the curve of the time cost of IPSUS and our GPU proposals using protein and DNA sequences of different sizes. (2) Recall that the implementations of both IPSUS and our GPU proposal use other third-party code for suffix array construction. In order to show the performance gain in the algorithmic part after the suffix array construction, we compared the time cost of our GPU proposal against IPSUS on the computation after the suffix array is constructed. Figures 2 and 3 also reports the results on this comparison, which shows that our GPU proposal is faster than IPSUS by a factor of 8.2 and 11.92 for protein and DNA sequences, respectively. (3) It is important to observe the speedup gained by our GPU proposal for the

---

[3] Our code: http://penguin.ewu.edu/~bojianxu/publications.

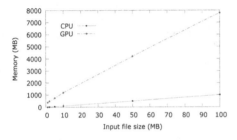
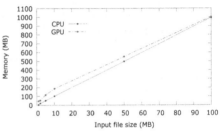

Fig. 4. Peak memory usage of exact SUS computation on both protein and DNA

Fig. 5. Peak memory usage of exact SUS computation on both protein and DNA (excluding SA construction)

computation excluding suffix array construction, because roughly 50% of the total computation time of IPSUS is spend on suffix array construction. Therefore, in the case where the suffix array of the string is given, we can simply copy the suffix array to the GPU memory and continue to finish the rest of the computation with a good speedup overall (and a significant memory space saving for GPU, which will be explained in the space cost comparison analysis next).

Space. Figure 4 shows the linear peak memory usage for the exact SUS computation observed in IPSUS and our GPU proposal. The space cost of each algorithm stays the same regardless of string type (protein or DNA). While both proposals show a linear space cost, our GPU proposal overall uses more memory space by a factor of 7.8 in the worst case in our experiments. However, the majority memory cost overhead in our GPU proposal is spent by CUDPP, the third party GPU suffix array construction program. This claim is demonstrated by Fig. 5, which shows that, after the suffix array construction, the memory usages of IPSUS and our GPU proposal are nearly the same. This gives practitioners a time and memory usage trade-off: If the suffix array is constructed in CPU and then shipped to GPU, the memory space cost can be significantly reduced, while a descend speedup is still maintained (see the previous time cost result analysis).

5.2 Experimental Results on Approximate SUS

We performed our experiments on approximate SUS computation on string size up to 200 KB, due to the intensive computation time cost by the algorithms. We observe that experiments with string size up to 200 KB already serve well enough to present the performance comparison between IPSUS and our GPU proposal. In experiments, we set $k \in \{1, 5, 10, 15\}$, the number of mismatches allowed in approximate SUS. In the implementation of our GPU proposal for approximate SUS computation, we have used dynamic parallelism to avoid the many CPU-GPU kernel launching and communication caused by the nature of the algorithm.

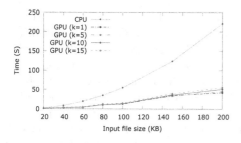

Fig. 6. Time cost of approx. SUS computation on protein

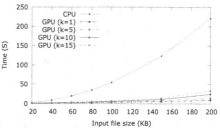

Fig. 7. Time cost of approx. SUS computation on DNA

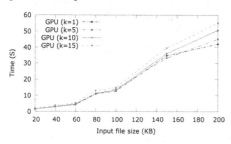

Fig. 8. Time cost of approximate SUS computation on protein (GPU only)

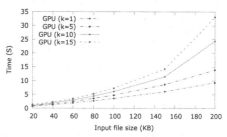

Fig. 9. Time cost of approximate SUS computation on DNA (GPU only)

Time. (1) Figures 6 and 7 clearly show that IPSUS has a quadratic time cost, while our GPU solution has a super-linear time cost but is much faster, for both string types, for any string size, and for any value of k. The speedup for the GPU solution is 23.557x and 5.254x for DNA and protein sequences of size 200KB, respectively. Different speedup for DNA and protein is caused by the fact that the average size of SUS in protein is longer and thus needs more comparisons in the GPU solu-

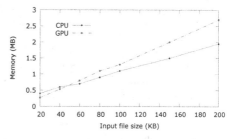

Fig. 10. Peak memory usage of approximate SUS computation on both DNA and protein

tion. (2) Figures 6 and 7 also show that, the longer the string is, the more speedup our GPU solution attains, since the GPU solution's time cost has a smaller growth rate than the quadratic time cost of IPSUS. (3) The time cost of IPSUS is irrelevant to the value of k (Figs. 6 and 7), where our GPU solution spend more computation time when the value of k increases (Figs. 8 and 9), because more comparisons are needed in GPU solution for the approximate SUS.

Space. Figure 10 shows the peak memory space cost for the approximate SUS observed in IPSUS and the GPU solution. The IPSUS and GPU solution both have space cost that is linear of string size. Our GPU solution use slightly more space, due to the additional constants and small sized arrays used for the approximate LSUS computation. It is consistent to the theoretical analysis that the space costs of both CPU and GPU solutions are irrelevant to string type and the value of k. It is worthing noting that our GPU proposal uses dynamic parallelism, which incurs a constant amount of memory usage overhead. That memory cost overhead is about 712 MB consistently, used by the GPU system for device runtimes execution tracking and management that are involved in dynamic parallelism. We have also tried the GPU implementation implementation without using dynamic parallelism. It turns out that the space cost of the GPU solution is exactly the same as what have in Fig. 10, but the speedup attained by the GPU proposal become much smaller—about 1.5x for protein and about 6x for DNA sequences.

References

1. A lightweight suffix-sorting library. https://code.google.com/p/libdivsufsort
2. Compressed indexes and their testbeds. http://pizzachili.dcc.uchile.cl/texts.html
3. Cuda data parallel primitives library. http://cudpp.github.io/
4. Haubold, B., Pierstorff, N., Möller, F., Wiehe, T.: Genome comparison without alignment using shortest unique substrings. BMC Bioinf. **6**(1), 123 (2005)
5. Hon, W.K., Thankachan, S.V., Xu, B.: In-place algorithms for exact and approximate shortest unique substring problems. Theor. Comput. Sci. **690**, 12–25 (2017)
6. Hu, X., Pei, J., Tao, Y.: Shortest Unique queries on strings. In: Moura, E., Crochemore, M. (eds.) SPIRE 2014. LNCS, vol. 8799, pp. 161–172. Springer, Cham (2014). https://doi.org/10.1007/978-3-319-11918-2_16
7. İleri, A.M., Külekci, M.O., Xu, B.: A simple yet time-optimal and linear-space algorithm for shortest unique substring queries. Theor. Comput. Sci. **562**, 621–633 (2015)
8. Kasai, T., Lee, G., Arimura, H., Arikawa, S., Park, K.: Linear-time longest-common-prefix computation in suffix arrays and its applications. In: Proceedings of the Symposium on Combinatorial Pattern Matching, pp. 181–192 (2001)
9. Pei, J., Wu, W.C.H., Yeh, M.Y.: On shortest unique substring queries. In: Proceedings of IEEE International Conference on Data Engineering (ICDE), pp. 937–948 (2013)
10. Tian, Y., Xu, B.: On longest repeat queries using GPU. In: Renz, M., Shahabi, C., Zhou, X., Cheema, M.A. (eds.) DASFAA 2015. LNCS, vol. 9049, pp. 316–333. Springer, Cham (2015). https://doi.org/10.1007/978-3-319-18120-2_19
11. Tsuruta, K., Inenaga, S., Bannai, H., Takeda, M.: Shortest unique substrings queries in optimal time. In: Geffert, V., Preneel, B., Rovan, B., Štuller, J., Tjoa, A.M. (eds.) SOFSEM 2014. LNCS, vol. 8327, pp. 503–513. Springer, Cham (2014). https://doi.org/10.1007/978-3-319-04298-5_44
12. Wang, L., Baxter, S., Owens, J.D.: Fast parallel suffix array on the GPU. In: Träff, J.L., Hunold, S., Versaci, F. (eds.) Euro-Par 2015. LNCS, vol. 9233, pp. 573–587. Springer, Heidelberg (2015). https://doi.org/10.1007/978-3-662-48096-0_44

Memory-Efficient and Stabilizing Management System and Parallel Methods for RELION Using CUDA and MPI

Jingrong Zhang[1,2], Zihao Wang[1,2], Yu Chen[1,2], Zhiyong Liu[1],
and Fa Zhang[1(✉)]

[1] High Performance Computer Research Center, Institute of Computing Technology,
Chinese Academy of Sciences, Beijing 100190, China
{zhangjingrong,wangzihao,chenyu,zyliu,zhangfa}@ict.ac.cn
[2] University of Chinese Academy of Sciences, Beijing, China

Abstract. In cryo-electron microscopy, RELION has been proven to be a powerful tool for high-resolution reconstruction and has quickly gained its popularity. However, as the data processed in cryoEM is large and the algorithm of RELION is computation-intensive, the refinement procedure of RELION appears quite time-consuming and memory-demanding. These two problems have become major bottlenecks for its usage. Even though there have been efforts on paralleling RELION, the global memory size still may not meet its requirement. Also as by now there is no automatic memory management system on GPU (Graphics Processing Unit), the fragmentation will increase with iteration. Eventually, it would crash the program. In our work, we designed a memory-efficient and stabilizing management system to guarantee the robustness of our program and the efficiency of GPU memory usage. To reduce the memory usage, we developed a novel RELION 2.0 data structure. Also, we proposed a weight calculation parallel algorithm to speedup the calculation. Experiments show that the memory system can avoid memory fragmentation and we can achieve better speedup ratio compared with RELION 2.0.

Keywords: cryoEM · RELION · CUDA · Performance tuning

1 Introduction

Single particle cryo-electron microscopy (CryoEM) uses images of randomly-oriented particles to reconstruct 3D density map [1]. Over the past decades, cryoEM is increasingly becoming a mainstream technology for studying the architecture of biological macromolecules [2]. RELION (REgularised LIkelihood OptimisatioN) [3,4] significantly increased possible resolution of reconstruction algorithm by adopting expectation maximization (EM) algorithm. By now, many cryoEM structures were determined to a resolution better than 5Å using RELION [2,5–7].

© Springer International Publishing AG, part of Springer Nature 2018
F. Zhang et al. (Eds.): ISBRA 2018, LNBI 10847, pp. 205–216, 2018.
https://doi.org/10.1007/978-3-319-94968-0_19

However, expectation maximization (EM) algorithms [8] need to integrate over the hidden variables in the expectation step, which requires large amounts of computing time [9] and memory. Taking the EM algorithm used in RELION as example, calculating one dataset usually takes days of time. In 3D refinement stage, the test data used in the tutorial of RELION [10] with 6,496 β-galactosidase 128 * 128 particles. It costs more than 2 GB memory per process and 40 CPU hours, even though this dataset is really small.

To accelerate RELION, we must accelerate the calculation of EM algorithm. In RELION, the calculation of weights takes a major part of the processing time. However, even though RELION 2.0 has been implemented on GPU, multithreads are launched. The calculation of weights still take up a large part of calculation time. In this step, a lot of global memory accessing occurrs. To solve this problem we developed a weight calculation parallel algorithm. By making use of the shared memory on GPU, we can dramatically reduce the time of global memory accessing thus shortening its processing time.

Like many other softwares used in cryoEM, in order to cope with the time-consuming problem, RELION adopts parallel technology on Graphics Processor Units (GPUs) to speedup the processing procedure [11]. However, the stability and performance of GPU relies heavily on its hardware characteristics. For memory-intensive applications like RELION, memory-related code may dramatically influence the performance. Also there is not a sophisticated memory management system to handle the fragmentation problem. Fragmentation is generated during the randomly allocating and de-allocating operation. Fragments divide the successive memory into several pieces of memory. When the program try to allocate a large piece of memory, even though the total available memory is sufficient, the allocation can still fall because of the fragmentation. However this severe problem has not caught enough attention. By now, there still is no stable memory management strategy introduced in RELION parallelization. For example, we use NVIDIA K20c to process one dataset EMPIAR-10028 with 360 * 360 particle size, "Out of Memory" error occurs during later iterations. Facing this condition, we developed a stabilizing memory manage system to ensure the robustness of program, especially for large dataset. In this system, we avoid fragmentation based on the characteristics of iterative methods. Also our method can be applied to other iterative methods.

As the total memory consumption of RELION exceeds the memory of GPU. Reducing memory consumption is of great importance. Weight array is the most memory demanding data structure. We analysed the data structure of weight array carefully and find out that we can reduce the memory requirements of weight array by redesign its structure delicately.

In summarize, in this work, we designed a memory-efficient and stabilizing management system to guarantee the robustness of our program and the efficiency of GPU memory usage. Then to reduce the memory usage, we developed a novel data structure. Also, we developed a weight calculation parallel algorithm to speedup the calculation.

Our paper is organized as follows: in Sect. 2, we introduce the related work; in Sect. 3, we introduce our stabilizing management system; in Sect. 4 a memory-efficient data structure is shown; in Sect. 5, we present the weight calculation parallel algorithm. The results of experiment are presented in Sect. 6. After that we conclude the paper.

2 Related Work

In this section we will introduce the basic algorithm flow of RELION and its accelerating work.

RELION adopts a modified adaptive expectation maximization (EM) algorithm based on the work of Tagare et al. [8]. Instead of assigning the best orientation for each particle, RELION generates a probability distribution over all orientations for each particle. We use $\Gamma_{i\phi}^{(n)}$ to indicate the posterior probability of orientation assignment ϕ for the ith image in the nth iteration ($i \in [0, I]$). In particular, the orientations $\phi \in [0, \Phi]$ is in a five dimension domain, comprising 3 rotations and 2 translations, i.e. $\Phi = (rRot, rTilt, rPsi, tX, tY)$.

The expectation-maximization algorithm iteratively optimizes the structure through a two-step procedure. The first step is "Expectation", in this step, computer-generated projections (here we also call them re-projections) of the structure are compared with the real particle images, resulting in relative similarity information about the relevant orientations of the images. The second step is "Maximization", the images are combined with the prior information into a smooth, 3D reconstruction. Meanwhile, the power of the noise and the signal in the data are updated.

In step "Expectation", RELION implements a modified version of the adaptive expectation-maximization algorithm. It owns two times finer procedure. For each particle image i, in the first pass ($oversampling = 0$), $\Gamma_i^{(n)}$ is evaluated over the entire domain using a relatively coarsely sampled grid. RELION selects the sub-domain of all orientations ϕ corresponding to the highest values of $\Gamma_{i\phi}^{(n)}$ that sum to 99.9% of the total probability mass on the coarse grid. Then, in the second pass ($oversampling = 1$), $\Gamma_i^{(n)}$ is evaluated only over the selected sub-domain using a finer grid which owns 32 times more sampling points than the coarse sampling grid [4].

Based on the analysis of GeRelion [12] and Relion 2.0 [11], the most time-consuming part is the step "Expection". And the most memory demanding data structure is weight array Γ.

GeRelion [12] is a GPU-accelerated version of RELION 1.4. Based on the high parallel ability of GPU, GeRelion adopts a four-level parallel model. Also to reduce the memory consumption, the sparse array is compacted to continuous vector and an aux vector was introduced to store the indexes of the weight array, which also will cause memory consumption. When the weight array is not sparse, this structure will take up more memory than original structure. Based on this consideration, we developed a new data structure which uses less memory.

RELION 2.0 [11] also accelerates the expectation-maximization algorithm on GPU. To reduce the memory consumption, it used single precision instead of double. Meanwhile, as the texture memory owns cache, RELION 2.0 stores the re-projection image on texture memory to accelerate the memory accessing. However, as the size of re-projection increase, one projection image will not fit in the texture memory cache. This will introduce frequently data swap in cache. The accelerating performance will be significantly weaken. Meanwhile, the calculation of weight array takes up majority time of program, in which accessing re-projection takes up a big part of the time. So we designed a different accelerate strategy to speedup this part.

3   Stack-Based Stabilizing Management System

RELION involved a lot of internal variables to perform its calculation. For many of these variables, their memory requirement may differ with sampling and particles, which means these variables require frequently and repeatedly allocation and de-allocation. Without memory management system on GPU, repeatedly allocation and de-allocation may introduce memory fragmentations. In later stage of iterations, these fragments can cause very serious consequence, i.e. application crash.

As mentioned above, there is no suitable memory management system by now. RELION 2.0 adopted linked list to collect memory fragments. The linked list can merge adjacent fragments. But it still can not prevent the generating of fragments.

Essentially EM algorithm is one of those iterative methods. Within each iteration, there is a loop for processing each particle. When processing one particle, there are variables allocating and de-allocating during the calculation for each orientation. So the lifetime of different variables can be classified into four types, global scope, iteration scope, during processing one particle (job scope) and during processing one orientation. As the lifetime is quite regular, we can control them using characteristic. The corresponding operation is shown Fig. 1. At the beginning of the whole program, we first allocate variables with global scope. While at the beginning of each iteration, we allocate memory for variables of iteration scope. Then, we allocate variable used in each job of processing one particle. After one job completed, the variables of job scope are released. At the end of one iteration, the variables of iteration scope are released. And at the end of program, variables of global scope are released. As the variables allocated last will alway be freed first. Based on this rule, we can build a stack based memory management system thoroughly to avoid memory fragments.

4   Memory-Efficient Data Structure

As mentioned above, the memory requirements of weight array are quite large, especially for the fine sampling pass. And, as mentioned, the global memory of

Allocate Memory Pool
Allocate Variables with Global Scope
Allocate Variables with Iteration Scope
Allocate Variables with Job Scope
Deallocate Variables with Job Scope
Deallocate Variables with Iteration Scope
Deallocate Variables with Global Scope

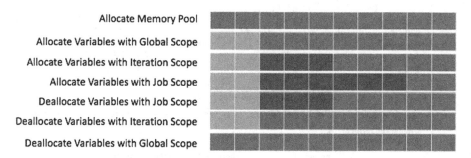

Fig. 1. Memory consumption on host.

GPU is quite limited. So more efficient data structure to store the weight array is of great importance.

By analyzing the data structure of weight array carefully, we find out that we can reduce the memory requirements of weight array by redesigning its structure delicately. In the first pass, all possible assignments are calculated. But before the second pass, RELION selects assignments with high probability. In the second pass, only assignments with high probability will be considered with fine sampling steps. Because the adoption of fine sampling, the size of weight array in the second pass is 32 times (the multiplication of 8 orientations and 4 translations) as large as that in the first pass. While because the selecting step, most of the assignments are neglected, only a few elements in weight array are non-zero.

Obviously, the storage of neglected assignments wastes a lot of memory. In our compressed structure, we can only store the non-zero value and their corresponding subscript. We observed that the assignments in first pass correspond to 32 assignments in the second pass. The weight of these 32 assignments must be stored in weight array successively. This trait can be used to reduce the size of aux index array. Instead of storing the column subscript of each non-zero element, only the first subscript of the 32 assignments will be stored.

The conversion of compressed data structure and original structure is shown in Fig. 2. The upper side of this figure shows the original data structure. The upper left array indicates the weight array, the upper right array is its corresponding array, this array indicates which assignments should be used in the second pass.

The bottom side of this figure shows our compressed data structure. The array in the middle stores all the non-zero elements. Below this array, we show the subscript. For each group of successive 32 elements, we only store the subscript of the first assignment. The array at the bottom stores the information about the number of non-zero elements for each particle. For convenient data accessing, we use the start point of each particle to show this information. For each particle, the number elements is N, the non-zero elements in matrix is n. For each particle, *size_original*, *size_conpressed* and *size_new* shows the number of bits cost in

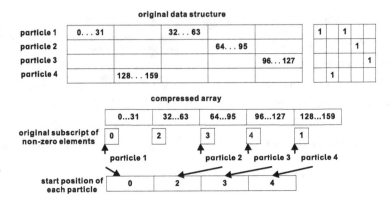

Fig. 2. Memory-efficient data structure.

original version of RELION, general compressed method and our method. We can easily know, when $n > 0.67N$, the memory consumption of general compressed method will exceed original version. However the memory consumption of our method won't exceed original version unless $n > 0.98N$.

$$size\_original = N * (32 * 8)$$
$$size\_conpressed = n * (32 * (8 + 4)) \qquad (4.1)$$
$$size\_new = n * (32 * 8 + 4)$$

5   Weight Calculation Parallel Algorithm

In step Get Squared Differences (GSD), it calculates the actual squared difference term of the gaussian probability function. Its main part can be written in Formula 5.1:

$$\Gamma_{i\phi}^{(n)} = \frac{1}{2} \sum_{k=1}^{N} ((Re(p_{\Phi k}) - Re(X_{ik}))^2 + (Im(p_{\Phi k}) - Im(X_{ik}))^2) \qquad (5.1)$$

where $p_{\phi k}$ is the kth component of the projection at orientation assignment ϕ. X_{ik} is the kth component of the particle at the ith image.

Different from later iterations, in the first iteration, this step calculates the normalized cross-correlation coefficient. Although the formulas of these two conditions are different. Their computational complexity is of the same order. As for each particle, we need to calculate the weight for each assignment. This means each projection image should compare with each particle image, accordingly. Its calculation can be formalized as following:

$$
\begin{bmatrix}
\Gamma_{i_1\phi_1} & \Gamma_{i_1\phi_2} & \cdots & \Gamma_{i_1\phi_m} \\
\Gamma_{i_2\phi_1} & \Gamma_{i_2\phi_2} & \cdots & \Gamma_{i_2\phi_m} \\
& \cdots & & \\
\Gamma_{i_k\phi_1} & \Gamma_{i_k\phi_2} & \cdots & \Gamma_{i_k\phi_m}
\end{bmatrix}
=
\begin{bmatrix}
\sum_{k=1}^{N} f_{i_1,\phi_1,k} & \cdots & \sum_{k=1}^{N} f_{i_1,\phi_m,k} \\
\sum_{k=1}^{N} f_{i_2,\phi_1,k} & \cdots & \sum_{k=1}^{N} f_{i_2,\phi_m,k} \\
& \cdots & \\
\sum_{k=1}^{N} f_{i_u,\phi_1,k} & \cdots & \sum_{k=1}^{N} f_{i_u,\phi_m,k}
\end{bmatrix}
$$

$$
=
\begin{bmatrix}
\sum_{k=1}^{n_1} f_{i_1,\phi_1,k} & \cdots & \sum_{k=1}^{n_1} f_{i_1,\phi_m,k} \\
\sum_{k=1}^{n_1} f_{i_2,\phi_1,k} & \cdots & \sum_{k=1}^{n_1} f_{i_2,\phi_m,k} \\
& \cdots & \\
\sum_{k=1}^{n_1} f_{i_u,\phi_1,k} & \cdots & \sum_{k=1}^{n_1} f_{i_u,\phi_m,k}
\end{bmatrix}
+ \cdots +
\begin{bmatrix}
\sum_{k=n_c}^{N} f_{i_1,\phi_1,k} & \cdots & \sum_{k=n_c}^{N} f_{i_1,\phi_m,k} \\
\sum_{k=n_c}^{N} f_{i_2,\phi_1,k} & \cdots & \sum_{k=n_c}^{N} f_{i_2,\phi_m,k} \\
& \cdots & \\
\sum_{k=n_c}^{N} f_{i_u,\phi_1,k} & \cdots & \sum_{k=n_c}^{N} f_{i_u,\phi_m,k}
\end{bmatrix}
\tag{5.2}
$$

$$
=
\begin{bmatrix}
\Gamma_{i_1\phi_1}(1,n_1) & \cdots & \Gamma_{i_1\phi_m}(1,n_1) \\
\Gamma_{i_2\phi_1}(1,n_1) & \cdots & \Gamma_{i_2\phi_m}(1,n_1) \\
\Gamma_{i_k\phi_1}(1,n_1) & \cdots & \Gamma_{i_k\phi_m}(1,n_1)
\end{bmatrix}
+ \cdots +
\begin{bmatrix}
\Gamma_{i_1\phi_1}(n_c,N) & \cdots & \Gamma_{i_1\phi_m}(n_c,N) \\
\Gamma_{i_2\phi_1}(n_c,N) & \cdots & \Gamma_{i_2\phi_m}(n_c,N) \\
\Gamma_{i_k\phi_1}(n_c,N) & \cdots & \Gamma_{i_k\phi_m}(n_c,N)
\end{bmatrix}
$$

In formula 5.2, the first line shows the common way of calculation. The calculation of weight $\Gamma_{i\phi}$ is independent of $\Gamma_{i'\phi'}$. Considering the structure of GPU card, accessing time to global memory can be 100 times greater than shared memory [13]. So reducing the global memory accessing can dramatically improve the time performance of applications. Here $\Gamma_{i_1\phi_1}(n,n') = \sum_{k=n}^{n'} f_{i_1,\phi_1,k}$ means the partial results of $\Gamma_{i_1\phi_1}$. The second and third lines of formula 5.2 show that we can calculate part of weights first and sum these intermediate results to get the final results. Generally speaking, instead of calculating $\Gamma_{i\phi}$ independently, we calculate a group of $\Gamma_{i\phi}(n,n')$ together. First, threads load a portion of projection $p_\phi(n,n')$ and particle $X_i(n,n')$ into shared memory, where we can access them much more quickly. Then, the partial results of $\Gamma_{i_1\phi_1}$ can be calculated using shared memory.

As shared memory are available for all threads of one block, we can reuse the data and reduce the data accessing time. Take a 32 * 32 block as an example, the global memory accessing time can be reduced by 32 times. As step "Get Squared Differences (GSD)" take majority time in expectation step, reducing global memory accessing can efficiently reduce the processing time of this step.

6   Experiments

6.1   Environment and Test Dataset

Generally, we adopts two dataset to perform our experiment. First, we use the dataset used in the tutorial of RELION, owning 6,496 β-galactosidase particles (EMPIAR-10017 [10]). The size of each particle is 128 * 128. Its experiments are

carried out on a machine running the Ubuntu operating system 64-bit with an
Intel(R) Xeon(R) CPU E5630 at 2.53 GHz. The GPU card is NVIDIA Tesla
K20c, with 2496 stream processors and 5 GB global memory.

Another dataset is "plasmodium falciparum 80S ribosome bound to the anti-
protozoan drug emetine", which owns 100,000 particles with pixels size 360 * 360
(EMPIAR-10028 [2,5]). Its experiments are carried out on a machine running
the Ubuntu operating system 14.04, 64-bit with an Intel(R) Xeon(R) CPU E5-
2680 v3 at 2.50 GHz. The GPU card is NVIDIA GeForce GTX Titan X, with
3072 stream processers and 12288 MB global memory. In our test, we use 5 MPI
processes (one thread for each process), each slave process owns one GPU.

6.2   Performance of Stack-Based Memory Management System

In this section, we show the detailed working process of stack-based memory
management system. To do so, we record the variation of used memory when
running the dataset EMPIAR-10028. As shown in Fig. 3, the top pointer indicates that memory before this pointer is used memory. Memory with larger
address is continuous free memory. In this method, the fragment are avoided. In
this figure, we can see that the memory operation indeed shows the characteristic
of iterative algorithm. But as marked with red rectangle, the memory operation
owns randomness. Also the randomness doesn't influent the working process of
stack-based memory.

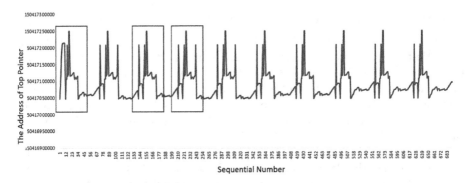

Fig. 3. The top pointer of used memory (Color figure online)

6.3   Memory Consumption Optimization

We modified the data structure of array $\Gamma^{(n)}$, which saves a lot of memory.
For dataset EMPIAR-10017, we show the memory consumption of double precision version. As discussed in RELION 2.0 the single precision version doesn't
adversely affect results. We still keeps the double version for users demanding
more accurate calculation. As the size of single-precision floating-point number

is half of double ones, the memory of single precision is just half of the double version.

We record the max memory consumption of array $\Gamma^{(n)}$ for one particle to show the performance. As shown in Fig. 4. The most memory demanding array appears in the middle of the iterations. In these iterations, our structure significantly reduce the memory requirements. The upper bound of the size of array $\Gamma^{(n)}$ is mainly the sampling number in first pass in both float mode and double mode. The memory consumption of array $\Gamma^{(n)}$ is generally 1/32 of the old version.

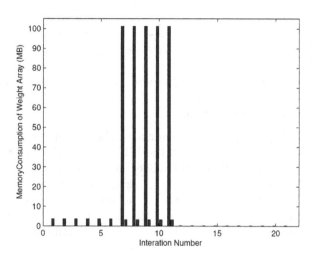

Fig. 4. Maximum memory consumption of array "$\Gamma^{(n)}$" for one particle for β-galactosidase in double mode.

6.4   The Speedup Ratio of Weight Calculation Parallel Algorithm

In this section, we test the performance of our weight calculation parallel algorithm. Different from the texture memory, the performance of our method is not restricted by the total size of re-projection. We compared our method with RELION 2.0, both methods are GPU parallelized. The speedup ratio shows the improvement by our method. We tested the methods on dataset EMPIAR-10028. In Fig. 5, the horizontal axis indicates the iteration number. As the size and number of re-projection image may vary during iterating. We use different iterations to indicate different processing size of data. As we can see, the performance of two methods is similar. However, in iterations of larger dataset, the advantage of reusing shared memory shows up.

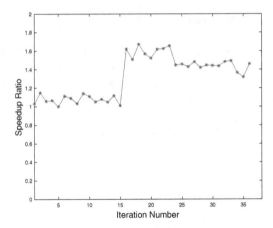

Fig. 5. Speedup ratio of weight calculation parallel algorithm.

6.5   The General Speedup Ratio

After analyzing the speedup ratio, we summarize the total speedup ratio for each iteration. As our program is based on the MPI version of RELION, we don't modify the master-slave MPI parallel model. In this experiment, we compare the CUDA+MPI version with MPI version using the same number of processes and threads. Different from the MPI version, each slave process owns one GPU. From Fig. 6, we can see that the double precision version can reach 80x speedup ratio for dataset EMPIAR-10028 in the most time-consuming iterations.

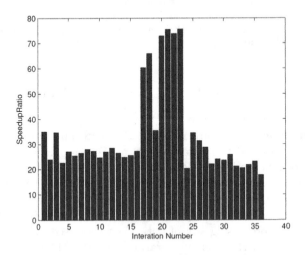

Fig. 6. The general speedup ratio.

To compare the performance of RELION 2.0 and our method, we used dataset EMPIAR-10017 with single precision version. We compare them with the original MPI version and record the speedup ratio. Figure 7 shows the speedup ratio respectively. Generally speaking, our version owns better speedup ratio. Especially in the most time consuming iteration, our version can reach 105x speedup ratio, while the speedup ratio of RELION 2.0 is 96x.

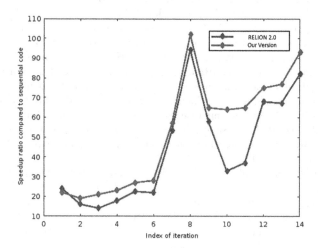

Fig. 7. Compare the performance of RELION 2.0 with our method.

7 Conclusion

In this work, we proposed stack-based memory management system, which can ensure the program to proceed without "Out of Memory" error and enable it to process dataset with large particle size. Then we introduced our compressed data structure which will dramatically reduce the memory consumption. After that, we developed the weight calculation parallel algorithm using shared memory. Its performance won't be affected by the size of re-projection. The results of experiment show that the memory system can avoid memory fragments. And we have achieved better speedup ratio compared to RELION 2.0.

Acknowledgments. This research is supported by the National Key Research and Development Program of China (2017YFA0504702), the NSFC projects Grant No. U1611263, U1611261, 61472397, 61502455, 61672493 and Special Program for Applied Research on Super Computation of the NSFC-Guangdong Joint Fund (the second phase).

References

1. Li, X., Grigorieff, N., Cheng, Y.: GPU-enabled FREALIGN: accelerating single particle 3D reconstruction and refinement in fourier space on graphics processors. J. Struct. Biol. **172**(3), 407–412 (2010)
2. Bai, X., McMullan, G., Scheres, S.H.: How cryo-EM is revolutionizing structural biology. Trends Biochem. Sci. **40**(1), 49–57 (2015)

3. Scheres, S.H.: A Bayesian view on cryo-EM structure determination. J. Mol. Biol. **415**(2), 406–418 (2012)
4. Scheres, S.H.: RELION: implementation of a Bayesian approach to cryo-EM structure determination. J. Struct. Biol. **180**(3), 519–530 (2012)
5. Wong, W., Bai, X., Brown, A., Fernandez, I.S., Hanssen, E., Condron, M., Tan, Y.H., Baum, J., Scheres, S.H.: Cryo-EM structure of the plasmodium falciparum 80S ribosome bound to the anti-protozoan drug emetine. Elife **3**, e03080 (2014)
6. Amunts, A., Brown, A., Bai, X., Llácer, J.L., Hussain, T., Emsley, P., Long, F., Murshudov, G., Scheres, S.H., Ramakrishnan, V.: Structure of the yeast mitochondrial large ribosomal subunit. Science **343**(6178), 1485–1489 (2014)
7. Liao, M., Cao, E., Julius, D., Cheng, Y.: Structure of the TRPV1 ion channel determined by electron cryo-microscopy. Nature **504**(7478), 107–112 (2013)
8. Tagare, H.D., Barthel, A., Sigworth, F.J.: An adaptive expectation-maximization algorithm with GPU implementation for electron cryomicroscopy. J. Struct. Biol. **171**(3), 256–265 (2010)
9. Sigworth, F.J., Doerschuk, P.C., Carazo, J., Scheres, S.H.W.: Maximum-likelihood methods in cryo-EM. Part i: theoretical basis and overview of existing approaches. Methods Enzymol. **482**, 263 (2010)
10. Scheres, S.H.: Single-particle processing in RELION-1.3 (2014)
11. Kimanius, D., Forsberg, B.O., Scheres, S.H., Lindahl, E.: Accelerated cryo-EM structure determination with parallelisation using GPUs in RELION-2. eLife **5**, e18722 (2016)
12. Su, H., Wen, W., Du, X., Lu, X., Liao, M., Li, D.: Gerelion: GPU-enhanced parallel implementation of single particle cryo-EM image processing. bioRxiv 075887 (2016)
13. Corporation N.: CUDA in C best practices guide. NVIDIA Corporation (2016)

GPU Accelerated Ray Tracing
for the Beta-Barrel Detection
from Three-Dimensional Cryo-EM Maps

Albert Ng, Adedayo Odesile, and Dong Si[(✉)]

Department of Computing and Software Systems,
University of Washington Bothell, Bothell, WA 98011, USA
dongsi@uw.edu

Abstract. Cryo-electron microscopy is a technique that is capable of producing high quality three-dimensional density maps of proteins. The identification of secondary structures from within these proteins is important to help understand the protein's overall structure and function. One of the more commonly found secondary structures is the β barrel. In previous papers, we presented a novel approach utilizing a genetic algorithm and ray tracing to identify and isolate β barrels from the density maps. However, one key limitation of that approach was the computational cost of ray tracing portion. In this paper, we applied parallel processing and graphical processing units (GPU) to increase the performance of the ray tracing. We tested this method on both experimental and simulated cryo-EM density maps. The results suggest that we were successful in speeding up our method significantly using parallelization and graphical processing units.

1 Introduction

Cryo-electron (Cryo-EM) microscopy is a laboratory technique used to produce three-dimensional electron density maps of large molecules, such as proteins [1]. Proteins are the fundamental building blocks of life on earth and the study of protein structures is vital to many fields. Cryo-EM works by freezing a molecule to extremely low temperatures to reduce the amount of movement from that molecule. Afterwards, using an electron microscope, a large number of two-dimensional images are taken of the frozen molecule. These two-dimensional images are then used to reconstruct a three-dimensional electron density map of the molecule being studied (See Fig. 1).

Cryo-EM technology has improved in recent years to atomic resolutions, where individual atoms within the larger molecules can be easily distinguished from each other [21]. However, there still exists a large number of cryo-EM density maps produced over the years that were resolved at medium resolutions (5–10 Å) [9]. One of the key advantages of these medium resolution density maps, over the more modern high resolution maps, is that secondary structures can still be identified [4].

© Springer International Publishing AG, part of Springer Nature 2018
F. Zhang et al. (Eds.): ISBRA 2018, LNBI 10847, pp. 217–226, 2018.
https://doi.org/10.1007/978-3-319-94968-0_20

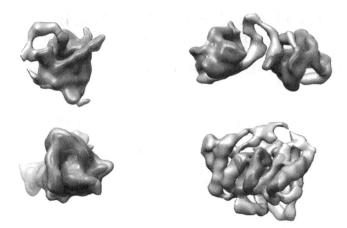

Fig. 1. Examples of cryo-electron (cryo-EM) microscopy density maps at medium resolutions. Highlighted in red within the density maps are β barrels. (Color figure online)

Secondary structures are substructures within proteins that combine with one another to help form the overall shape and structure of the protein. There exists two types of secondary structures: α-helices and β sheets. Due to the regular nature of α-helices, there have been many successful attempts in developing methods to automatically detect α-helices directly from the cryo-EM density map [4,5,15,20]. However, because β sheets are capable of folding and twisting into numerous different geometries and shapes, it is far more difficult to achieve automatic detection for β sheets. Some methods are capable of automatically detecting some β sheets structures [16–19]. One such β sheet structure is the β barrel, which is commonly found in membrane and transport proteins.

In [11,12], we proposed a method to isolate and detect β barrels from medium resolution cryo-EM density maps. The method used a combination of a genetic algorithm and ray tracing to detect and isolate β barrel density from was shown to be capable of detecting both experimental and simulated density. One of the key limitations of that proposed method was that the ray tracing portion was relatively computationally expensive. This limited the number of rays that could be casted and we felt that this limited how much detail the ray tracing method could capture. In order to speed up the ray tracing, parallelizing the ray tracing portion of the method represents a potential solution to this limitation.

In this paper, we explored a solution to this by parallelizing the ray tracing algorithm using accelerators such as graphical processing units (GPU). Additionally, we explored the effect of number rays casted on the accuracy of β barrel detection.

2   Method

The method described in this paper is separated into two main parts: a genetic algorithm component and a ray tracing component (see Fig. 2). The genetic algorithm component attempts to fit a model cylinder to the general β barrel region. The ray tracing component uses the fitted cylinder as a base to isolate and detect the true shape of the β barrel.

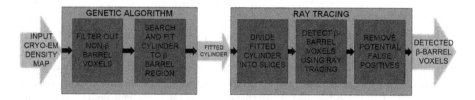

Fig. 2. Flowchart of the major steps in this method.

2.1 Genetic Algorithm

The purpose of the genetic algorithm is to search the entire density map to identify the general β barrel region and to fit a cylinder to the region as a basic approximation.

Preprocessing. Before the genetic algorithm is initiated, unnecessary voxels are filtered out from the cryo-EM density maps beforehand. First, Chimera [13] is used to select a density threshold to filter out background voxels. Second, using *Gorgon* [3] and *SSETracer* [20], α-helix voxels are removed.

The removal of density in the previous preprocessing step tends to leave behind small isolated clusters of non-β barrel voxels. The density map is then divided into clusters, defined as a group of voxels that is at least 2.0 Å away from any other cluster. Any cluster with a voxel count below a manually defined threshold is removed. Additionally, this clustering allows for the ability to detect multiple β barrels in the same density map. The following steps are applied to each remaining cluster.

Genetic Algorithm. The genetic algorithm is an optimization algorithm that is well suited to searching across large search spaces that attempts to mimic how natural selection works [8]. The genetic algorithm searches the entire density map to fit an ideal cylinder to the general β barrel region. Afterwards, a second algorithm is used to accurately detect and isolate the β barrel from the density map.

The candidate used in this genetic algorithm is an open-ended ideal cylinder. The cylinder is represented by two points, that represent the center of the two circles at the ends, and the radius of the cylinder. An initial population of candidates is randomly generated in order to seed the first generation for the genetic algorithm.

Fitness Score. The fitness equation represents the equation that we wish to optimize using the genetic algorithm. In this method, the fitness equation, F, is given by Eq. 1.

$$F = \sum_i^N \frac{(d_i)^2}{N} + \sum_j^M (d_j)^3 \tag{1}$$

The first portion of the fitness equation represents the mean-squared error. N is the total number of voxels and d_i is the distance between the voxel and the cylinder. The second portion of the fitness equation represents a penalty for when voxels are located inside the cylinder. M represents the number of voxels located inside the cylinder and d_j is the distance between the voxel inside the cylinder and the surface of the cylinder.

Crossover. The crossover function is used to create the next generation of candidates. In this method, crossover is done by selecting the top 50% of the current candidates as parents for the next generation and the bottom 50% are discarded. The parents are then randomly paired up with one another to produce two children candidates. Each children candidate randomly selects one cylinder endpoint from each parent and its radius is set to the average of the radius of its parents.

Mutation. The mutation function is used to add some randomness to the genetic algorithm to help avoid it become trapped in local minima. Every child candidate has a chance that one of its seven parameters may be modified by a mutation factor. This method uses a non-uniform mutation strategy [21] to control the size of the mutation factor.

2.2   Ray Tracing

Ray tracing is a computer graphics technique that attempts to mimic light rays [7]. This method uses the ray tracing algorithm to find the β barrel from the rest of the density map. Once the genetic algorithm has fit a cylinder to the β barrel region, the cylinder is used as the reference point from where the ray tracing is done.

Ray Tracing Algorithm. The cylinder is first split up into evenly divided circular slices stacked on top of each other. Within each slice, rays are shot out from the center of the slice out towards the edges of the circle. Because voxels are infinitesimally small, the casted ray is given a thickness of 0.5 Å. The closest voxel to the cylinder slice that each ray intersects with is then found. This is added to the set of solution β barrel voxels for each set. If the ray intersected with no voxels, then it is assumed that the ray was shot into empty space and no voxel is added to the solution set.

The number of slices that the cylinder is divided into and the rays shot from each slice is defined by the user. The higher the number of slices and rays used results in more rays casted in general, resulting in potentially more detail captured from the β barrel surrounding the cylinder. However, increasing the number of slices and rays increases the computational cost of the ray tracing and can greatly increase the time needed to finish.

GPU Accelerated Ray Tracing. One of the key issues with ray tracing is its relative high computational costs. In [11,12], the ray tracing algorithm was done on a single CPU thread and this resulted in it taking up the vast majority of the runtime. Because of this, ways to speed up the ray tracing algorithm is necessary. One such way is to parallelize the ray tracing algorithm and use graphical processing units to help accelerate the computations [14].

The ray tracing algorithm naturally tends itself to parallelization, as it consists of the tracing of numerous independent rays on the same data set [6]. The algorithm does not have any of the classic problems that prevent effective parallelization, such as data dependency between parallelized elements. In this method, each ray is designed to be executed in parallel. This should drastically reduce the total amount of time necessary to complete the ray tracing.

However, the average CPU only has two to four cores and only two to four rays can be run in parallel. Due to Amdahl's Law [2], the absolute theoretical maximum speedup from parallelizing the ray tracing algorithm on average CPUs is limited to two to four times faster, and far slower in real conditions. Accelerators, like GPUs, allow for the simultaneous execution of numerous threads at a given time. To take advantage of accelerators, the ray tracing algorithm was modified such that every ray trace would be executed on its own GPU thread to maximize parallelism.

In order to utilize the graphical processing units, the C++ Accelerated Massive Parallelism (AMP) library was used. Previous code written by Adedayo Odesile was used as the basis for this[1].

Extending and Shrinking the Cylinder. For each slice, the ratio of rays that intersected with a voxel over the total number of rays casted is calculated. Starting from the slices at the edges, the ratio is checked to make sure that it is over a threshold. For this paper, a ratio of 0.6 was used for all tests. Slice and its voxels are removed if this threshold is not met and the next slice is then checked. This continues until a slice that reaches the threshold is reached. If all slices are removed, then this method would suggest that no β barrel exists. However, if the ratios at the initial edges match the threshold, then an additional slice is then added to that end. This continues until an added slice does not meet the threshold ratio.

Removal of Outlier Voxels. Due to the irregular shape of β barrels, there are occasionally gaps in the β barrel walls. Because this method assumes an ideal cylinder shape, these gaps allow for some of the rays casted to shoot past the β barrel and intersect with non β barrel voxels that should be blocked by

---

[1] https://github.com/oddyloops/RayTracer.

the β barrel walls. This method removes these voxels by automatically removing voxels that are two standard deviations away from the mean distance between the detected β barrel voxels and the fitted cylinder.

Results Metrics

Sensitivity. Sensitivity (*Sens*) indicates the ratio of β barrel α carbons that were correctly detected by this method. An α carbon is deemed to be detected if there are voxels labeled as β barrel voxels within 2.0 Å from it. Sensitivity is calculated using Eq. 2.

$$Sensitivity = \frac{\#\ of\ \beta\ barrel\ \alpha\ carbons\ detected}{total\ \#\ of\ \beta\ barrel\ \alpha\ carbons} \tag{2}$$

Specificity. Specificity (*Spec*) indicates the ratio of non-β barrel α carbons the were correctly detected by this method. An α carbon is deemed to be detected if there are no voxels labels as β-barrel voxels within 2.0 Å from it. Specificity is calculated using Eq. 3.

$$Specificity = 1 - \frac{\#\ of\ non\text{-}\beta\ barrel\ \alpha\ carbons\ detected}{total\ \#\ of\ non\text{-}\beta\ barrel\ \alpha\ carbons} \tag{3}$$

Execution Time. Execution time (t) is used to measure performance between the GPU-accelerated method and the original non-parallelized methods. The execution time is defined as starting from when the method starts and stops when ray tracing is complete.

3   Results

Simulated Data. Eleven simulated density maps were tested using the original non-parallelized method and the new GPU-accelerated method. These density maps were generated using the base Protein Data Bank (PDB) file and the *EMAN2* program *pdb2mrc* at a resolution of 9 Å and 1 Å per pixel. In Fig. 3, an example of the results is shown.

Fig. 3. Example of results using protein 1AJZ. (**Left**) Simulated density map of protein 1AJZ at 9 Å. (**Center**) Detected β barrel surface (in red) superimposed onto the density map (**Right**) Detected β barrel surface (in red) superimposed over the true PDB structure. (Color figure online)

Table 1. Accuracy and execution time (in seconds) on simulated density maps

PDB ID	Voxels	Original			GPU_1			GPU_2			GPU_3		
		Sens	Spec	t(sec)	Sens	Spec	t(sec)	Sens	Spec	t(sec)	Sens	Spec	t(sec)
1AJZ_A	20242	1.00	0.80	137	1.00	0.80	36	1.00	0.79	46	1.00	0.80	36
1AL7_A	26189	0.97	0.82	184	0.98	0.84	37	1.00	0.83	36	1.00	0.83	37
1JB3_A	11808	0.97	0.79	68	0.98	0.79	20	0.98	0.75	23	0.98	0.74	22
1NNX_A	6938	1.00	0.88	49	1.00	0.85	20	1.00	0.81	24	1.00	0.85	21
1TIM_A	16999	0.99	0.75	127	0.98	0.75	29	0.99	0.75	31	0.99	0.76	29
1Y0Y_A	25434	0.97	0.73	196	0.97	0.72	56	0.99	0.72	60	0.99	0.74	57
2DYI_A	12879	0.93	0.82	83	0.93	0.82	35	0.92	0.83	35	0.93	0.80	35
2F01_A	8963	0.95	0.88	65	0.95	0.85	26	0.95	0.91	27	0.96	0.92	26
2VDF_A	16051	0.92	0.89	104	0.92	0.91	24	0.92	0.86	23	0.92	0.84	23
3GP6_A	12820	0.93	0.86	75	0.93	0.86	37	0.94	0.84	38	0.93	0.86	38
3ULJ_A	6849	0.93	0.87	41	0.93	0.88	18	0.95	0.83	18	0.95	0.80	18
Average		0.96	0.83	102.64	0.96	0.82	30.73	0.97	0.81	32.18	0.97	0.81	31.09

In Table 1, the results using both methods are shown. Original represents the results generated using the non-parallelized method, dividing the cylinder into 40 slices, and casting a ray every 1° around the slice. GPU_1 represents results generated using the GPU accelerated method, dividing the cylinders into 40 slices, and casting a ray every 1°. GPU_2 also uses the GPU accelerated method and dividing the cylinders into 40 slices, but instead casts a ray every 0.1°. Lastly, GPU_3 uses the GPU accelerated method and casts a ray every 1°, but uses 80 slices instead.

As shown in Table 1, the difference between the original method and the GPU accelerated method is significant in terms of performance. By using the GPU, the ray tracing algorithm averages approximately three times faster compared to the original version of the method. This improvement is even more pronounced for density maps with very high voxel counts, such as 1AL7, as this improvement increases to nearly five times faster than the original.

However, the results also demonstrate that when the number of rays and slices is increased, the accuracy is not generally affected. In terms of sensitivity, although increasing the number of rays and slices suggests a slight increase in sensitivity. In terms of specificity, increasing the number of rays and slices seems to show a minor decrease in specificity. However, in both cases, the differences are less than a percent. These results suggest that the default number of rays and slices that we used in the original method is sufficient to accurately detect and isolate the β barrel and the number of rays and slices was not a limiting factor in previous results.

Experimental Data. Eight experimental density maps were tested using both the original non-parallelized method and the GPU-accelerated method. Unlike the simulated density maps, however, these were obtained from the EM Data

Table 2. Accuracy and execution time (in seconds) on experimental density maps

EMDB ID(Res)	Voxels	Original			GPU_1			GPU_2			GPU_3		
		$Sens$	$Spec$	$t(sec)$	$Sens$	$Spec$	$t(sec)$	$Sens$	$Spec$	$t(sec)$	$Sens$	$Spec$	$t(sec)$
1657 (5.8 Å)	3720	0.73	0.78	7	0.73	0.78	2	0.73	0.77	2	0.73	0.73	2
1780 (5.5 Å)	4570	0.82	0.76	11	0.82	0.76	3	0.86	0.74	3	0.86	0.74	3
1849_L (8.25 Å)	2592	0.81	0.91	4	0.81	0.91	1.07	0.81	0.91	1.12	0.84	0.91	1.09
1849_W (8.25 Å)	2509	0.93	0.81	4	0.93	0.81	1.03	0.93	0.78	1.05	0.90	0.80	1.07
2169 (8.1 Å)	1800	0.85	0.73	4	0.85	0.73	1.10	0.85	0.72	1.03	0.89	0.75	1.03
2605 (5.5 Å)	867	0.93	0.93	2	0.96	0.93	1.08	0.93	0.93	1.13	0.89	0.93	1.12
5036 (6.7 Å)	3524	0.83	0.83	8	0.83	0.83	3	0.83	0.85	3	0.83	0.83	3
6396 (6.4 Å)	3051	0.93	0.74	6	0.93	0.73	2	0.97	0.77	2	0.67	0.78	2
Average		0.85	0.81	5.75	0.86	0.81	1.77	0.86	0.81	1.78	0.86	0.81	1.78

Bank (EMDB). Like with the simulated results, the same four ray/slice configurations were used to test the experimental density maps.

Table 2 demonstrates similar results to the simulated density maps. The GPU-accelerated method generally reduced the computation time by around three times compared to the original method. Like with the simulated density maps, the sensitivity generally increased slightly and specificity decreased slightly as rays and slices increased. However, the differences in sensitivity and specificity between the configurations averaged less than a percent difference.

Testing Environment. All tests were performed using a desktop computer with an Intel i7-4790k @ 4.0 GHz processor, 16 GB of RAM, and a nVidia GeForce 980TI.

4 Conclusion

In this paper, one way to improve the previous method of β barrel detection was to focus on the performance of the ray tracing. The previous ray tracing algorithm was parallelized and graphics processing units were used to accelerate the calculations. The results suggest that utilizing GPUs to accelerate the ray tracing algorithm generally resulted in a significant decrease in computation time. Additionally, the results suggest that the number of rays and cylinder slices used by the previous method was sufficient to accurately detect β barrels from density maps. However, this should be further explored to confirm this and to potentially find an optimal number of rays/slices.

Future Work. Future work should focus on working on improving the genetic algorithm portion of this method. The performance of the ray tracing algorithm no longer seems to be the major bottleneck of the method and work should be done on improving the preprocessing step of the genetic algorithm. Additionally, the use of other machine learning techniques, such as convolutional neural networks [10], should be explored.

Acknowledgments. This work was supported by the Graduate Research Award from the Computing and Software Systems division of University of Washington Bothell and the startup fund 74-0525. We would like to thank Dr. Kelvin Sung for his assistance on this paper.

References

1. Adrian, M., Dubochet, J., Lepault, J., McDowall, A.W.: Cryo-electron microscopy of viruses. Nature **308**(5954), 32–36 (1984). https://www.nature.com/articles/308032a0
2. Amdahl, G.M.: Validity of the single processor approach to achieving large scale computing capabilities. In: Proceedings of the April 18–20, Spring Joint Computer Conference, pp. 483–485. ACM (1967). https://doi.org/10.1145/1465482.1465560
3. Baker, M., Baker, M., Hryc, C., Ju, T., Chiu, W.: Gorgon and pathwalking: macromolecular modeling tools for subnanometer resolution density maps. Biopolymers **97**, 655–668 (2012)
4. Baker, M.L., Ju, T., Chiu, W.: Identification of secondary structure elements in intermediate-resolution density maps. Structure **15**(1), 7–19 (2007). http://www.sciencedirect.com/science/article/pii/S0969212606004722
5. Dal Palù, A., He, J., Pontelli, E., Lu, Y.: Identification of -Helices From Low Resolution Protein Density Maps, pp. 89–98
6. Glassner, A.S.: Space subdivision for fast ray tracing. IEEE Comput. Graph. Appl. **4**(10), 15–24 (1984)
7. Goldstein, R.A., Nagel, R.: 3-D visual simulation. Trans. Soc. Comput. Simul. **16**(1), 25–31 (1971). https://doi.org/10.1177/003754977101600104
8. Holland, J.H.: Genetic algorithms. Sci. Am. **267**(1), 66–72 (1992). https://doi.org/10.1038/scientificamerican0792-66
9. Lawson, C.L., Baker, M.L., Best, C., Bi, C., Dougherty, M., Feng, P., van Ginkel, G., Devkota, B., Lagerstedt, I., Ludtke, S.J., Newman, R.H., Oldfield, T.J., Rees, I., Sahni, G., Sala, R., Velankar, S., Warren, J., Westbrook, J.D., Henrick, K., Kleywegt, G.J., Berman, H.M., Chiu, W.: EMDataBank.org: unified data resource for CryoEM. Nucleic Acids Res. **39**, D456–D464 (2011). https://www.ncbi.nlm.nih.gov/pmc/articles/PMC3013769/
10. Li, R., Si, D., Zeng, T., Ji, S., He, J.: Deep convolutional neural networks for detecting secondary structures in protein density maps from cryo-electron microscopy. In: 2016 IEEE International Conference on Bioinformatics and Biomedicine (BIBM), pp. 41–46 (2016)
11. Ng, A., Si, D.: Beta-barrel detection for medium resolution cryo-electron microscopy density maps using genetic algorithms and ray tracing
12. Ng, A., Si, D.: Genetic algorithm based beta-barrel detection for medium resolution cryo-EM density maps. In: Cai, Z., Daescu, O., Li, M. (eds.) ISBRA 2017. LNCS, vol. 10330, pp. 174–185. Springer, Cham (2017). https://doi.org/10.1007/978-3-319-59575-7_16
13. Pettersen, E.F., Goddard, T.D., Huang, C.C., Couch, G.S., Greenblatt, D.M., Meng, E.C., Ferrin, T.E.: UCSF chimera-a visualization system for exploratory research and analysis. J. Comput. Chem. **25**(13), 1605–1612 (2004). http://online library.wiley.com/doi/10.1002/jcc.20084/full
14. Reinhard, E., Jansen, F.W.: Rendering large scenes using parallel ray tracing. Parallel Comput. **23**(7), 873–885 (1997). http://www.sciencedirect.com/science/article/pii/S0167819197000318

15. Rusu, M., Wriggers, W.: Evolutionary bidirectional expansion for the tracing of alpha helices in cryo-electron microscopy reconstructions. J. Struct. Biol. **177**(2), 410–419 (2012). http://www.sciencedirect.com/science/article/pii/S1047847711003455

16. Si, D., He, J.: Combining image processing and modeling to generate traces of beta-strands from cryo-EM density images of beta-barrels. In: 2014 36th Annual International Conference of the IEEE Engineering in Medicine and Biology Society, pp. 3941–3944 (2014)

17. Si, D.: Automatic detection of beta-barrel from medium resolution cryo-EM density maps. In: Proceedings of the 7th ACM International Conference on Bioinformatics, Computational Biology, and Health Informatics, pp. 156–164. ACM (2016). https://doi.org/10.1145/2975167.2975183

18. Si, D., He, J.: Beta-sheet detection and representation from medium resolution cryo-EM density maps. In: Proceedings of the International Conference on Bioinformatics, Computational Biology and Biomedical Informatics, pp. 764:764–764:770. ACM (2013). https://doi.org/10.1145/2506583.2506707

19. Si, D., He, J.: Modeling beta-traces for beta-barrels from cryo-EM density maps. **2017**, 1–9 (2017). https://doi.org/10.1155/2017/1793213. https://www.hindawi.com/journals/bmri/2017/1793213/

20. Si, D., Ji, S., Nasr, K.A., He, J.: A machine learning approach for the identification of protein secondary structure elements from electron cryo-microscopy density maps. Biopolymers **97**(9), 698–708 (2012). http://onlinelibrary.wiley.com/doi/10.1002/bip.22063/abstract

21. Zhou, Z.H.: Atomic resolution cryo electron microscopy of macromolecular complexes. In: Ludtke, S.J., Venkataram Prasad, B.V. (eds.) Advances in Protein Chemistry and Structural Biology, vol. 82, pp. 1–35. Academic Press, Cambridge (2011). https://doi.org/10.1016/B978-0-12-386507-6.00001-4. http://www.sciencedirect.com/science/article/pii/B9780123865076000014

A Fast Genome Sequence Aligner Based on Minimal Perfect Hash Algorithm Realized with FPGA Based Heterogeneous Computing Platform

Ke Huang[1,2], Shubo Yang[1], Zhaojian Luo[1], Ke Yang[1], Menghan Chen[1], Guopeng Wei[1], and Jian Huang[2(✉)]

[1] Nanjing Gezhi Genomics Bioscience, Inc., Chengdu 610051, China
{huangke,yangshubo,luozhaojian,yangke,
chenmenghan,guopeng}@gezhigene.com
[2] Center for Informational Biology,
University of Electronic Science and Technology of China,
Chengdu 610054, China
hj@uestc.edu.cn

Abstract. A fast genome sequence aligner is proposed in this paper. The alignment algorithm is based on minimal perfect hash, reducing the memory occupation and improving memory access efficiency. Several strategies and techniques are adopted to improve the speed and accuracy of the aligner. Realized with a field programmable gate array (FPGA) based heterogeneous computing platform, the aligner achieves similar accuracy compared with BWA-MEM while the speed is around 10 times faster than BWA-MEM.

Keywords: FPGA · Minimal perfect hash · Fast alignment · Smith-Waterman

1 Introduction

As the cost of next generation sequencing (NGS) technology continues to decrease, high throughput NGS technology has made its way into a variety of fields, from plant biology to human infectious disease, cancer research, and clinical medicine. With the advent of newest Illumina NovaSeq sequencer, the sequencing throughput has soared to 6T base pairs every two days [1], making the analysis of massive gene data a bottleneck.

The sequence alignment is a very key and time-consuming step of genetic data analysis. BWA [2] is an aligner which has been widely accepted for NGS data alignment. To further improve the speed and accuracy, researchers have kept studying new algorithms and optimal realization [3, 4]. Among these researches, aligners realized with FPGA offer very good speed and accuracy performance. An exact matcher based on minimal perfect hash realized with FPGA is reported in [3]. However, the reference index build and algorithm design is quite simple, limiting its application only

© Springer International Publishing AG, part of Springer Nature 2018
F. Zhang et al. (Eds.): ISBRA 2018, LNBI 10847, pp. 227–232, 2018.
https://doi.org/10.1007/978-3-319-94968-0_21

in sequence exact match. A FPGA based aligner DRAGEN is reported in [4], which is the fastest sequence aligner reported by now. However, very few technical detail of the aligner is reported.

This paper proposes a fast sequence aligner based on minimal perfect hash algorithm, reducing the memory occupation and improving memory access efficiency. Several strategies and techniques are adopted to improve the speed and accuracy of the aligner. Realized with a FPGA based heterogeneous computing platform,the aligner achieves similar accuracy compared with BWA-MEM while the speed is around 10 times faster than BWA-MEM.

2 Algorithm and Realization

The aligner basically employs the classical seed-and-extend strategy. The seed-and-extend method was pioneered by BLAST, which builds hash indexes of small sub-strings (seeds) of the reference genome and checks seeds from the reads against it for exact matches. It then uses a local extension process at each seed location to find good alignments. BWA-MEM also employs seed-and-extend strategy, while its reference index is Burrows-Wheeler transform (BWT) based, not hash-based.

Compared with BWT-based reference index, hash-based index needs less memory access but requires more memory occupation. If the memory space is enough, hash-based aligner should achieve better speed performance theoretically. However, memory space is not infinite in most computing systems. Moreover, hash collision and reference duplication greatly influence the speed performance of hash-based aligner. Thus proper organization of reference index is crucial to hash-based aligner. A hash-based reference genome index method is introduced in this paper to properly organize reference index to 16 GB memory.

A. Minimal Perfect Hash Introduction

A perfect hash function for a set S is a hash function that maps distinct elements in S to a set of integers, with no collisions. In mathematical terms, it is an injective function. A minimal perfect hash function is a perfect hash function that maps n keys to n consecutive integers. A more formal way of expressing this is: Let j and k be elements of some finite set S. F is a minimal perfect hash function if and only if $F(j) = F(k)$ implies $j = k$ (injectivity) and there exists an integer X such that the range of F is from X to $X + |S| - 1$. Figure 1 shows the definition of perfect hash and minimal perfect hash.

Minimal perfect hash algorithm eliminates hash collision and builds compact reference index with no vacant memory space. These advantages greatly reduce computation complexity and alleviate space limitation, accelerating hash table look up and saving memory space.

B. Aligner Strategies and Techniques

This paper will not discuss the minimal perfect hash construction as details can be found in both [5, 6]. With a set of known keys, we can construct a minimal perfect hash using a generalized method called hash and displace. In the aligner, Jenkin's Spooky

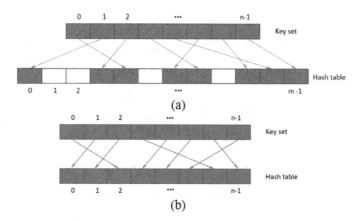

Fig. 1. (a) Perfect hash (b) Minimal perfect hash

Hash [7] was chosen because it is both fast in software and easy to implement in hardware due to its reliance on only shifts, adds, and XOR operations. And the seed input of Jenkin's Spooky Hash function facilitates minimal perfect hash realization. Figure 2 illustrates the basic idea of reference index organization. The reference genome index is composed of 37 K-mer tables, ranging from 19 bp to 127 bp with a interval of 3 bp. Each K-mer table consists of an intermediate table, an address table and a duplicate table. The intermediate table is employed to realize minimal perfect hash [3], and this table stores an information bit indicating whether the table value is an offset or rehash seed. If a K-mer is unique in the reference, the address table records the K-mer address in the reference and K-mer occurrence number in the reference. If a K-mer is not unique in the reference, the duplicate table records all the locations of the K-mer, while the address table records the access address of duplicate table and the K-mer occurrence number in the reference. Meanwhile, the address table also stores 2 bytes reference sequence at each location as verification bytes. When looking up a K-mer, the aligner compares K-mer with reference sequence at the candidate location, greatly reducing false hash hits. Noting that each K-mer table only records the K-mers whose occurrence number is smaller than 16. If the occurrence number is not smaller than 16, we record the occurrence number as -1 and keep extending the K-mer in the reference by 3 bp until the occurrence number smaller than 16. The maximum length of K-mer is 127 bp in the aligner. If a 127 bp K-mer occurs more than 15 times in the reference, the aligner drops this K-mer. By sampling both forward and reverse complemented human reference genome by every 4 bp, we successfully construct the 37 K-mer tables within 15 GB. Considering another 1 GB for human reference genome, total reference index size is smaller than 16 GB. The pseudocode of K-mer looking up in is shown in Fig. 3.

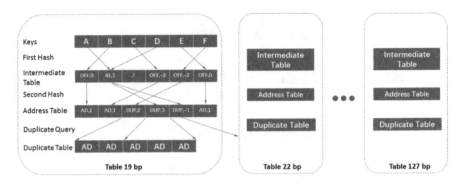

Fig. 2. Basic idea of reference index organization

Algorithm 1 Pseudocode of the K-mer looking up
1: **function** K-mer_lookup(K-mer, K , IntermediateTable, AddressTable, DuplicateTable)
2:      IntermediateTable_add ← hash (K-mer, seed = 0);
3:      inter_value ← IntermediateTable[K][IntermediateTable_add];
4:      **if** (inter_value is an offset) **then**
5:          address_table_add ← IntermediateTable_add + inter_value;
6:      **else**
7:          address_table_add ← hash (K-mer, seed = inter_value);
8:      **end if**
9:      address_value ← AddressTable[K][address_table_add];
10:     **if** (address_value indicates 1 occurrence in the reference) **then**
11:         results ← address information in the address_value ;
12:     **else if** (address_value indicates more than 1 and less than 16 occurrence in the reference) **then**
13:         results ← get addresses from DuplicateTable[K] according to address information in the address_value;
14:     **else if** (address_value indicates more than 15 occurrence in the reference) **then**
15:         results ← K-mer_lookup((K+3)-mer, K+3, IntermediateTable, AddressTable, DuplicateTable);
16:     **else**
17:         results ← NULL;
18:     **end if**
19:     **return** results;
20: **end function**

Fig. 3. K-mer looking up pseudocode

C. Aligner Realization

When aligning a read, the aligner extracts overlapping K-mers from a read starting at each base pair. These K-mers are mapped by DRAM hash table queries, each to zero or more reference positions. Matches along similar alignment diagonals are grouped into seed chains, which are sent to a voting module to conservatively filter out low quality seed chains. Each seed chain is then extended by Smith-Waterman gapped local alignment, permitting mismatch, clipping and indels. Tracing from the maximum score, a back-trace module generates alignment score and CIGAR string. All gapped local alignment results are compared to obtain best and second-best scores to generate the estimated mapping quality. All aligner modules, including K-mer mapping, seed chain filtering, Smith-Waterman local aligner and back-trace, are realized with FPGA and designed to be highly parallel and pipelined.

3   Results

We choose several dataset including whole exome sequencing (WES) and whole-genome sequencing (WGS) data to test the aligner speed and accuracy under various situation. In the test pipeline, the proposed aligner maps raw FASTQ to generate BAM files. BWA-MEM is employed to map these FASTQ files to generate BAM files at the same time. Subsequently, these BAM files are processed by GATK pipeline to call variants. The VCF files are compared to check calling consistency. All these tests are run on a server with 1 Xillinx FPGA card, 256 GB memory, 1 TB SSD hard disk and dual 10-core 2.3 GHz Intel Xeon E5 CPU supporting 40 logical threads. Figure 4 shows the run time of BWA-MEM and the proposed aligner. Table 1 illustrates the mapping rate and calling consistency of proposed aligner and BWA-MEM. It is shown that the FPGA aligner achieves similar accuracy compared with BWA-MEM while the aligner speed is around 10 times faster than BWA-MEM.

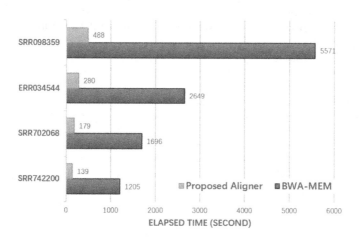

Fig. 4. Run time of the proposed aligner and BWA-MEM

Table 1. Mapping rate and calling consistency

Test data	BWA-MEM mapping rate	Proposed aligner mapping rate	GATK calling consistency
ERR034544	99.54%	99.54%	98.9%
SRR098359	93.36%	94.42%	97.7%
SRR702068	99.64%	99.67%	99.1%
SRR742200	99.12%	99.07%	98.5%

References

1. https://emea.illumina.com/content/dam/illumine-marketing/documents/products/datasheets/novaseq-6000-system-specification-sheet-770-2016-025.pdf
2. Li, H., Durbin, R.: Fast and accurate short read alignment with Burrows-Wheeler transform. Bioinformatics **25**, 1754–1760 (2009)
3. Nelson, C., Townsend, K., Rao, B.S., Jones, P., Zambreno, J.: Shepard: a fast exact match short read aligner. In: Tenth ACM/IEEE International Conference on Formal Methods and Models for Codesign, pp. 91–94 (2012)
4. Miller, N.A., Farrow, E.G., Gibson, M., Willig, L.K., Twist, G., Yoo, B., Marrs, T., Corder, S., Krivohlavek, L., Walter, A., Petrikin, J.E., Saunders, C.J., Thiffault, I., Soden, S.E., Smith, L.D., Dinwiddie, D.L., Herd, S., Cakici, J.A., Catreux, S., Ruehle, M., Kingsmore, S.F.: A 26-hour system of highly sensitive whole genome sequencing for emergency management of genetic diseases. Genome Med. **7**, 100 (2015)
5. Belazzougui, D., Botelho, F.C., Dietzfelbinger, M.: Hash, displace, and compress. In: Proceedings of the European Symposium on Algorithms, pp. 682–693 (2009)
6. Hanov, S.: Throw away the keys: easy, minimal perfect hashing (2011). http://stevehanov.ca/blog/index.php?id=119
7. Jenkins, B.: Spookyhash: a 128-bit noncryptographic hash (2012). http://burtleburtle.net/bob/hash/spooky.html

A Pattern Recognition Tool for Medium-Resolution Cryo-EM Density Maps and Low-Resolution Cryo-ET Density Maps

Devin Haslam[1], Salim Sazzed[1], Willy Wriggers[3], Julio Kovcas[3], Junha Song[2], Manfred Auer[2], and Jing He[1(✉)]

[1] Department of Computer Science, Old Dominion University, Norfolk, VA 23529, USA
jhe@cs.odu.edu
[2] Cell and Tissue Imaging, Molecular Biophysics and Integrated Bioimaging Division, Lawrence Berkeley National Laboratory, Berkeley, CA 94720, USA
[3] Department of Mechanical and Aerospace Engineering, Old Dominion University, Norfolk, VA 23529, USA

Abstract. Cryo-electron microscopy (Cryo-EM) and cryo-electron tomography (cryo-ET) produce 3-D density maps of biological molecules at a range of resolution levels. Pattern recognition tools are important in distinguishing biological components from volumetric maps with the available resolutions. One of the most distinct characters in density maps at medium (5–10 Å) resolution is the visibility of protein secondary structures. Although computational methods have been developed, the accurate detection of helices and β-strands from cryo-EM density maps is still an active research area. We have developed a tool for protein secondary structure detection and evaluation of medium resolution 3-D cryo-EM density maps that combines three computational methods (SSETracer, StrandTwister, and AxisComparison). The program was integrated in UCSF Chimera, a popular visualization software in the cryo-EM community. In related work, we have developed BundleTrac, a computational method to trace filaments in a bundle from lower resolution cryo-ET density maps. It has been applied to actin filament tracing in stereocilia with good accuracy and can be potentially added as a tool in Chimera.

Keywords: Pattern recognition · Cryo-electron microscopy · Density map
Helix · Beta-strands · Filament · Stereocilia

1 Introduction

The use of a transmission electron microscope to determine the 3-D volumetric image of biological molecules is a powerful approach to study the structure and function of macromolecules, recognized by the 2017 Nobel Prize in Chemistry [1]. Many large molecular machines that had been difficult to visualize in detail can now been resolved to near-atomic resolution [2]. While the atomic structure of some macromolecules can now be solved directly, the number of available density maps at medium resolution (5–10 Å) has also increased steadily as seen in Electron Microscopy Data Bank (EMDB) [3].

© Springer International Publishing AG, part of Springer Nature 2018
F. Zhang et al. (Eds.): ISBRA 2018, LNBI 10847, pp. 233–238, 2018.
https://doi.org/10.1007/978-3-319-94968-0_22

Due to the quality of the density map at such a resolution range, it is challenging to detect structure information with high accuracy; hence, it is challenging to obtain accurate atomic models that correspond to the medium-resolution density.

While atomic-level details may not be directly available, secondary structure elements such as helices and β-sheets are often visible in cryo-EM density maps at medium resolution. In general, a helix appears approximately as a cylinder and a β-sheet appears as a thin layer of density. Various computational methods have been developed to detect helices and β-sheets using their shape patterns [4–9], including HelixTracer, SSEhunter, SSELearner, SSETracer and VolTrac [5, 6, 9–11]. There are very few tools available to detect α-helices from medium-resolution density maps, and none to model β-strands in the density region of a β-sheet. In order to develop an accurate method for secondary structure modeling, it is essential to evaluate the accuracy of the detection. In this paper, we introduce a tool for the detection of α-helices, β-sheets, β-strands, and quantitative evaluation of accuracy.

2 Integrated Interface for Protein Secondary Structure Detection and Evaluation in UCSF Chimera

UCSF Chimera is a comprehensive visualization and analysis software widely used in the cryo-EM community for volumetric data and atomic the structure of molecules [12]. Chimera uses Python at the highest layer to organize individual functional components. The universal Python framework makes it convenient to package user-developed methods into Chimera. We have integrated three computational methods with Chimera so that they utilize existing capabilities without reinventing them (Fig. 1). SSETracer is a method for the detection of the locations of helices (redlines in Fig. 2A and β-sheets (blue regions in Fig. 2B) from a 3D density map at a medium resolution [5]. StrandTwister is a method to predict the location of β-strands (lines in Fig. 2C) from an isolated density region of a β-sheet [13]. AxisComparison is a method that uses an idea of arc-length association to evaluate the accuracy of detected helices and β-strands [14]. It compares the detected traces of helices and β-strands (red and black lines in Fig. 2D) with the axes derived from the atomic structure (green lines in Fig. 2D). The cross and longitudinal discrepancies are quantified for each line trace.

Integration of the three secondary structure analysis methods in UCSF Chimera allows a user to perform our three methods interactively so that a user may use various manipulation and visualization options in Chimera for an individual helix or β-strand and for the subsequent quantitative evaluation. This makes it convenient to scan through secondary structure detection and evaluation, particularly when there is a large number of secondary structure elements detected in the density map. The Chimera plugin is downloadable from http://www.cs.odu.edu/~jhe.

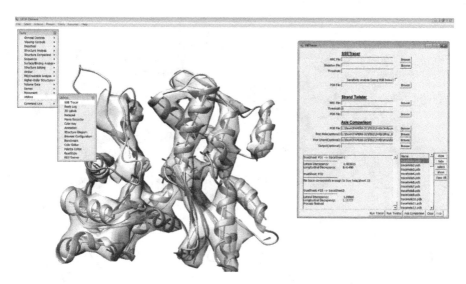

Fig. 1. Integrated interface in UCSF Chimera for SSETracer, StrandTwister, and AxisComparison.

3 Filament Tracing in a Bundle

Cryo-electron tomography (Cryo-ET) is a technique to obtain 3-D images of much larger cellular targets such as organelles, which are often complex and thus do not lend themselves to averaging. The data collection process in cryo-ET differs from that of cryo-EM. Instead of taking single images of particles in a random orientation and averaging the different views and reconstructing the object given the different orientations, cryo-ET collects multiple images from the same object, with the specimen being tilted at different angles. However, due to tilt angle limitations and limits on the acceptable total radiation dose, tomograms often display reconstruction artifacts, anisotropic resolution and a high level of noise. In particular, orientational (missing-wedge) artifacts are prominent in the reconstructed 3-D volume due to limitations in the data collection geometry. Furthermore, cryo-ET density maps usually show a much lower resolution (30–50 Å) than those obtained using single-particle cryo-EM or other implicit averaging approaches, which makes the direct modeling and interpretation of structural features difficult.

We have developed a computational method, BundleTrac, to trace filaments in a bundle and applied it to stereocilia density maps obtained using cryo-ET (Fig. 3) [15]. BundleTrac is a semi-automated method that starts with user-defined seed points in UCSF Chimera on a cross-section of the bundle. It traces the rest of the filaments using the geometric pattern of a bundle of filaments.

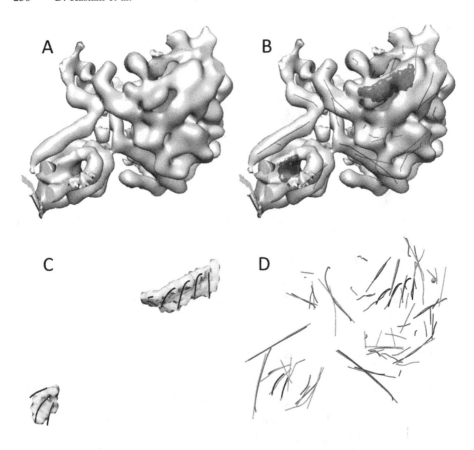

Fig. 2. Three computational methods SSETracer, StrandTwister, and AxisComparison. (A) An isolated density region of cryo-EM map EMD-3204 (EMDB ID) that corresponds to a single chain of the protein; (B) SSETracer detected helices (red lines) and β-sheets (blue) superimposed on the density map; (C) StrandTwister detected β-strands (black lines) from the β-sheets (cyan or blue in (B)); (D) AxisComparison quantifies the error between detected helices (red lines)/β-strands (black lines) and the true axes obtained from atomic structure of 5FKX (PDB ID). UCSF Chimera was used as a platform to develop an integrated interface for the three methods. (Color figure online)

4   Summary

We present, in this paper, a pattern recognition tool for protein secondary structure detection and evaluation of cryo-EM density maps at medium resolutions. We show that the three computational methods in secondary structure detection and evaluation can be combined and inserted in the framework of Chimera to utilize existing resources in Chimera. BundleTrac potentially can be inserted in Chimera using a similar approach in the future to benefit pattern recognition needs in cryo-ET.

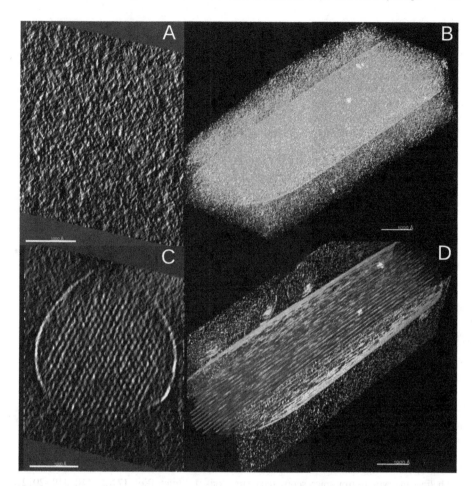

Fig. 3. Detection of actin filaments from a bundle using BundleTrac. (A) a cross-section of the bundle in a stereocilium density map obtained using cryo-ET technique; (B) the density map of the stereocilium; (C) The cross-section at the same location of the bundle as in (A) after an average along the calculated bundle axis; (D) a subset of filaments (red lines) detected using BundleTrac superimposed with the image of a stereocilium. Visualization was performed in Chimera. (Color figure online)

Acknowledgements and Author Contributions. The work in this article was supported, in part, by NSF DBI-1356621 (to J.H), NIH R01-GM062968 (to W.W.) and NIH P01-GM051487 (to M.A.). D.H. and J.H. designed the plugin of Chimera; D.H. implemented it. We would like to thank Taylor Gerpheide and Stephanie Zeil for initial work in the tool development when they worked as undergraduate research assistants. S.S., J.S., J.K., W.W., M.A. and J.H. worked together in the development of BundleTrac. J.H.. W.W. and M.A. wrote the paper. We thank Peter Barr-Gillespie for sample preparation and imaging of stereocilia.

References

1. Nobelprize.org: The Nobel Prize in Chemistry 2017: Nobel Media AB 2014. https://www. nobelprize.org/nobel_prizes/chemistry/laureates/2017/. Accessed 11 Apr 2018
2. Shalev-Benami, M., Zhang, Y., Rozenberg, H., Nobe, Y., Taoka, M., Matzov, D., et al.: Atomic resolution snapshot of Leishmania ribosome inhibition by the aminoglycoside paromomycin. Nat. Commun. **8**(1), 1589 (2017)
3. Europe PDBi: EMDB statistics release (2018). http://www.ebi.ac.uk/pdbe/emdb/statistics_ releases.html/
4. Jiang, W., Baker, M.L., Ludtke, S.J., Chiu, W.: Bridging the information gap: computational tools for intermediate resolution structure interpretation. J. Mol. Biol. **308**(5), 1033–1044 (2001)
5. Si, D., He, J.: Beta-sheet detection and representation from medium resolution Cryo-EM density maps. In: Proceedings of the International Conference on Bioinformatics, Computational Biology and Biomedical Informatics, Washington DC, USA, pp. 764–770. ACM (2013). https://doi.org/10.1145/2506583.2506707
6. Baker, M.L., Ju, T., Chiu, W.: Identification of secondary structure elements in intermediate-resolution density maps. Structure **15**(1), 7–19 (2007)
7. Kong, Y., Zhang, X., Baker, T.S., Ma, J.: A Structural-informatics approach for tracing beta-sheets: building pseudo-C(alpha) traces for beta-strands in intermediate-resolution density maps. J. Mol. Biol. **339**(1), 117–130 (2004)
8. Zeyun, Y., Bajaj, C.: Computational approaches for automatic structural analysis of large biomolecular complexes. IEEE/ACM Trans. Comput. Biol. Bioinform. **5**(4), 568–582 (2008)
9. Dal Palu, A., He, J., Pontelli, E., Lu, Y.: Identification of Alpha-Helices from low resolution protein density maps. In: Proceeding of Computational Systems Bioinformatics Conference, CSB, pp. 89–98 (2006)
10. Si, D., Ji, S., Nasr, K.A., He, J.: A machine learning approach for the identification of protein secondary structure elements from electron cryo-microscopy density maps. Biopolymers **97** (9), 698–708 (2012)
11. Rusu, M., Wriggers, W.: Evolutionary bidirectional expansion for the tracing of alpha helices in cryo-electron microscopy reconstructions. J. Struct. Biol. **177**(2), 410–419 (2012)
12. Pettersen, E.F., Goddard, T.D., Huang, C.C., Couch, G.S., Greenblatt, D.M., Meng, E.C., et al.: UCSF Chimera–a visualization system for exploratory research and analysis. J. Comput. Chem. **25**(13), 1605–1612 (2004)
13. Si, D., He, J.: Tracing beta strands using StrandTwister from cryo-EM density maps at medium resolutions. Structure **22**(11), 1665–1676 (2014)
14. Stephanie, Z., Julio, K., Willy, W., Jing, H.: Comparing an atomic model or structure to a corresponding cryo-electron microscopy image at the central axis of a helix. J. Comput. Biol. **24**(1), 52–67 (2017)
15. Sazzed, S., Song, J., Kovacs, A.J., Wriggers, W., Auer, M., He, J.: Tracing actin filament bundles in three-dimensional electron tomography density maps of hair cell stereocilia. Molecules **23**(4), 882 (2018)

Machine and Deep Learning

Combining Sequence and Epigenomic Data to Predict Transcription Factor Binding Sites Using Deep Learning

Fang Jing[1], Shao-Wu Zhang[1(✉)], Zhen Cao[2], and Shihua Zhang[2,3(✉)]

[1] Key Laboratory of Information Fusion Technology of Ministry of Education, College of Automation, Northwestern Polytechnical University, Xi'an 710072, China
zhangsw@nwpu.edu.cn

[2] NCMIS, CEMS, RCSDS, Academy of Mathematics and Systems Science, Chinese Academy of Sciences, Beijing 100190, China
zsh@amss.ac.cn

[3] School of Mathematical Sciences, University of Chinese Academy of Sciences, Beijing 100049, China

Abstract. Knowing the transcription factor binding sites (TFBSs) is essential for modeling the underlying binding mechanisms and follow-up cellular functions. Convolutional neural networks (CNNs) have outperformed methods in predicting TFBSs from the primary DNA sequence. In addition to DNA sequences, histone modifications and chromatin accessibility are also important factors influencing their activity. They have been explored to predict TFBSs recently. However, current methods rarely take into account histone modifications and chromatin accessibility using CNN in an integrative framework. To this end, we developed a general CNN model to integrate these data for predicting TFBSs. We systematically benchmarked a series of architecture variants by changing network structure in terms of width and depth, and explored the effects of sample length at flanking regions. We evaluated the performance of the three types of data and their combinations using 256 ChIP-seq experiments and also compared it with competing machine learning methods. We find that contributions from these three types of data are complementary to each other. Moreover, the integrative CNN framework is superior to traditional machine learning methods with significant improvements.

Keywords: Bioinformatics · Machine learning
Transcription factors binding sites · Convolutional neural networks
DNA accessibility · Histone modification

1 Introduction

It has been well known that transcription factors (TFs) are key proteins decoding the information in the genome to express a precise and unique set of proteins and RNAs in each cell type in the cellular system [1]. How TFs bind to specific DNA-regulatory sequences (known as TF binding site, or TFBS for short) to cooperatively modulate the

© Springer International Publishing AG, part of Springer Nature 2018
F. Zhang et al. (Eds.): ISBRA 2018, LNBI 10847, pp. 241–252, 2018.
https://doi.org/10.1007/978-3-319-94968-0_23

gene transcription and protein synthesis is an essential procedure, which plays key roles in many biological processes [2, 3]. Moreover, it has been reported that some genomic variants in such TFBSs are associated with serious diseases including cancer and so on [4]. In the past decade, large amount of immunoprecipitation followed by high throughput sequencing (ChIP-seq) data have been generated and profiled to study the mechanisms behind these regulatory processes [5]. However, the ChIP-seq experiment can only profile one TF binding map in a given cell type one time [6, 7]. Hence it is not possible to profile every TF binding maps in all cell types due to the large number of TF-cell combinations and the high experimental cost [6, 7]. Thus, accurate computational methods are desired to decode the underlying binding rules under different circumstances. Naturally, how to predict TFBSs in DNA sequences is a basic problem in bioinformatics.

In this background, using primary DNA sequences to predict the TFBSs has become a direct and promising paradigm. At first position weight matrices (PWMs) based methods achieved great success in modeling the DNA binding protein process [8]. Later, gkm-SVM (i.e., gapped k-mers along with support vector machine) shows great superiority over the PWM-based methods [9]. More recently, convolutional neural networks [10], coupled with the one-hot coding format of DNA sequences [11–20], attracted great interest in predicting TFBSs. However, prediction or imputation of TFBSs using solely primary DNA sequences lacks the ability of dealing with cell type-specific binding events.

As a result, more and more methods turn to using cell type-specific information for addressing this issue. In addition to primary DNA sequences, other local chromatin information such as chromatin accessibility and histone modifications also have great impact to the binding of TFs to their target sites [21]. Their analysis suggested models learned from one TF was transferable across diverse TFs. Xin and Rohs [22] built a L2-regularized multiple linear regression (MLR) model to analyze histone modification patterns associated with TFBSs and showed that histone modification patterns contribute to TF binding specificities. Their results suggested that adding histone modification or chromatin accessibility information could increase the prediction performance of a classifier. However, there still exist limitations to be addressed when integrating data from different sources.

In the last few years, the fast development of deep learning or deep neural networks such as the convolutional neural networks (CNNs) attracts great attentions for the predicting of TFBSs. First, the convolution filters fitting in well with the one-hot coding format of DNA sequence can mimic the characteristics of DNA motifs [12–15, 23, 24]. Meanwhile, the learning procedure of CNN automatically extract features, which may overcome the information loss of handcrafted features. Second, the deep learning framework is flexible enough to integrate different sources of data. In addition to DNA sequence data, other data sources can be put as inputs using a computational graph, which is a directed acyclic graph representing the arbitrary information flow [25]. Third, the use of graphics processing unit (GPU) makes the training process of deep learning and especially CNNs extremely faster than before. This enables the CNN models to be applicable to deal with large amount of biological samples. However, all the existing CNN based models use solely primary DNA sequence to predict TFBSs.

Currently, it is not clear how to effectively integrate DNA sequence information with other local chromatin information (e.g., DNase and histone modification) using CNN.

To this end, we disentangled the contributions of DNA sequence and DNase I hypersensitivity (DHS for short) and histone modifications (HMS for short) in distinguishing TFBSs from background based on a CNN model (Fig. 1). To explore how to use DHS and HMS to train the neural networks, we first benchmarked a series of architecture variants by changing network structure in terms of width and depth. We also explored the effects of sample length at flanking regions 5' and 3' of the motif binding sites ranging from 5 to 101 bp of DHS and HMS data. Based on detailed experimental setup, we evaluated the performance of the three types of data and their combinations using 256 ChIP-seq experiments [15]. We find that contributions from these three types of data are complementary to each other. Moreover, the results show distinct superiority of the integrative framework over traditional machine learning methods. We expect to see wide applications of integrating multiple types of data with deep learning methods not only for TFBSs prediction, but also for other genomic studies in near future.

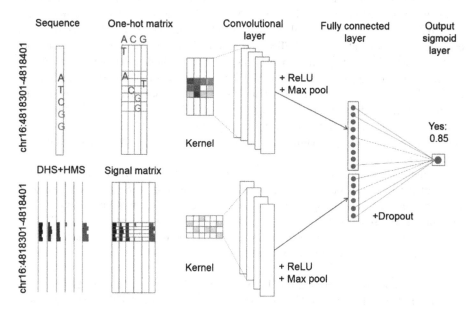

Fig. 1. Overview of the unified framework for predicting TFBSs using CNN.

2   Materials and Methods

2.1   Datasets

We downloaded 256 TF ChIP-seq experiments for 15 cell types from [15]. Each experiment includes training and testing datasets in fastq format. In the datasets, DNA sequences and its location in the reference genome (hg19) and labels are given.

The positive and negative samples have matched GC-content and sequence length (101 bp). Then we downloaded normalized DNase-seq (DHS) and five core histone modifications (HMS) ChIP-seq data (H3K4me3, H3K4me1, H3K36me3, H3K9me3, H3K27me3) for the 15 cell types from the REMC database [26]. The DHS and HMS data are genome-wide –log10 (p-value) signal coverage tracks in bigwig format.

According to the location of the sample in the sequence datasets, we extracted the signal values of the corresponding positions from the DHS and HMS signal coverage tracks. The DNase-seq or each histone modification data was represented in a feature vector (where each nucleotide position has a value). Thus, TFBSs and non-TFBSs were described as three types of features: (1) a one-hot vector for a DNA sequence; (2) a vector for DHS at each nucleotide position; (3) a vector for each HMS at each nucleotide position. For each dataset, we used 70% samples for training, 10% samples for validating and 20% for testing.

2.2   Neural Network Setup

For a DNA sequence, TFBSs and non-TFBSs were described as one dimensional image with four channels. Each base pair (A, C, T, G) was denoted as a one-hot vector [1, 0, 0, 0], [0, 1, 0, 0], [0, 0, 1, 0] and [0, 0, 0, 1] respectively. For DNase-seq data and each histone modification data, TFBSs and non-TFBSs were described as one channel vector at each nucleotide position, For HMS, existing methods calculated the statistical values (such average reads number in each base pair) within the range of hundreds or thousands nucleotide. However, such a simplistic approach may not fully use the information in HMS data. So we used histone modification data of single base resolution in our study. HMS and DHS are contiguous attributes describing surrounding epigenetic marks and chromatin accessibility that may be related to the binding of specific TFs [27].

From the viewpoint of data, to examine how these models perform quantitatively in terms of the length of flanking regions used in calculating DHS and HMS, we tried different length scales ranging from 5 to 101 bp centered on the motif binding sites. For example, if we used DNase-seq data with 101 bp, the vector was of size 1×101 for a sample; if we used five histone modifications data with 71 bp, the dimension of a vector was of size 1×71, and they were combined as matrix with size of 5×71 for a sample.

For the purpose of combining DHS, HMS and sequence in the unified deep learning framework, after collecting DNA sequence, HMS, DHS, labels data and encoding features for each sample, we first implemented five different models: sequence CNN model, using DNA sequence as features; DHS CNN model, using DHS as features; DHS Deep Neural Networks (DNN) model, using DHS as features; HMS CNN model, using HMS as features; HMS DNN model, using HMS as features. We used CNN and DNN models to compare which one was more suitable for DHS and HMS data. The CNN consists of a convolutional layer, a max-pooling layer, a fully connected layer, a dropout layer [28] and an output layer. DNN consists of one or two full connection layers, a dropout layer after each full connection layer and an output layer. For CNN models, we vary the number of kernels, the size of kernel window, and the number of neurons in the full connection layer. For DNN models, we vary the number of layers, and the number of neural in each full connection layer.

After determining an appropriate model, hyper-parameters and sample length for each data, we then studied the combinations performances of two types of data implementing three different models: sequence + HMS model, using a combination of DNA sequence and HMS as features, sequence + DHS model, using a combination of DNA sequence and DHS as features, DHS + HMS model, using a combination of DHS and HMS as features. We suggest an integrative model combing all three types of data (sequence + HMS + DHS model) as features at last.

For training, we used the cross-entropy as the loss function. Given this loss function and different hyper-parameters (see below), the models were trained using the standard error back-propagation algorithm and AdaDetla method [29]. Passing all the training data through the model once is an epoch. We set each model for 100 epochs and 128 mini-batch size and validated the model after each epoch. Then the early-stop trick was used to stop training as the error on validation set is higher than the last four epochs. The best model was chosen according to the accuracy on the validation set.

2.3 Leave-One-Feature-Out of the HMS Model

To determine the importance of each histone modification feature in the classification models by combining five core histone modification features, we implemented CNN models where we left out one of the features at a time. We recorded the AUC for each model compared to the model that used all five histone modification features.

2.4 Comparison with Conventional Learning Methods with HMS and DHS Data

We evaluated whether conventional learning methods can get comparable predictions compared with CNN. We predicted the TFBSs using k-Nearest Neighbor (kNN), Logistic Regression (LR), Random Forest (RF) classifiers. For KNN, LR, RF, we implemented these baselines using the python based scikit-learn package.

For the kNN classifier implementation, this model was trained on varying hyper-parameter values of n_neighbors: 1, 3, or 5, weights: 'uniform', or 'distance', the algorithm was 'auto'. The n_neighbors parameter defines the number of neighbors to be used for prediction. The weights define weight function used in prediction. Uniform means all points in each neighborhood are weighted equally. Distance means weight points by the inverse of their distance. In this case, closer neighbors of a query point will have a greater influence than neighbors which are a little far away.

For the LR classifier implementation, the model was trained on varying hyper-parameter values of penalty: 'l1' or 'l2', C: 0.1, 1, or 10. The penalty is used to specify the norm used in the penalization. C is the inverse of regularization strength, smaller values specify stronger regularization.

For the RF Classifier implementation, we varied the number of trees in the forest, n_estimators: 10, 20, 30, ..., 100, 200, 300, used to train each model.

All the above models were trained on the training set, and evaluated on the corresponding testing set. For kNN, we selected n_neighbors = 5, weights = 'distance'. For RF, we selected n_estimators = 100. For LR, we selected penalty = 12, C = 1.

2.5 Implementation

We used python and Keras framework to train neural networks. We used python and skcikit-learn to train conventional machine learning methods [30]. All the source codes are available at http://page.amss.ac.cn/shihua.zhang/.

3 Results

3.1 Long Sample Length and CNN Architecture Improve TFBSs Prediction Based on Histone Modification Profiles

For predicting TFBSs, we considered several practical aspects to make full use of HMS data. We first tested the effects of using different sample lengths. We used different sample lengths to train the CNN models and different hyper-parameters for each length. For each length, we selected the results of best hyper-parameters. As expected, the longer the sequence length was, the better the model performs (Fig. 2A). The improvement may come from the extra context information contained in the longer samples.

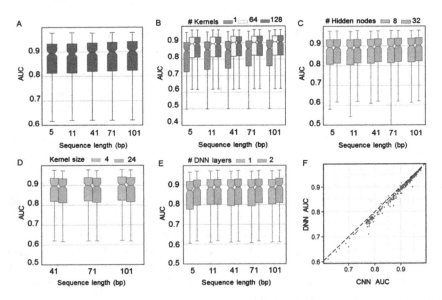

Fig. 2. Performance evaluation of CNN with respect to sample length and model structure using HMS data in terms of the distribution of AUCs across 256 experiments. (A) The effect of sequence length. (B) The effect of kernel number. (C) The effect of neuron number. (D) The effect of kernel window size. (E) The effect of sample length and DNN model structure. (F) The performance comparison of DNN versus CNN.

In addition to different sequence lengths, proper model architecture was also needed. First, more convolutional kernels could also improve the prediction performance (Fig. 2B). This observation shows additional kernels add power in extracting features. However, when more than 64 kernels were used, the improvement seemed to be saturated for the 256 experiments (Fig. 2B). Second, more neurons in the full connection layer of CNN could improve the prediction performance (Fig. 2C). And adding more neurons could improve the results. We observe that small kernel window size achieves better performance than using large ones (Fig. 2D) while big kernel window size usually used in sequence-based CNN models. This suggests that HMS features is different from sequence, and big window size may lose some information. Since the small window size is good, we are wondering how DNN performs. For comparison, we trained DNN with HMS data. We find that deeper neural networks and longer sample length work better too for DNN (Fig. 2E). As model with more neurons and layers could represent more abstract features, this observation emphasizes sufficient neurons and layers are needed to extract abstract features. However, the performance of DNN is still slightly worse than that of CNN, indicating the importance of combining convolution operation with HMS data (Fig. 2F).

3.2 Different Histone Modification Features Contribute Diversely

How each individual histone modification feature contribute relative to all five features together? We conducted leave-one-feature-out feature selection experiments to train the CNN models by using merely four histone modifications data with the same hyperparameters in previous section. Our results suggest that H3K4me3 mark is the most important mark and H3K4me1 is the second most important one (Fig. 3). We also known that H3K4me3 denotes a specific chemical modification of proteins used to package DNA in eukaryotic cells, which is commonly associated with active transcription of nearby genes [26]. While H3K4me1 has been shown distinct enrichment at active and primed enhancers, indicating its underlying strong connections with enhancer activity and function. However, the remaining three marks H3K27me3, H3K36me3, H3K9me3 play limited impacts on the prediction performance. This is very consistent with their well-known characteristics that H3K27me3, H3K36me3, H3K9me3 are found in facultatively repressed genes, actively transcribed gene bodies, and constitutively repressed genes respectively. Thus, this is reasonable that H3K9me3 shows the worst prediction ability to TFBSs. In summary, the histone modification importance observations are in consistent with their general functions and might provide further insights into the importance of different types of data in a similar way.

3.3 TFBSs Prediction Results Based on DNase-seq Profiles

Similar to HMS data, we also considered several practical aspects to make full use of DNase-seq data. We first tested the effects of using different sample lengths. As expected, the longer the sequence length is, the better the model performs (Fig. 4A). This indicates that the improvement may also come from the extra context information contained in the longer samples. For model architectures, more convolutional kernels could also improve the prediction performance (Fig. 4B). Thus, no matter what the data

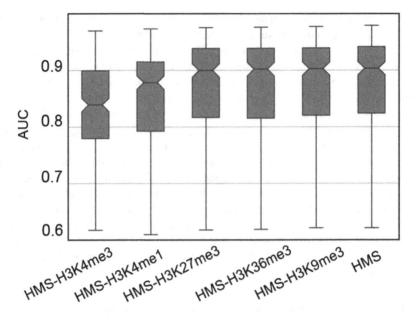

Fig. 3. Performance comparison of different HMS combinations in terms of distribution of AUCs across 256 experiments. HMS means using all five histone modification marks. HMS-H3K4me3 means using other four histone modification marks except H3K4me3.

type is, the additional kernels are beneficial to enhance power in extracting features and improve model performance. By changing the number of neurons in the last dense layer of CNN, we can see that models with more hidden neurons achieve better performance (Fig. 4C). This observation was similar with that of HMS data. We also see that CNN models with small and large kernel window sizes (4 and 24) achieve almost the same performance for different sample lengths (Fig. 4D). This suggests that kernel window sizes (4 and 24) could not distinctly influence DHS data information. For comparison with CNN, we also trained DNN using different sequence lengths and hyper-parameters for the DHS data. Similarly, the deeper neural networks and longer sample length also work better based on DHS data (Fig. 4E). Moreover, the performance of DNN is slightly worse than that of CNN, indicating the importance of combining convolution operation with DHS data (Fig. 4F).

3.4 Comparison of CNN with Conventional Learning Methods with HMS and DHS Data

We have shown that CNN models with HMS and DHS data could make very promising predictions for diverse TFs. In this section, we evaluated whether conventional learning methods can get such predictions compared to CNN. As we showed that for DHS and HMS, the longer the sequence length was, the better the model performed. Here all sample lengths used were set as 101 bp. We adopted the popular k-Nearest Neighbor (kNN), Logistic Regression (LR) and Random Forest (RF) for this task. The

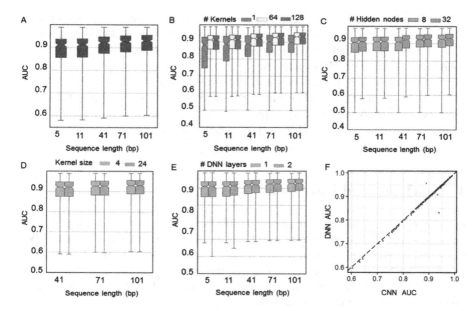

Fig. 4. Performance evaluation of CNN with respect to sample length and model structure using DHS data in terms of the distribution of AUCs across 256 experiments. (A) The effect of sequence length. (B) The effect of kernel number. (C) The effect of neuron number. (D) The effect of kernel window size. (E) The effect of sample length and DNN model structure. (F) The performance comparison of DNN versus CNN.

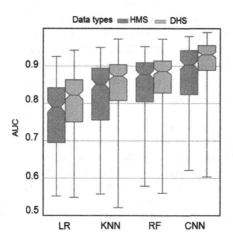

Fig. 5. Comparison of CNN with conventional learning methods in terms of the distribution of AUCs across 256 experiments. KNN: k-Nearest Neighbor; LR: Logistic Regression; RF: Random Forest.

best hyper-parameters of these methods were also chosen according to the performance on testing set (Methods and Supplementary Information). In both HMS and DHS cases, CNN perform significantly better than conventional classifiers in term of the distribution of AUCs across 256 experiments (Fig. 5). This was not surprisingly, as deep learning models could automatically extract high-level features in the DHS or HMS data due to its elaborate architectures. We note that most conventional learning methods are shallow models, which limited their performance. Taken together, our study suggests that CNN model is a more reliable tool for predicting the TFBSs by integrating these three types of data.

4   Conclusion and Discussion

In this work, we systematically explored the effects of epigenomic information from the chromatin accessibility and histone modifications data on the basis of a series of CNN architectures. We suggest an integrative CNN framework to combine primary DNA sequence, DHS and HMS data to predict cell type-specific TFBSs. Thorough evaluation demonstrate that the integrative framework show much better performance than using primary DNA sequence data only.

Chromatin accessibility and histone modifications are critical factors enabling the binding of TFs to their target genes. Chromatin accessibility has been widely used in conventional methods. But conventional methods required a lot of time for large input data and they used low resolution canonical features. Thus, we expect to improve discrimination ability through deep learning approach by automatically extracting efficient features. Histone modifications data is less used in TFBSs prediction than chromatin accessibility. The reason is that DNase-seq can give base pair resolution whereas DNA sequence was nicked, histone modification ChIP-seq gives a region where protein interacting with DNA sequence, so it only gives low resolution information compared to DNase-seq data. Besides DNA sequence and DHS data, we suggest that the HMS data can also provide extra context information despite of the low experimental resolution. In short, our work suggests combining more data in deep learning model may be beneficial.

Acknowledgement. Fang Jing would like to thank the support of the National Center for Mathematics and Interdisciplinary Sciences, Academy of Mathematics and Systems Science, CAS, during his visit. The work was supported by the National Natural Science Foundation of China [No. 61473232 and 91430111 to SWZ; No. 61621003 and 11661141019 to SZ]; the Strategic Priority Research Program of the Chinese Academy of Sciences (CAS) [No. XDB13040600], the Key Research Program of the Chinese Academy of Sciences, [No. KFZD-SW-219] and CAS Frontier Science Research Key Project for Top Young Scientist [No. QYZDB-SSW-SYS008].

References

1. Mitchell, P.J., Tjian, R.: Transcriptional regulation in mammalian cells by sequence-specific DNA binding proteins. Science **245**, 371–378 (1989)
2. Junion, G., Spivakov, M., Girardot, C., Braun, M., Gustafson, E.H., Birney, E., Furlong, E. E.: A transcription factor collective defines cardiac cell fate and reflects lineage history. Cell **148**, 473–486 (2012)
3. Vaquerizas, J.M., Kummerfeld, S.K., Teichmann, S.A., Luscombe, N.M.: A census of human transcription factors: function, expression and evolution. Nature Rev. Genet. **10**, 252–263 (2009)
4. Lee, T.I., Young, R.A.: Transcriptional regulation and its misregulation in disease. Cell **152**, 1237–1251 (2013)
5. Neph, S., Vierstra, J., Stergachis, A.B., Reynolds, A.P., Haugen, E., Vernot, B., Thurman, R. E., John, S., Sandstrom, R., Johnson, A.K.: An expansive human regulatory lexicon encoded in transcription factor footprints. Nature **489**, 83–90 (2012)
6. Gilfillan, G.D., Hughes, T., Sheng, Y., Hjorthaug, H.S., Straub, T., Gervin, K., Harris, J.R., Undlien, D.E., Lyle, R.: Limitations and possibilities of low cell number ChIP-seq. BMC Genom. **13**, 645 (2012)
7. Park, P.J.: ChIP–seq: advantages and challenges of a maturing technology. Nature Rev. Genet. **10**, 669–680 (2009)
8. Warner, J.B., Philippakis, A.A., Jaeger, S.A., He, F.S., Lin, J., Bulyk, M.L.: Systematic identification of mammalian regulatory motifs' target genes and functions. Nat. Methods **5**, 347–353 (2008)
9. Ghandi, M., Lee, D., Mohammad-Noori, M., Beer, M.A.: Enhanced regulatory sequence prediction using gapped k-mer features. PLoS Comput. Biol. **10**, e1003711 (2014)
10. LeCun, Y., Bengio, Y., Hinton, G.: Deep learning. Nature **521**, 436–444 (2015)
11. Angermueller, C., Lee, H.J., Reik, W., Stegle, O.: DeepCpG: accurate prediction of single-cell DNA methylation states using deep learning. Genome Biol. **18**, 67 (2017)
12. Qin, Q., Feng, J.: Imputation for transcription factor binding predictions based on deep learning. PLoS Comput. Biol. **13**, e1005403 (2017)
13. Yang, B., Liu, F., Ren, C., Ouyang, Z., Xie, Z., Bo, X., Shu, W.: BiRen: predicting enhancers with a deep-learning-based model using the DNA sequence alone. Bioinformatics **33**, 1930–1936 (2017)
14. Kelley, D.R., Snoek, J., Rinn, J.L.: Basset: learning the regulatory code of the accessible genome with deep convolutional neural networks. Genome Res. **26**(7), 990–999 (2016)
15. Zeng, H., Edwards, M.D., Liu, G., Gifford, D.K.: Convolutional neural network architectures for predicting DNA–protein binding. Bioinformatics **32**, i121–i127 (2016)
16. Jurtz, V.I., Johansen, A.R., Nielsen, M., Almagro Armenteros, J.J., Nielsen, H., Sønderby, C.K., Winther, O., Sønderby, S.K.: An introduction to deep learning on biological sequence data: examples and solutions. Bioinformatics **33**, 3685–3690 (2017)
17. Liu, Q., Xia, F., Yin, Q., Jiang, R.: Chromatin accessibility prediction via a hybrid deep convolutional neural network. Bioinformatics **34**(5), 732–738 (2017). https://doi.org/10.1093/bioinformatics/btx679
18. Min, X., Zeng, W., Chen, N., Chen, T., Jiang, R.: Chromatin accessibility prediction via convolutional long short-term memory networks with k-mer embedding. Bioinformatics **33**, i92–i101 (2017)
19. Bu, H., Gan, Y., Wang, Y., Zhou, S., Guan, J.: A new method for enhancer prediction based on deep belief network. BMC Bioinform. **18**, 418 (2017)

20. Zhang, J., Peng, W., Wang, L.: LeNup: learning nucleosome positioning from DNA sequences with improved convolutional neural networks. Bioinformatics **34**(10), 1705–1712 (2018). https://doi.org/10.1093/bioinformatics/bty003
21. Piqueregi, R., Degner, J.F., Pai, A.A., Gaffney, D.J., Gilad, Y., Pritchard, J.K.: Accurate inference of transcription factor binding from DNA sequence and chromatin accessibility data. Genome Res. **21**, 447–455 (2011)
22. Xin, B., Rohs, R.: Relationship between histone modifications and transcription factor binding is protein family specific. Genome Res. (2018). https://doi.org/10.1101/gr.220079.116
23. Min, X., Zeng, W., Chen, S., Chen, N., Chen, T., Jiang, R.: Predicting enhancers with deep convolutional neural networks. BMC Bioinform. **18**, 478 (2017)
24. Zhou, J., Troyanskaya, O.G.: Predicting effects of noncoding variants with deep learning–based sequence model. Nat. Methods **12**, 931–934 (2015)
25. Abadi, M., Barham, P., Chen, J., Chen, Z., Davis, A., Dean, J., Devin, M., Ghemawat, S., Irving, G., Isard, M.: TensorFlow: a system for large-scale machine learning. In: OSDI 2016, pp. 265–283 (2016)
26. Kundaje, A., Meuleman, W., Ernst, J., Bilenky, M., Yen, A., Kheradpour, P., Zhang, Z., Heravi-Moussavi, A., Liu, Y., Amin, V.: Integrative analysis of 111 reference human epigenomes. Nature **518**, 317–330 (2015)
27. Ziller, M.J., Edri, R., Yaffe, Y., Donaghey, J., Pop, R., Mallard, W., Issner, R., Gifford, C.A., Goren, A., Xing, J.: Dissecting neural differentiation regulatory networks through epigenetic footprinting. Nature **518**, 355–359 (2015)
28. Srivastava, N., Hinton, G.E., Krizhevsky, A., Sutskever, I., Salakhutdinov, R.: Dropout: a simple way to prevent neural networks from overfitting. J. Mach. Learn. Res. **15**, 1929–1958 (2014)
29. Zeiler, M.D.: ADADELTA: an adaptive learning rate method. arXiv preprint arXiv:1212.5701 (2012)
30. Pedregosa, F., Varoquaux, G., Gramfort, A., Michel, V., Thirion, B., Grisel, O., Blondel, M., Prettenhofer, P., Weiss, R., Dubourg, V.: Scikit-learn: machine learning in Python. J. Mach. Learn. Res. **12**, 2825–2830 (2011)

A Deep Learning Method for Prediction of Benign Epilepsy with Centrotemporal Spikes

Ming Yan[1,2], Ling Liu[3(✉)], Sihan Chen[3], and Yi Pan[2(✉)]

[1] College of Computer Science, Sichuan University,
Chengdu 610065, People's Republic of China
[2] Department of Computer Science, Georgia State University, Atlanta 30302, USA
yipan@gsu.edu
[3] West China Hospital, Sichuan University,
Chengdu 610041, People's Republic of China
zjllxx1968@163.com

Abstract. Benign epilepsy with centrotemporal spikes (BECT) is the most common epilepsy in the children. The research of BECT mainly focuses on the comparative analysis of the BECT patients and the healthy controls. Different from the existing methods, we proposed a 3D convolution neural network (3DCNN) that directly predicts the disease of BECT from raw magnetic resonance imaging (MRI). The experiment shows our 3DCNN model get an 89.80% accuracy in the five-fold cross-validation evaluation which is over a large margin than the benchmark method.

Keywords: BECT · 3DCNN · MRI

1 Introduction

The disease of benign epilepsy with centrotemporal spikes (BECT) is first reported in the 1950s and is now the most common epilepsy syndrome in children between the age of 4 and 13 years old [9]. For a long-term, doctors diagnosis the BECT is a "benign" case, and there is a dispute on drug treatment to the young children [6]. However, a series of review works reveal some BECT patients have verbal dysfunction, attention deficit, and language impairment [11]. Recent research reveals that children with benign epilepsy with centrotemporal spikes may even arise sudden unexpected death in epilepsy (SUDEP) [7]. Therefore, the International League Against Epilepsy suggests the term of "benign" is improper in the denomination [16]. Unfortunately, the precise reason and mechanism causing the BECT are still unclear. The researchers want to get more profound knowledge from the comparison of the BECT patients and the healthy controls.

Supported by the science and technology department of Sichuan province (No. 18MZGC0127).

In a study of 200 BECT patients, BECT instance always has a spike-waves in their sleeping electroencephalograph (EEG), and it also is a localization-related epilepsy [3]. Moreover, from 98 consecutive BECT patients investigation in 15 years, Gelisse confirms that the BECT patient has an abnormal brain lesion in their MRI images [10]. As we know, MRI is a static reflection of the anatomical structure, while functional MRI (fMRI) views the metabolic function. Regional homogeneity (ReHo) reveals that the BECT patients have a high value in their fMRI [17]. Moreover, another researcher points out that the fMRI has a spike-related activation region related to EEG expression in the BECT patients [4]. All the works research around the reasons and the related differences in the BECT from different methods and data sources. But the doctor diagnoses BECT by clinical behavior and EEG expression.

Our manuscript presents a data-driven deep learning method, termed 3D convolutional neural network (3DCNN), to predict BECT. Different from the previous methods which aim to explore the differences and relations between the BECT patients and the healthy controls. 3DCNN classifies the BECT patients from the healthy controls. Moreover, the model of 3DCNN does BECT prediction only depending on the raw MRI image data, which is much more straightforward than the clinical diagnosis with EEG and clinical behaviors. Furthermore, the clinical diagnosis is after the event of epilepsy seizure, while 3DCNN model could predict epilepsy in advance.

2   Related Works

2.1   BECT

BECT researchers are mostly focusing on revealing the cause of BECT and the relationship between the BECT and the health. For this cause, some work supports the BECT a genetic disease. A study of DNA linkage shows that there are some regions on chromosome $15q15$ have strong linkage in the BECT families [12]. For the relation analysis between BECT and healthy children, the children with BECT have an expression on their EEG and MRI [3]. In a 2-years tracking of 9 BECT children shows that the BECT patients have an interictal epileptic discharge during their sleeping [2]. In a further study of the EEG signal, the θ EEG frequency activity overall cortical regions in the analysis of 21 BECT children [1]. Similarly, another research shows the frontal, temporal, and occipital regions of BECT patients are thinner than the healthy controls. Conversely, the subcortical volumes are larger in the BECT patients [8].

2.2   Deep Neural Networks

The deep neural networks have achieved great success in the biomedical area [13]. 3D U-Net extends 2D convolution neural network to the 3D, which gets good performance on the kidney segmentation [5]. In the detection of lymph node, the classifier of deep convolution neural network predicts the candidate region with a high accuracy which beyond the previous works [15]. DeepMeSH learns semantic representations and assigns the appropriate indexings to citations automatically [14]. All the biomedical methods embrace the deep neural network.

3   Method

A 3-dimensional convolution neural network is proposed to predict the disease of BECT from the raw MRI data and learns the data representation from original data. As the MRI data is a 3D data, 3DCNN is fit for the process raw MRI data in an end-to-end way. With 3D convolution kernel in 3DCNN could learn more spatial information than the standard 2D convolutional neural networks with the 2D kernels. In Fig. 1, the children's whole skull high-resolution 3D MRI data is acquired from the MRI machine. Then, the data is sent to the 3DCNN model to train and prediction.

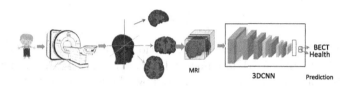

Fig. 1. The workflow of BECT prediction.

3.1   3DCNN

Our 3DCNN is a seven layers deep neural network consists the layers of 3D convolution, batch normalization, and Softmax. As the input is a high-resolution 3D MRI data, 3DCNN is applied deep convolution layers with small kernel numbers to overcome the overfitting problem. Meanwhile, 3DMaxPooling is introduced to dimension reduction. Figure 2 shows the architecture of 3DCNN.

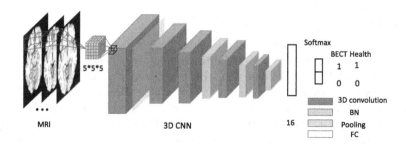

Fig. 2. The architecture of 3DCNN.

Except for the output layer, all the layers are 3D convolution layers, which have a much stronger nonlinear mapping power and feature extraction capacity than traditional 2D convolution neural network. 3D convolution operation is an extension of 2D convolution which reformulate in Eq. 1:

$$y_{i'',j'',k'',d''} = \sum_{i=1}^{I}\sum_{j=1}^{J}\sum_{d=1}^{D}\sum_{c=1}^{C} f(w_{i,j,d}x_{i',j',d',c'}) \tag{1}$$

The 3D convolution computation is similar to 2D which adds the third dimension of d. The variable i, j, d notate the index of convolution width, height, and depth. c notates the kernel or channel. The capital words are the corresponding total number. The superscript of x and y denotes the index of feature map or activation. The backpropagation computation of 3D convolution follows the chain-rule which multiples the sensitivity (θ) to the transposition of convolution kernel (W^T) in Eq. 2. Equation 3 denotes the gradient computation ($\frac{\partial J}{\partial W^l}$) from the higher layer's sensitivity (θ^{l+1}) and activations (a^l). The variable J denotes the cost function.

$$\theta^l = \theta^{l+1}W^T \tag{2}$$

$$\frac{\partial J}{\partial W^l} = \theta^{l+1}a^{lT} \tag{3}$$

Our method is different from the traditional clinical diagnosis of BECT, which only uses the raw MRI data material instead of EEG and clinical behavior. Moreover, 3D convolution has a much more powerful feature extraction capacity than the traditional 2D convolution. At last, due to the 3D high dimension in MRI data, all the parameters and convolution kernels in 3DCNN are small to prevent the overfitting.

4   Experiments

4.1   Data

There are 80 samples in our MRI dataset which consists 40 BECT instances and 40 control instances. All the data acquired from a SIEMENS TrioTim 3T with the same resolution of $256 \times 256 \times 200$ in each case. The only data processing is scaling down to the range of (0–1).

4.2   Baseline

As far as we know, no existing deep-learning based methods in BECT prediction. The research works of the BECT disease mainly concerning to explore the connection and relations between the BECT patients and the healthy controls. Therefore, we construct a traditional 2D convolution neural network (2DCNN) as the baseline model for BECT prediction. The convolution kernel sizes of 2DCNN are set the same 3DCNN. And the stride of 2DCNN is modified to get a full convolution architecture which reduces the number of weights and overcomes the overfitting problem. So, the third dimension of MRI in 2DCNN is computed as the channel dimension. However, 3DCNN extends one more dimension as the data channel dimension in the input layer. The detailed comparison of 2DCNN and 3DCNN are shown in Table 1.

Table 1. Comparison of 2DCNN and 3DCNN.

Layer name	Input	Layer1	Layer2	Layer3	Layer4	Layer5	Layer6
2DCNN	$256 \times 256 \times 200 \times 1$	$5 \times 5 \times 5$	$3 \times 3 \times 3$	$2 \times 4 \times 4$	$1 \times 1 \times 1$	16	2
3DCNN	$256 \times 256 \times 200$	5×5	3×3	3×3	1×1	16	2

4.3 Result

We evaluate 3DCNN on our MRI dataset in five-fold cross-validation. The whole 80-cases dataset is divided into training dataset (54 instances) and testing dataset (16 instances), and the evaluation criterion is the prediction accuracy. The learning rate is set to $1e-4$ under Adam optimizer. Table 2 shows the performance of 3DCNN and 2DCNN on the test dataset.

Table 2. Comparison of BECT prediction accuracy.

Model name	Accuracy-1	Accuracy-2	Accuracy-3	Accuracy-4	Accuracy-5	Total
2DCNN	87.50%	81.25%	81.25%	75.00%	87.50%	82.50%
3DCNN	93.25%	87.50%	93.25%	87.50%	87.50%	89.80%

The result shows 3DCNN achieving a prediction accuracy of 89.80% on the five-fold cross-validation evaluation, which outperforms the 2DCNN baseline model (82.50%). 3DCNN performs a consistently high performance on the whole testing dataset which shows 3D convolution neural network has a good capacity of data presentation and feature extraction in 3D MRI images.

5   Conclusion

A 3DCNN is proposed to BECT prediction, which is a data-driven method on the BECT prediction. Different to the verification research work, 3DCNN directly predict the disease from raw MRI dataset. In the further work, we will introduce more data source and dataset to our deep convolution model (Eg: hand design feature, and fMRI data, etc.).

References

1. Adebimpe, A., Aarabi, A., Bourel-Ponchel, E., Mahmoudzadeh, M., Wallois, F.: EEG resting state functional connectivity analysis in children with benign epilepsy with centrotemporal spikes. Front. Neurosci. **10**, 143 (2016)
2. Baglietto, M.G., Battaglia, F.M., Nobili, L., Tortorelli, S., De Negri, E., Calevo, M.G., Veneselli, E., De Negri, M.: Neuropsychological disorders related to interictal epileptic discharges during sleep in benign epilepsy of childhood with centrotemporal or rolandic spikes. Dev. Med. Child Neurol. **43**(6), 407–412 (2001)

3. Beaussart, M.: Benign epilepsy of children with rolandic (centro-temporal) paroxysmal foci a clinical entity. Study of 221 cases. Epilepsia **13**(6), 795–811 (1972)
4. Boor, S., Vucurevic, G., Pfleiderer, C., Stoeter, P., Kutschke, G., Boor, R.: EEG-related functional MRI in benign childhood epilepsy with centrotemporal spikes. Epilepsia **44**(5), 688–692 (2003)
5. Çiçek, Ö., Abdulkadir, A., Lienkamp, S.S., Brox, T., Ronneberger, O.: 3D U-Net: learning dense volumetric segmentation from sparse annotation. In: Ourselin, S., Joskowicz, L., Sabuncu, M.R., Unal, G., Wells, W. (eds.) MICCAI 2016. LNCS, vol. 9901, pp. 424–432. Springer, Cham (2016). https://doi.org/10.1007/978-3-319-46723-8_49
6. Croona, C., Kihlgren, M., Lundberg, S., Eeg-Olofsson, O., Eeg-Olofsson, K.E.: Neuropsychological findings in children with benign childhood epilepsy with centrotemporal spikes. Dev. Med. Child Neurol. **41**(12), 813–818 (1999)
7. Doumlele, K., Friedman, D., Buchhalter, J., Donner, E.J., Louik, J., Devinsky, O.: Sudden unexpected death in epilepsy among patients with benign childhood epilepsy with centrotemporal spikes. JAMA Neurol. **74**(6), 645–649 (2017)
8. Garcia-Ramos, C., Jackson, D.C., Lin, J.J., Dabbs, K., Jones, J.E., Hsu, D.A., Stafstrom, C.E., Zawadzki, L., Seidenberg, M., Prabhakaran, V., et al.: Cognition and brain development in children with benign epilepsy with centrotemporal spikes. Epilepsia **56**(10), 1615–1622 (2015)
9. Gastaut, Y.: Un element deroutant de la semeiologie electroencephalographique: les pointes prerolandique sans signification focale. Rev. Neurol. **87**, 408–490 (1952)
10. Gelisse, P., Corda, D., Raybaud, C., Dravet, C., Bureau, M., Genton, P.: Abnormal neuroimaging in patients with benign epilepsy with centrotemporal spikes. Epilepsia **44**(3), 372–378 (2003)
11. Liasis, A., Bamiou, D., Boyd, S., Towell, A.: Evidence for a neurophysiologic auditory deficit in children with benign epilepsy with centro-temporal spikes. J. Neural Transm. **113**(7), 939–949 (2006)
12. Neubauer, B., Fiedler, B., Himmelein, B., Kämpfer, F., Lässker, U., Schwabe, G., Spanier, I., Tams, D., Bretscher, C., Moldenhauer, K., et al.: Centrotemporal spikes in families with rolandic epilepsy linkage to chromosome 15q14. Neurology **51**(6), 1608–1612 (1998)
13. Yu, N., Li, Z., Yu, Z.: A survey on encoding schemes for genomic data representation and feature learning? From signal processing to machine learning. Big Data Min. Anal. **1**(3), 23–40 (2018)
14. Peng, S., You, R., Wang, H., Zhai, C., Mamitsuka, H., Zhu, S.: DeepMeSH: deep semantic representation for improving large-scale MeSH indexing. Bioinformatics **32**(12), i70–i79 (2016)
15. Roth, H.R., et al.: A new 2.5D representation for lymph node detection using random sets of deep convolutional neural network observations. In: Golland, P., Hata, N., Barillot, C., Hornegger, J., Howe, R. (eds.) MICCAI 2014. LNCS, vol. 8673, pp. 520–527. Springer, Cham (2014). https://doi.org/10.1007/978-3-319-10404-1_65
16. Uliel-Sibony, S., Kramer, U.: Benign childhood epilepsy with centro-temporal spikes (BCECTSs), electrical status epilepticus in sleep (ESES), and academic decline? How aggressive should we be? Epilepsy Behav. **44**, 117–120 (2015)
17. Zeng, H., Ramos, C.G., Nair, V.A., Hu, Y., Liao, J., La, C., Chen, L., Gan, Y., Wen, F., Hermann, B., et al.: Regional homogeneity (ReHo) changes in new onset versus chronic benign epilepsy of childhood with centrotemporal spikes (BECTs): a resting state fMRI study. Epilepsy Res. **116**, 79–85 (2015)

LSTM Recurrent Neural Networks for Influenza Trends Prediction

Liyuan Liu, Meng Han$^{(\boxtimes)}$, Yiyun Zhou, and Yan Wang

Kennesaw State University, Kennesaw, GA, USA
{lliyuan,yzhou20,ywang63}@students.kennesaw.edu, menghan@kennesaw.edu

Abstract. Influenza-like illness (ILI) is an acute respiratory infection causes substantial mortality and morbidity. Predict Influenza trends and response to a health disease rapidly is crucial to diminish the loss of life. In this paper, we employ the long short term memory (LSTM) recurrent neural networks to forecast the influenza trends. We are the first one to use multiple and novel data sources including virologic surveillance, influenza geographic spread, Google trends, climate and air pollution to predict influenza trends. Moreover, We find there are several environmental and climatic factors have the significant correlation with ILI rate.

Keywords: Influenza-like illness · Influenza trends · Google trends
Climate change · Air pollution · Long short term memory

1 Introduction

Influenza-like illness (ILI) is an acute respiratory infection causes substantial mortality and morbidity. Influenza leads to the hospitalization of more than 200,000 people yearly and results in 36,000 deaths from flu or flu-related complications in the United States [1]. The New York Times reported 2017–2018 influenza season is the worst in nearly a decade in the United States, the cumulative amount of ILI in January and February 2018 is 573,622, higher than the yearly cumulative amount of ILI in 2011 [2]. Besides, a total of 142 influenza-associated pediatric deaths for the 2017–2018 flu season have been reported by the United States Centers for Disease Control and Prevention (CDC) [3]. Therefore, effective influenza prediction and early outbreak detection are valuable in bioinformatics research. There are numerous studies related to the influenza prediction [4]. Google Flu Trends operated a linear model to provided estimates of influenza activity for more than 25 countries but is now no longer publishing current estimates [5]. Dugas *et al.* employed generalized linear models with Google trends data [6] and Paul *et al.* applied the linear model with Google trends and twitter data [7] to predict influenza. However, predict influenza trends with social media data is a post-verify scheme [8,9]. People usually post influenza-related tweets after the seasonal outbreak. Furthermore, in traditional time series models, they mainly used linear models that cannot describe the uncertainty and non-linearity relationship very well, the models' accuracy will decrease when

© Springer International Publishing AG, part of Springer Nature 2018
F. Zhang et al. (Eds.): ISBRA 2018, LNBI 10847, pp. 259–264, 2018.
https://doi.org/10.1007/978-3-319-94968-0_25

the datasets have numerous variables. Since the limitation of conventional time series models, there are several novel methods emerged for analyzing time series data, Guo *et al.* employed Bayesian-inference-based methods to analysis time series data, Dietterich *et al.* introduced several machine learning methods for analysis sequential data [10, 11]. Deep learning is a family member of machine learning methods which based on learning data representations and handle the uncertainty and non-linear problems. The architecture could deploy various hidden layers with non-linear processing units to extract data features and transfer to a different dimension. It is worth to mention, deep learning algorithms are rarely implemented in influenza prediction. In this paper, we assemble CDC data, Google trends data, climatic data, air pollution data to forecast influenza trends which few researchers focused before. We employ pearson correlation to prove $PM_{2.5}$ and carbon monoxide (CO) are significant impact the rate of ILI. The Lofgren *et al.*'s research shows that variations of temperature associated with high levels of seasonal influenza. There exist substantial evidence to add climatic data as the predictor variables [12]. Based on the recent research and advanced deep learning algorithm, we combine various dataset and deploy long short-term memory (LSTM) recurrent neural networks to forecast st influenza trends. The dataset includes virologic surveillance, influenza geographic spread, Google trends, climate and air pollution. The experiment results indicate our approach can predict the influenza trend very well.

2   Data Source

In this research, we focus on Georgia state in the United States. However, our approach could also apply in global regions. We collect related datasets based on Georgia state area from 2012 to 2018. The virologic surveillance and influenza geographic spread data collected from Centers for Disease Control and Prevention (CDC) based on FluView system. The terms highly correlated with influenza geographic spread [7] adopt to collect Google trends data. Since it is hard to assemble state's climate and air pollution data by week, we use data from the six most representative cities in Georgia: Athens, Atlanta, Augusta, Columbus, Macon, Savannah. The climate data reached from Climate Data Online (CDO) and the air pollution data collected from the United States Environmental Protection Agency. The weather data includes daily information of minimum temperature, maximum temperature, and average temperature. The pollution data includes the variables: $PM_{2.5}$ concentration, Carbon Monoxide (CO) concentration, Lead (Pb) concentration, Nitrogen Dioxide (NO_2) concentration, Ozone concentration, PM_{10} concentration, Sulfur Dioxide (SO_2) concentration. We preprocess all the data we mention above and organize them by weekly. Finally, the dataset for model contains 32 independent variables, 1 dependent variable, and 1 time index.

3   Method

Correlation analysis: Pearson correlation coefficient is the method employed to measure the linear dependence between two variables. It has the value between positive 1 and negative 1, where 1 is the total positive linear correlation and 0 is no linear correlation, and negative 1 is total negative linear correlation. Figure 1 is the pearson parametric correlation matrix of the relationship between air pollution, climate change and the rate of unweighted ILI. From the pearson correlation matrix, we can observe $PM_{2.5}$ and CO have the significant correlation with the rate of unweighted ILI as their p-values are smaller than 0.05.

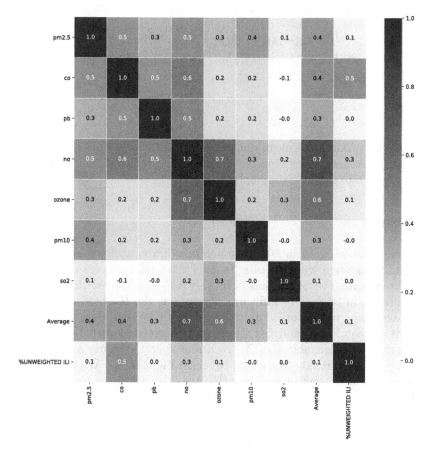

Fig. 1. Variables correlation between air pollution, temperature change and Influenza-like illness

Long Short-term Memory: It is the recurrent neural network (RNN) architecture that was designed by Hochreiter and Schmidhuber to address the vanishing and exploding gradient problems of traditional RNNs [13,14]. The LSTM is composed of some memory blocks. Memory block contains memory cells and gates. Memory cells able to remembering the temporal state of the network by self-connections and the gates control the flow of information. Each memory block contains an input gate to control the flow of input activations into the memory cell, an output gate to control the output flow of cell activations into the rest of the network and a forget gate [14]. The compact form of the equations for an LSTM unit with a forget gate shown follows, where the x_t denotes the input vector to the LSTM unit, f_t denotes the forget gate's activation vector, i_t denotes the input gate's activation vector, o_t denotes the output gate's activation vector and h_t denotes the output vector of LSTM unite, the initial value $c_0 = 0$ and $h_0 = 0$, the operator $\odot$ denotes the Hadamard product [13,15].

$$i_t = \sigma_g(W_i x_t + U_i h_{t-1} + b_i) \tag{1}$$

$$f_t = \sigma_g(W_f x_t + U_f h_{t-1} + b_f) \tag{2}$$

$$o_t = \sigma_g(W_o x_t + U_o h_{t-1} + b_o) \tag{3}$$

$$c_t = f_t \odot c_{t-1} + i_t \odot \sigma_c(W_c x_t + U_c h_{t-1} + b_c) \tag{4}$$

$$h_t = o_t \odot \sigma_h(c_t) \tag{5}$$

Model evaluation: Root mean squared error (RMSE) captures the square root of the difference between ground truth values and predicted values. We employ RMSE to evaluate the LSTM models.

4   Results

From week 1 to week 10 in 2018, temperature dramatically changes have the significant correlation with ILI rate. In 2017–2018 influenza season, the google search interest of "Flu Shot" and "Flu Vaccine" has 2 times than the google search interest in 2016–2017 influenza season, but the ILI rate still soaring in this season. Using the historical air pollution and CDC data from 2012 to 2017, we observed that $PM_{2.5}$ and CO significantly correlated with ILI rate. Using the long short-term memory (LSTM) recurrent neural networks algorithms, our predict trends are strongly fit the real trends as Fig. 2 shows. We test 4 different LSTM models: large/small sample size that the predictor variables come from air pollution, temperature change, and google trends datasets; large/small sample size that only use google trends data as predictor variables. The outcome variable of these 4 models are the rate of ILI collected from CDC. From Fig. 3, we can observe the multivariant LSTM models' RMSEs are smaller than the models' only use Google trends data. It proves the method of using the multivariant data sources to predict influenza trends is better than the method of only using google trends data. The extra related data could provide more insights through LSTM. In addition, we find the large sample size can improve the influenza prediction model performance evidently.

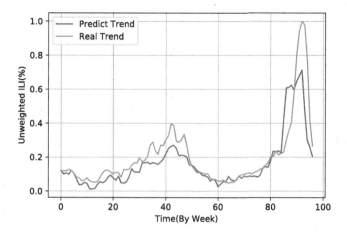

Fig. 2. Long short-term memory predict trends and real trends

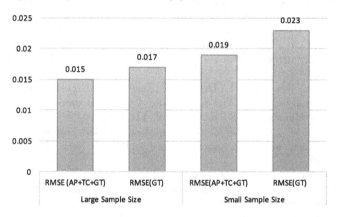

Fig. 3. RMSE of long short-term memory models

5   Conclusion

In this paper, we have three crucial contributions. First, we are the first one combine the air pollution data, climate data, google trends data, and CDC data to improve the model of influenza prediction. Second, the applicability of the LTSM method which is shown to outperform in other influenza prediction models. The multiple data sources LSTM models' performances are better than the LSTM models only rely the data from Google trends. Third, the data-driven approach is incorporated into our work to further confirm that the air pollution factors significantly correlated with ILI rate.

In the future work, we will incorporate the framework with multiple difference data sources and the improved deep learning algorithm to leverage the performance of the prediction and try to make the framework and the prediction could benefit the real life applications.

References

1. Lofgren, E., Fefferman, N.H., Naumov, Y.N., Gorski, J., Naumova, E.N.: Influenza seasonality: underlying causes and modeling theories. J. Virol. **81**(11), 5429–5436 (2007)
2. Mcneil, D.G.: This Flu Season Is the Worst in Nearly a Decade (2018). https://www.nytimes.com/2018/01/26/health/flu-rates-deaths.html. Accessed 13 Mar 2018
3. Centers for Disease Control and Prevention: Situation Update: Summary of Weekly FluView Report (2018). https://www.cdc.gov/flu/weekly/summary.htm. Accessed 5 Apr 2018
4. Han, M., Yan, M., Li, J., Ji, S., Li, Y.: Generating uncertain networks based on historical network snapshots. In: Du, D.-Z., Zhang, G. (eds.) COCOON 2013. LNCS, vol. 7936, pp. 747–758. Springer, Heidelberg (2013). https://doi.org/10.1007/978-3-642-38768-5_68
5. O'Connor, F.: Google Flu Trends calls out sick, indefinitely (2015). https://www.pcworld.com/article/2974153/websites/google-flu-trends-calls-out-sick-indefinitely.html. Accessed 13 Mar 2018
6. Dugas, A.F., Hsieh, Y.H., Levin, S.R., Pines, J.M., Mareiniss, D.P., Mohareb, A., Gaydos, C.A., Perl, T.M., Rothman, R.E.: Google flu trends: correlation with emergency department influenza rates and crowding metrics. Clin. Infect. Dis. **54**(4), 463–469 (2012)
7. Santillana, M., Nguyen, A.T., Dredze, M., Paul, M.J., Nsoesie, E.O., Brownstein, J.S.: Combining search, social media, and traditional data sources to improve influenza surveillance. PLoS Comput. Biol. **11**(10), e1004513 (2015)
8. Han, M., Yan, M., Cai, Z., Li, Y.: An exploration of broader influence maximization in timeliness networks with opportunistic selection. J. Netw. Comput. Appl. **63**, 39–49 (2016)
9. Albinali, H., Han, M., Wang, J., Gao, H., Li, Y.: The roles of social network mavens. In: 2016 12th International Conference on Mobile Ad-Hoc and Sensor Networks, MSN, pp. 1–8. IEEE (2016)
10. Guo, X., Liu, B., Chen, L., Chen, G., Pan, Y., Zhang, J.: Bayesian inference for functional dynamics exploring in fMRI data. Comput. Math. Methods Med. **2016**, 1–9 (2016)
11. Dieterich, T.G.: Machine learning for sequential data: a review. In: Caelli, T., Amin, A., Duin, R.P.W., de Ridder, D., Kamel, M. (eds.) SSPR /SPR 2002. LNCS, vol. 2396, pp. 15–30. Springer, Heidelberg (2002). https://doi.org/10.1007/3-540-70659-3_2
12. Georgia Health News: Georgia's flu death toll now at 51; season's peak is still ahead (online)
13. Hochreiter, S., Schmidhuber, J.: Long short-term memory. Neural Comput. **9**(8), 1735–1780 (1997)
14. Azzouni, A., Pujolle, G.: A long short-term memory recurrent neural network framework for network traffic matrix prediction. arXiv preprint arXiv:1705.05690 (2017)
15. Gers, F.A., Schmidhuber, J., Cummins, F.: Learning to forget: continual prediction with LSTM (1999)

Predicting Gene-Disease Associations with Manifold Learning

Ping Luo[1], Li-Ping Tian[2], Bolin Chen[3], Qianghua Xiao[4],
and Fang-Xiang Wu[1,5,6](✉)

[1] Division of Biomedical Engineering, University of Saskatchewan, Sakatoon, Canada
[2] School of Information, Beijing Wuzi University, Beijing, China
[3] School of Computer Science and Technology,
Northwestern Polytechnical University, Xi'an, China
[4] School of Mathematics and Physics, University of South China, Hengyang, China
[5] School of Mathematical Sciences, Nankai University, Tianjin, China
[6] Department of Mechanical Engineering, University of Saskatchewan,
Saskatoon, Canada
faw341@mail.usask.ca

Abstract. In this study, we propose a manifold learning-based method for predicting disease genes by assuming that a disease and its associated genes should be consistent in some lower dimensional manifold. The 10-fold cross-validation experiments show that the area under of the receiver operating characteristic (ROC) curve (AUC) generated by our approach is 0.7452 with high-quality gene-disease associations in OMIM dataset, which is greater that of the competing method PBCF (0.5700). 9 out of top 10 predicted gene-disease associations can be supported by existing literature, which is better than the result (6 out of top 10 predicted association) of the PBCF. All these results illustrate that our method outperforms the competing method.

1 Introduction

Complex diseases are usually caused by a group of disease genes. Uncovering gene-disease associations is critical for diagnosis, treatment and prevention of diseases. With advances in high-throughput techniques, a large amount of potential disease causative mutations have been generated. Further validation of these data is time-consuming and expensive. Thus, computationally predict disease genes is necessary to effectively translate the experimental data into legible disease-gene associations.

Current computational methods can be basically classified as two categories: the machine learning-based approaches, which focus on learning gene-disease relations [1–3], and the network-based techniques, which are based on the assumption that genes closed related in a network are associated with the same diseases [4–9]. As one kind of machine learning methods, matrix completion is also used to predict gene-disease associations [10,11]. However, matrix completion methods generally do not have the global optimal solutions and could take very long time to converge to even a local optimal solution [12,13].

© Springer International Publishing AG, part of Springer Nature 2018
F. Zhang et al. (Eds.): ISBRA 2018, LNBI 10847, pp. 265–271, 2018.
https://doi.org/10.1007/978-3-319-94968-0_26

In this study, we propose a manifold learning-based method which assumes that the geodetic distance between any disease and the genes associated with it are shorter than other non associated gene-disease pairs in a lower dimensional manifold [14, 15]. The experiments show that our method performs better than competing method in terms of the area under of the receiver operating characteristic (ROC) curve (AUC) and *de novo* validation.

2   Dataset and Method

2.1   Dataset

The disease-gene association data are downloaded from the Online Mendelian Inheritance in Man (OMIM) database [16]. To get the most reliable associations, three steps are adopted to preprocess the data. First, associations of disorders with tag '(3)' are selected since the molecular basis of these diseases is known, which means the associations are reliable. Second, disease terms are classified into distinct diseases by merging disease subtypes based on their given disorder names. Third, 475 diseases are removed because each of them is associated with only one gene which is not associated with any other diseases. As a result, the final dataset consists of 4257 associations between 1146 diseases and 2809 genes.

2.2   Method

The final dataset can be presented by a $n_d \times n_g$ gene-disease association matrix A, where $a_{ij} = 1$ if disease i is associated with gene j and $a_{ij} = 0$ otherwise, for $i = 1, \ldots, n_d$ and $j = 1, \ldots, n_g$, where $n_d = 1146$ and $n_g = 2809$. Intuitively, each disease can be represented by a n_g-dimensional 0–1 row vector while each gene can be represented by a n_d-dimensional 0–1 column vector. However, in higher dimensional different spaces, it is hard to learn the distance between a disease and a gene. In this study, we map diseases and genes into the same manifold with the lower dimensionality while assuming that the distance between a disease and its associated genes should be as short as possible on this manifold. This problem can be mathematically formulated as: finding k-dimensional representatives of genes $g_1, \ldots, g_{n_g}$ and k-dimensional representatives of diseases $d_1, \ldots, d_{n_g}$ such that the following objective function is minimized

$$Q_k = \sum_{i=1}^{n_d} \sum_{j=1}^{n_g} a_{ij} \left\| d_i - g_j \right\|^2. \tag{1}$$

All $d_1, \ldots, d_{n_d}$ and $g_1, \ldots, g_{n_g}$ are considered as k-dimensional column vectors. To make (1) well defined, we add the following constraints

$$\sum_{i=1}^{n_d} w_{r,i} d_i d_i^T = I_k \quad \text{and} \quad \sum_{j=1}^{n_g} w_{c,j} g_j g_j^T = I_k. \tag{2}$$

where I_k is the $k \times k$ identify matrix and

$$w_{r,i} = \sum_{j=1}^{n_g} a_{ij}, i = 1, \ldots, n_d \quad \text{and} \quad w_{c,j} = \sum_{i=1}^{n_d} a_{ij}, j = 1, \ldots, n_g. \tag{3}$$

Now the optimization problem with the objective function (1) and constraints (2) is well defined. To solve this optimization problem, we notice that

$$\begin{aligned} Q_k &= \sum_{i=1}^{n_d} \sum_{j=1}^{n_g} a_{ij} \left\| d_i - g_j \right\|^2 \\ &= \sum_{i=1}^{n_d} \sum_{j=1}^{n_g} a_{ij} (d_i - g_j)^T (d_i - g_j) \\ &= \sum_{i=1}^{n_d} \sum_{j=1}^{n_g} a_{ij} (d_i^T d_i - d_i^T g_j - g_j^T d_i + g_j^T g_j) \\ &= \sum_{i=1}^{n_d} w_{r,i} d_i^T d_i - 2 \sum_{i=1}^{n_d} \sum_{j=1}^{n_g} a_{ij} d_i^T g_j + \sum_{j=1}^{n_g} w_{c,j} g_j^T g_j \end{aligned} \tag{4}$$

Note that $d_i^T d_i = tr(d_i d_i^T)$ and $g_j^T g_j = tr(g_j g_j^T)$, we can have $\sum_{i=1}^{n_d} w_{r,i} d_i^T d_i = k$ and $\sum_{j=1}^{n_g} w_{c,j} g_j^T g_j = k$. Furthermore, let $W_r = diag[w_{r,1}, \ldots, w_{r,n_d}]$, $W_c = diag[w_{c,1}, \ldots, w_{r,n_g}]$, $X = W_r^{\frac{1}{2}} [d_1^T, \ldots, d_{n_d}^T]^T$, $Y = W_c^{\frac{1}{2}} [g_1^T, \ldots, g_{n_g}^T]^T$, and $C = W_r^{-\frac{1}{2}} A W_c^{-\frac{1}{2}}$, then we can obtain

$$Q_k = 2k - 2tr(X^T C Y) \tag{5}$$

From (5), minimizing the objective function (1) with constraints (2) is equivalent to maximizing the following objective function

$$P_k = tr(X^T C Y) \tag{6}$$

with the following constraints

$$X^T X = I_k \quad \text{and} \quad Y^T Y = I_k \tag{7}$$

From the matrix analysis [17], matrix C has the singular value decomposition as follows

$$C = \sum_{k=0}^{r-1} s_k v_k u_k. \tag{8}$$

where $r \leq min[n_g, n_d]$ is the rank of matrix C. Actually, it can be verified that matrix C is a so-called correspondence matrix [18]. Therefore, the maximum singular value of matrix C is 1, that is, $s_0 = 1 \geq s_1 \geq \cdots \geq s_{r-1} > 0$ are the non-zero singular values of matrix C. In addition, 1 is a single singular value if matrix C is non-decomposable and its corresponding singular vector pair is

$v_0 = (\sqrt{w_{r,1}}, \ldots, \sqrt{w_{r,n_d}})$ and $u_0 = (\sqrt{w_{c,1}}, \ldots, \sqrt{w_{c,n_g}})$. Furthermore, according to Eq. (8), the optimization problem of maximizing the objective function (6) with constraints (7) can be solved by

$$X^* = W_r^{\frac{1}{2}}(v_0, v_1, \ldots, v_k) \quad \text{and} \quad Y^* = W_c^{\frac{1}{2}}(u_0, u_1, \ldots, u_k). \tag{9}$$

From (9), if matrix C is non-decomposable, the first columns of X^* and Y^* are $W_r^{\frac{1}{2}}v_0$ and $W_c^{\frac{1}{2}}u_0$, which are the constant vectors of R^{n_d} and R^{n_g} with all components of 1, respectively. Therefore, they do not contribute to describe a gene or a disease in a manifold. Let $\hat{X}$ and $\hat{Y}$ denote the matrices by removing the first columns of matrices X^* and Y^*, respectively. As a result, each row vector of matrices $\hat{X}$ and $\hat{Y}$ is the representation of the corresponding disease and gene in a k-dimensional manifold. Then, the geodetic distance $gdist_{ij}$ between disease i and gene j can be calculated as follows

$$gdist_{ij} = \left\| \hat{X}(i,:) - \hat{Y}(j:,) \right\|^2. \tag{10}$$

for $i = 1, \ldots, n_d$ and $j = 1, \ldots, n_g$. The smaller the value of $gdist_{ij}$, the more possibly disease i associates with gene j.

In our dataset, matrix A is decomposable although no diagonal elements of matrices W_r or W_c are zeros. Therefore, matrix C is still decomposable. In addition, as matrix A is too sparse $(4257/(1146 * 2809))$, matrix C is also too sparse, which causes the singular value decomposition diverging. To solve these issues, we add a small positive number α to each element of matrix A, which make matrix $A + \alpha$ non-decomposable and its singular value decomposition converged. Furthermore, if we understand the value of a_{ij} in A as the probability that disease i associates with gene j, adding a value α can be understood as giving a small chance for every disease-gene pairs to be associated even if their association has not been verified yet. The value of α is discussed in the next section.

3   Experiments and Results

3.1   10 Fold Cross Validation

To evaluate our method, ROC curve and AUC value are computed based on 10-fold cross validation. In each round of the validation, we randomly select 10% of known associations as positive testing samples, and set their values in A to be zeros. The resultant matrix is denoted by $\hat{A}$. We also randomly select the same number of unknown associations in A as negative testing samples. Then we apply our method to $\hat{A}$ and calculate the geodetic distance between the disease-gene pairs in the testing set. This process is repeated 20 times.

In our method, the values of two parameters k and α can affect its performance. Thus, we use grid search to find the optimal values of k among $k = 50, \ldots, 950$ with the step increase of 50 and α among $\alpha = 0.01, 0.005, \ldots, 0.00001, 0.00005$. From the experiments, for different values of α,

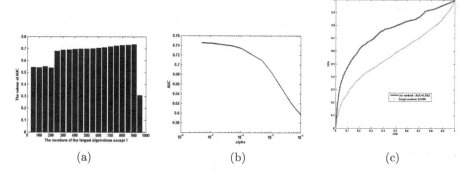

Fig. 1. The values of AUC in our experiments. (Color figure online)

our method achieves the best performance at $k = 900$ consistently. Figure 1(a) shows the AUC values over different values of k when $\alpha = 0.00005$. Figure 1(b) shows the AUC values over different values of α when $k = 900$. In summary, our method achieves the best performance with $\alpha = 0.00005$ and $k = 900$.

We also implement the best matrix completion-based method PBCF in [11] and compare it with our method on the same dataset. Figure 1(c) shows the comparison of ROC and AUC values. One can see that the ROC curve of our method (red) is always above that of PBCF (green) and the AUC value of our method (0.7452) is larger than that of PBCF (0.5700), which implies our method performs better.

3.2 *De Novo* Validation

To further evaluate our method, we also conduct the *de novo* validation and compare the results with PBCF. Out of the top 10 unknown gene-disease associations predicted by our method, 9 associations are found to be supported by existing literature. While in the ranked top 10 associations of PBCF, only 6 can be supported by existing studies.

4 Conclusions

In this study, we propose a novel method based on manifold learning to predict gene-disease associations. Evaluations performed on OMIM data have shown that our method outperforms the competing method PBCF in terms of both AUC with 10-fold cross validation and *de novo* validation. As can be seen, our proposed method only uses the experimentally validated gene-disease associations. However, it is believed that gene-gene similarity and disease-disease similarity can be used to improve the prediction of gene-disease associations. One of our future studies should be to integrate such data into our method to improve the prediction performance.

Acknowledgments. This work is supported in part by Natural Science and Engineering Research Council of Canada (NSERC), China Scholarship Council (CSC) and by the National Natural Science Foundation of China under Grant No. 61772552 and 61571052.

References

1. Glaab, E., Bacardit, J., Garibaldi, J.M., Natalio, K.: Using rule-based machine learning for candidate disease gene prioritization and sample classification of cancer gene expression data. PLoS ONE **7**(7), e39932 (2012)
2. Isakov, O., Dotan, I., Ben-Shachar, S.: Machine learningbased gene prioritization identifies novel candidate risk genes for inflammatory bowel disease. Inflamm. Bowel Dis. **23**(9), 15161523 (2017)
3. Mordelet, F., Vert, J.P.: Prodige: prioritization of disease genes with multitask machine learning from positive and unlabeled examples. BMC Bioinf. **12**, 389 (2011)
4. Chen, B., Wu, F.X.: Identifying protein complexes based on multiple topological structures in PPI networks. IEEE Trans. Nanobiosci. **12**(3), 165–172 (2016)
5. Chen, B., Li, M., Wang, J., Wu, F.X.: Disease gene identification by using graph kernels and Markov random fields. Sci. China Life Sci. **57**(11), 1054–1063 (2014)
6. Chen, B., Wang, J., Li, M., Wu, F.X.: Identifying disease genes by integrating multiple data sources. BMC Med. Genomics **7**(2), S2 (2014)
7. Chen, B., Li, M., Wang, J., Shang, X., Wu, F.X.: A fast and high performance multiple data integration algorithm for identifying human disease genes. BMC Med. Genomicse **8**(3), S2 (2015)
8. Chen, B., Shang, X., Li, M., Wang, J., Wu, F.X.: Identifying individual-cancer-related genes by rebalancing the training samples. IEEE Trans. Nanobiosci. **15**(4), 309–315 (2016)
9. Luo, P., Tian, L.P., Ruan, J., Wu, F.X.: Disease gene prediction by integrating PPI networks, clinical RNA-Seq data and OMIM data. IEEE/ACM Trans. Comput. Biol. Bioinf. (2017, in press). https://doi.org/10.1109/TCBB.2017.2770120
10. Natarajan, N., Dhillon, I.S.: Inductive matrix completion for predicting genedisease associations. Bioinformatics **30**(12), i60–i68 (2014)
11. Zeng, X., Ding, N., Rodrguez-Patn, A., Zou, Q.: Probability-based collaborative filtering model for predicting genedisease associations. BMC Med. Genomics **10**(S5), 76 (2017)
12. Li, L., Wu, L., Zhang, H., Wu, F.X.: A fast algorithm for nonnegative matrix factorization and its convergence. IEEE Trans. Neural Netw. Learn. Syst. **25**(10), 1855–1863 (2014)
13. Li-Ping, T., Luo, P., Wang, H., Huiru, Z., Wu, F.X.: CASNMF: a converged algorithm for symmetrical nonnegative matrix factorization. Neurocomputing **275**, 2031–2040 (2018)
14. Belkin, M., Niyogi, P.: Laplacian eigenmaps for dimensionality reduction and data representation. Neural Comput. **15**(6), 1373–1396 (2003)
15. Ham, J., Lee, D.D., Saul, L.K.: Semisupervised alignment of manifolds. In: AISTATS, pp. 120–127 (2005)

16. Amberger, J.S., Bocchini, C.A., Schiettecatte, F., Scott, A.F., Hamosh, A.: OMIM. org: online mendelian inheritance in man (OMIM®), an online catalog of human genes and genetic disorders. Nucleic Acids Res. **43**(D1), D789–D798 (2014)
17. Jolliffe, I.T.: Principal Component Analysis. Springer, New York (2002). https:// doi.org/10.1007/b98835
18. Greenacre, M.J.: Theory and Applications of Correspondence Analysis. Academic Press, New York (1984)

Data Analysis and Methodology

On Approaching the One-Sided Exemplar Adjacency Number Problem

Letu Qingge, Killian Smith, Sean Jungst, and Binhai Zhu[⊠]

Gianforte School of Computing, Montana State University,
Bozeman, MT 59717, USA
{letu.qingge,killian.smith,sean.jungst}@msu.montana.edu, bhz@montana.edu

Abstract. Given one generic linear genome $\mathcal{G}$ with gene duplications (over n gene families), an *exemplar* genome G is a permutation obtained from $\mathcal{G}$ by deleting duplicated genes such that G contains exactly one gene from each gene family (i.e., G is a permutation of length n). If we relax the constraint such that G^+ is obtained in the same way but has length at least k, then we call G^+ a *pseudo-exemplar* genome. Given $\mathcal{G}$ and one exemplar genome H over the same set of n gene families, the One-sided Exemplar Adjacency Number problem (One-sided EAN) is defined as follows: delete duplicated genes from the genome $\mathcal{G}$ to obtain an exemplar genome G of length n, such that the number of adjacencies between G and H is maximized. It is known that the problem is NP-hard; in fact, almost as hard to approximate as Independent Set, even when each gene (from the same gene family) appears at most twice in the generic genome $\mathcal{G}$. To overcome the constraint on the length of G, we define a slightly more general problem (One-sided EAN+) where we only need to obtain a pseudo-exemplar genome G^+ from $\mathcal{G}$ (by deleting duplicated genes) such that the number of adjacencies in H and G^+ is maximized. While One-sided EAN+ contains One-sided EAN as a special case, it does give us some flexibility in designing an algorithm. Firstly, we reformulate and relax the One-sided EAN+ problem as the maximum independent set (MIS) on a colored interval graph and hence reduce the appearance of each gene to at most two times. We show that this new relaxation is still NP-complete, though a simple factor-2 approximation algorithm can be designed; moreover, we also prove that the problem cannot be approximated within $2 - \varepsilon$ by a local search technique. Secondly, we use integer linear programming (ILP) to solve this relaxed problem exactly. Finally, we compare our results with the up-to-date software GREDU, with various simulation data. It turns out that our algorithm is more stable and can process genomes of length up to 12,000 (while GREDU sometimes can falter on such a large dataset).

1 Introduction

Comparing two generic genomes (with gene duplications) using different similarity measures, such as the exemplar genomic distance and the exemplar adjacency number, have been well studied in the past two decades. Sankoff first formulated

© Springer International Publishing AG, part of Springer Nature 2018
F. Zhang et al. (Eds.): ISBRA 2018, LNBI 10847, pp. 275–286, 2018.
https://doi.org/10.1007/978-3-319-94968-0_27

the exemplar breakpoint distance problem as selecting exactly one gene from each gene family, such that the breakpoint distance between the two resulting exemplar genomes is minimized [10]. The exemplar adjacency number problem is the complement of the exemplar breakpoint distance problem [5,6], which we will focus on in this paper. Formally, the Exemplar Adjacency Number (EAN) problem is defined as follows: given two generic linear genomes $\mathcal{G}$ and $\mathcal{H}$ over the same set of n gene families, delete duplicated genes from them to obtain two exemplar genomes G and H each of length n, such that the number of adjacencies between G and H is maximized. The One-Sided EAN problem is the one when $\mathcal{H}(= H)$ is given exemplar.

In theory, not only computing the exemplar breakpoint distance is NP-hard [3], even computing any approximation is NP-hard when each gene appears in $\mathcal{G}$ and $\mathcal{H}$ at most three times [4] or twice [2,8]. (They all showed that deciding whether the optimal exemplar breakpoint distance is zero, i.e, whether $G = H$, is NP-complete. Hence, computing any approximation solution is NP-hard.) Nonetheless, several algorithms have been proposed. Sankoff presented a branch-and-bound approach for the exemplar breakpoint distance problem [10]. Nguyen et al. gave a more efficient divide-and-conquer algorithm for the exemplar breakpoint distance problem [9]. Angibaud et al. developed an integer linear programming method for computing the exact breakpoint distance [1]. Shao and Moret gave a slight different formulation for the exemplar breakpoint distance problem where not all genes need to appear in the resulting (reduced) genomes, and they also proposed a fast and exact algorithm using integer linear programming [11].

The Exemplar Adjacency Number problem (EAN) is the complement of the breakpoint distance problem. Chen et al. showed that if one of the genomes $\mathcal{G}$ and $\mathcal{H}$ is exemplar and the other is 2-repetitive (each gene from each gene family appears at most twice), then this One-sided EAN problem admits neither a polynomial-time factor-$n^{0.5-\varepsilon}$ approximation (where n is the size of the gene family) unless $P = NP$, nor an FPT algorithm unless FPT=W[1] [5,6]. Chen et al. also designed a factor-$n^{0.5}$ approximation algorithm for the EAN problem when each gene appears at most twice in both $\mathcal{G}$ and $\mathcal{H}$ [6]. Because of the hardness result of the EAN problem (especially that a gene in each gene family must appear in G), it is not convenient to design efficient algorithms.

In this paper, we focus on the One-sided EAN problem whose input is $\mathcal{G}$ and H, over the same set of n gene families. We first relax the problem by allowing the reduced (pseudo-exemplar) genome G^+, obtained from $\mathcal{G}$ by deleting duplicated genes, to have a length less than n (but each gene can still appear most once), the objective is to maximize the number of adjacencies between G^+ and H. We call this problem One-sided EAN+, which is a general version of One-sided EAN and at least as hard as the latter. (The two problems are not exactly the same, e.g., $\mathcal{G} = 153642$ and $H = 123456$, with the latter there is no adjacency while in the former we could delete 3 to obtain an adjacency 56.)

On the other hand, One-sided EAN+ does allow a new relaxation of the problem as the maximum independent set (MIS) on new variants of interval graph for linear genomes (where the intervals correspond to the potential adjacency

ab or $-b - a$ in $\mathcal{G}$ are given the same color). While MIS on interval graphs are known to be polynomial-time solvable [7], in this paper, we show that the MIS problem on interval graphs becomes NP-complete when the vertices are colored. On the other hand, a simple factor-2 approximation can be obtained.

The above MIS approximation gives us a reduced/relaxed instance for the One-sided EAN+ problem. Let $\mathcal{G}'$ be the reduced instance for $\mathcal{G}$, obtained through the approximate MIS solution. $\mathcal{G}'$ has the property that each gene appears at most twice in $\mathcal{G}'$ (but some gene might not occur in $\mathcal{G}'$ at all). The next step is to use integer linear programming (ILP) to compute the exemplar adjacency between the pseudo-exemplar genome G^+ (obtained from $\mathcal{G}'$) and H. (Note that we never alter H, which is different from that of [11].)

We implement the algorithm and use simulated data (generated in the same way as in [11]. The comparison with GREDU indicates that our algorithm is more stable and generate comparable number of adjacencies (though the definitions of the adjacencies differ a little). The details are in Sect. 4.

This paper is organized as follows. In Sect. 2, we make necessary definitions. In Sect. 3, we approach the One-sided EAN+ problem by first reformulating it as the maximum independent set on colored interval graph, provide the NP-hardness proof of the MIS-problem and then design a simple 2-approximation algorithm (to obtain a reduced instance). Finally we use integer linear programming to solve the reduced instance exactly. In Sect. 4, we report our test results in comparison with GREDU on simulated data datasets. In Sect. 5, we conclude the paper and discuss the future work.

2 Preliminaries

Given n gene families (alphabet) $\mathcal{F}$, a genome $\mathcal{G}$ is a sequence of elements of $\mathcal{F}$ such that each element is with a sign ($+$ or $-$). Given a genome with a gene in each family appears exactly once (which is called *exemplar*) $G = g_1 g_2 \cdots g_n$, we say that gene g_i immediately precedes g_j, if $j = i + 1$. Given two exemplar genomes G, H, if gene a immediately precedes b in G and neither a immediately precedes b nor $-b$ immediately precedes $-a$ in H, then they constitute a break-point in G. The breakpoint distance is the number of breakpoints in G or H, denoted as $bd(G, H)$. Similarly, given genomes G, H with no duplications, if gene a immediately precedes b in G and either a immediately precedes b or $-b$ immediately precedes $-a$ in H, then they constitute an *adjacency* in G. The adjacency number is the number of adjacencies in G or H, denoted as $an(G, H)$. In the exemplar genomes G, H, we have $bd(G, H) + an(G, H) = n - 1$. For example, let $G = 12345, H = -5 - 4312$, then there are two adjacencies and two breakpoints between G and H.

Given generic genomes $\mathcal{G}$ and $\mathcal{H}$ the exemplar breakpoint distance between them, denoted as $ebd(\mathcal{G}, \mathcal{H})$, is the minimum breakpoint distance $bd(G, H)$ where G, H are exemplar genomes obtained from $\mathcal{G}$ and $\mathcal{H}$ respectively. Similarly, the exemplar adjacency number between $\mathcal{G}$ and $\mathcal{H}$, denoted as $ean(\mathcal{G}, \mathcal{H})$, is the maximum adjacency number $an(G, H)$, where G, H are exemplar genomes obtained from $\mathcal{G}$ and $\mathcal{H}$ respectively. Again, we have $ebd(\mathcal{G}, \mathcal{H}) + ean(\mathcal{G}, \mathcal{H}) = n - 1$.

The problem of computing $ean(\mathcal{G}, \mathcal{H})$ is formally defined as the *Exemplar Adjacency Number* (*EAN*) problem. When one of $\mathcal{G}, \mathcal{H}$ is given exemplar, we call the corresponding problem *One-sided EAN* problem. Throughout this paper, we assume $\mathcal{H}$ is given exemplar and we use H instead of $\mathcal{H}$ henceforth. A genome G^+ with length at least k ($k \leq n$), obtained from $\mathcal{G}$ by deleting duplicated genes (but each gene cannot appear more than once in G^+), is called a *pseudo-exemplar* genome. (Note that the definition of breakpoints and adjacencies between pseudo-exemplar genomes are the same as between exemplar genomes.) Given k, the problem of computing G^+ from $\mathcal{G}$ such that the number of adjacencies between G^+ and H is maximized, is hence called the *One-sided EAN+* problem. (One-sided EAN+ contains One-sided EAN as a special case when $k = n$; hence it is at least as hard as the latter.)

Finally, an interval graph $\mathcal{I} = (V, E)$ is a graph whose vertices have an one-to-one correspondence to a set of intervals on a line. An interval $u = (l(u), r(u))$ is represented by its left endpoint $l(u)$ and the right endpoint $r(u)$. There is an edge between two vertices $u, v \in V$ iff the intervals have a non-empty intersection. See Fig. 1 for an example.

Starting in the next section, we present a new method for the One-sided EAN+ problem. We start with a different relaxation/formulation of the problem.

3   A New Formulation of the Problem

Assume that we are given a generic linear genome $\mathcal{G}$ and an exemplar linear genome H, over the same gene family. Firstly, we try to identify some disjoint intervals in $\mathcal{G}$, one for each color (corresponding to some 2-substring in H). These intervals are the vertices of the corresponding colored interval graph $\mathcal{I}$. Then we try to identify the maximum number of disjoint intervals in $\mathcal{G}$, each of a different color. Clearly they correspond to a maximum independent set in $\mathcal{I}$. These intervals are formally constructed as follows.

For each 2-substring $a_i a_{i+1}$ in H, we list all the minimal intervals $a_i \beta a_{i+1}$ or $-a_{i+1} \beta - a_i$ (for unsigned genomes $a_{i+1} \beta a_i$) in $\mathcal{G}$ such that the contents β could be deleted to have a potential adjacency $a_i a_{i+1}$ or $-a_{i+1} - a_i$ ($a_{i+1} a_i$ for unsigned genomes). Note that a substring $x \beta y$ is *minimal* if the substring β does not contain x, or y, or a subsequence (potential adjacency) $-y - x$. All these minimal intervals in $\mathcal{G}$ corresponding to $a_i a_{i+1}$ in H will be given the same color. See Fig. 1 for an example of this construction with a colored interval graph where an unsigned genome is used.

We define the Maximum Independent Set (MIS) problem in a Colored Interval Graph (MIS-CIG for short) as follows.

Problem: MIS-CIG

Instance: A set $\mathcal{I}$ of m intervals on a line, each is in one of the $k(k < m)$ colors, and a parameter $k_1 \leq k$.

Question: Are there k_1 disjoint intervals of different colors?

We show next that MIS-CIG is NP-complete.

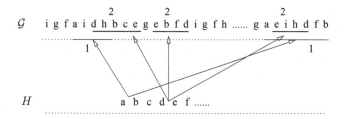

Fig. 1. Formulation of the one-sided EAN problem potentially as MIS in a colored interval graph. Note that in $\mathcal{G}$ the two intervals labeled with color 1 correspond to the adjacency ab, while the three intervals labeled with color 2 correspond to the adjacency de.

3.1 Hardness of MIS-CIG

Theorem 1. *The decision version of the MIS-CIG problem is NP-complete.*

Proof. It is not hard to see that the decision version of MIS-CIG is in NP. We reduce the classic 3SAT problem to MIS-CIG to prove that it is NP-complete. Given a Boolean formula ϕ in 3-conjunctive normal form, $\phi = F_1 \wedge F_2 \wedge \cdots \wedge F_\ell$, where each of the ℓ clauses F_i has three distinct literals from a set of m boolean variables $x_1, x_2, \cdots, x_m$, the problem is to decide whether ϕ is satisfiable.

For each variable x_i and its negation $\bar{x}_i$, we construct two copies of interleaving intervals of four different colors $4i - 3, 4i - 2, 4i - 1, 4i$ (See Fig. 2). If x_i is assigned TRUE, we select $4i - 3, 4i - 1$ at the top-right corner and $4i - 2, 4i$ at the bottom-left corner. If x_i is assigned FALSE, we select $4i - 3, 4i - 1$ at the top-left corner and $4i - 2, 4i$ at the bottom-right corner. For each clause F_j, we create three very small intervals of the same color (F_j). We put the interval with color F_j under the interval $4i - 3$ corresponding to x_i if x_i appears in F_j, and we put the interval with color F_j under the interval $4i - 3$ of $\bar{x}_i$ if $\bar{x}_j$ appears in F_j. We give an example to illustrate the construction. Assume we have a 3SAT formula $\phi = F_1 \wedge F_2 \wedge F_3 \wedge F_4$, with $F_1 = (x_1 \vee \bar{x}_2 \vee x_3), F_2 = (\bar{x}_1 \vee x_2 \vee \bar{x}_4), F_3 = (\bar{x}_2 \vee \bar{x}_3 \vee x_4)$, and $F_4 = (x_1 \vee \bar{x}_3 \vee \bar{x}_4)$. In Fig. 2, we show the construction on x_1, x_2 as the remaining construction for x_3, x_4 is similar.

We show that ϕ is satisfiable if and only if the colored interval graph has an IS of $4m + l$ intervals (all with different colors).

"$\rightarrow$" Suppose that the 3SAT instance ϕ is satisfiable. If x_i is assigned TRUE, we select the top-right intervals $4i - 3, 4i - 1$, the bottom-left intervals $4i - 2, 4i$ and all other F_j intervals in the bottom row under the unselected interval $4i - 3$ as part of the MIS. (If multiple F_j intervals are selected, we just arbitrarily keep one of them.) If x_i is assigned FALSE, we select the top-left intervals $4i - 3, 4i - 1$, the bottom-right intervals $4i - 2, 4i$ and all other F_j intervals in the bottom row under the unselected interval $4i - 3$ as part of the MIS. Consequently, we obtain an MIS with $4m + \ell$ intervals (all with different colors).

"←" Suppose that the colored interval graph contains an independent set of $4m + \ell$ intervals, all with different colors. For the two groups of interleaving intervals $4i - 3, 4i - 2, 4i - 1$ and $4i$ (corresponding to x_i and $\bar{x}_i$ respectively), clearly we could only select them in two different ways. We could either select the top-right $4i - 3, 4i - 1$ and the bottom-left $4i - 2, 4i$ intervals, we could select the top-left $4i - 3, 4i - 1$ and the bottom-right $4i - 2, 4i$ intervals. In the former case, we assign x_i TRUE, and then the F_j intervals containing x_i will be in the independent set as the top-left interval $4i - 3$, which intersects all the F_j intervals containing x_i, is not selected. In this case, F_j evaluates to TRUE and is satisfied. Similarly, we could show F_j containing $\bar{x}_i$ is also satisfied if we select the top-left $4i - 3, 4i - 1$ and the bottom-right $4i - 2, 4i$ intervals, which corresponds to assigning x_i FALSE. Then, we have a truth assignment for all the variables such that each F_j in ϕ is satisfied.

The reduction takes $O(m + \ell)$ time. Hence the theorem is proven.   □

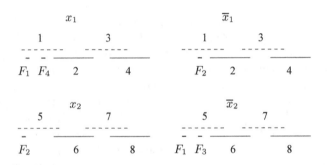

Fig. 2. Illustration for the reduction from the 3SAT instance ϕ. To save space, only the construction for $x_1, \bar{x}_1$ and $x_2, \bar{x}_2$ are shown.

4   Algorithms

In this section, we design an efficient factor-2 approximation algorithm for the MIS-CIG problem. Moreover, we prove that the approximation factor cannot be improved to be less than 2 by some local search technique. This gives us a reduced instance $\mathcal{G}'$ where each gene appears at most twice. On top of this, we use integer linear programming to efficiently delete extra gene duplications while computing the maximum number of adjacencies.

4.1   A Factor 2-Approximation

A factor-2 approximation for the MIS-CIG can be obtained using the well-known greedy method. Firstly, the intervals are ordered by their monotonically increasing right endpoints. Secondly, we scan the intervals from left to right and put

Algorithm 1. Greedy algorithm

1. $m \leftarrow$ size of $\mathcal{I}$.
2. $S \leftarrow \emptyset$.
3. Sort all intervals in $\mathcal{I}$ according to their right endpoints as $I_1, \ldots, I_m$.
4. Delete from $\mathcal{I}$ all the intervals of the same color with I_1 or overlapping with I_1.
5. Delete I_1 from $\mathcal{I}$.
6. $S \leftarrow S \cup \{I_1\}$.
7. **while** $\mathcal{I} \neq \emptyset$
8. $I_t \leftarrow$ the first interval in $\mathcal{I}$.
9. $S \leftarrow S \cup \{I_t\}$.
10.   Delete all the intervals of the same color with I_t or overlapping with I_t.
11.   Delete I_t from $\mathcal{I}$.
12. Return S.

the first interval I_1 in the solution. Then we delete all the intervals of the same color or overlapping with I_1 and repeat this process.

It is straightforward to see that the algorithm returns a factor-2 approximation for MIS-CIG. Let $I_i \in S$ and let I_j be some interval which intersects $r(I_i)$ (the right endpoint of I_i). As the algorithm scans intervals from left to right, any optimal solution not containing I_i must either contain an interval of the same color with I_i, or contain an interval of the same color with I_j (inclusive of I_j), or both. Hence the approximation algorithm would return at least a half of the optimal solution associated with $r(I_i)$. Applying this argument recursively would give us a factor-2 approximation.

4.2   A Local Search Improvement?

After obtaining the greedy algorithm, it is easy to come up with an improvement using a local search method. Let $\mathcal{I}$ be the input intervals for the MIS-CIG problem. We search for a subset of c intervals in S, S_c, such that putting S_c back to $\mathcal{I} \backslash S$ enables us to find locally a subset S'_c of more than c independent intervals with different colors and all the colors in S'_c do not appear in $S \backslash S_c$. Then $(S \backslash S_c) \cup S'_c$ would give us a better solution than S and we update $S \leftarrow (S \backslash S_c) \cup S'_c$. This process will continue until no S_c can be found. For a constant c, we call the corresponding local search procedure c-*local search*. Unfortunately, for $c = 1, 2$, this local search method cannot improve the result in the worst case. We summarize the result as follows.

Theorem 2. *Algorithm 1 returns a 2-approximation for MIS-CIG. For $c = 1, 2$ and some constant ε, the solution obtained by Algorithm 1 cannot be improved to have an approximation factor smaller than $2 - \varepsilon$ using c-local search.*

Proof. Assume to the contrary that with c-local search, $c = 1, 2$, we could obtain a $2 - \varepsilon$ approximation for the MIS-CIG problem. We construct an instance with $3n + 2$ intervals, with $2n$ colors $\{0, 1, \ldots, 2n - 1\}$. The greedy algorithm would select the intervals at the bottom row with colors $0, 1, 2, \ldots, n$. The optimal solution is to select all the intervals in the top row (except the last interval with color 0), with colors $0, 1, \ldots, 2n - 2, 2n - 1$. When context is clear, we just call an interval with color i interval i (Fig. 3).

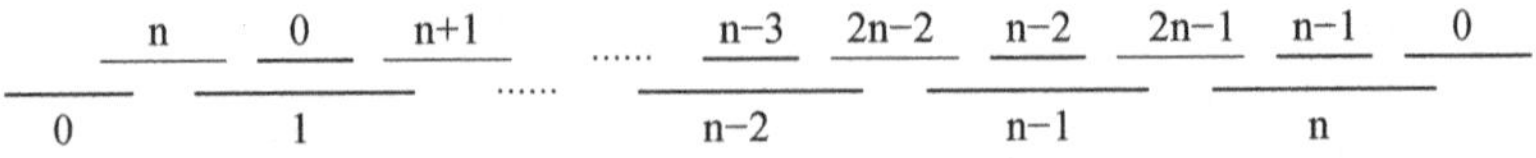

Fig. 3. Illustration for the proof of Theorem 2.

To see that c-local search does not incur any improvement for $c = 1, 2$, it is important to notice that any interval i ($i > 0$) at the bottom row contains an interval $i - 1$ at the top row; moreover, the two neighboring intervals of interval $i - 1$ at the top row, $n + i - 1$ and $(n + i)$ mod $2n$, are both intersecting with the corresponding neighboring intervals of i at the bottom row. (For interval 0 at the bottom row, it intersects interval n at the top row.) Hence, for 1-local search, it is impossible to obtain a better result by swapping interval i at the bottom row with two intervals in new colors at the top row. Similarly, for 2-local search, it is impossible to obtain a better result by swapping two intervals i, j at the bottom row with three intervals in new colors at the top row.

For this instance, the approximation factor is

$$\frac{2n}{n + 1} = 2 - \frac{2}{n + 1} > 2 - \varepsilon,$$

when n is sufficiently large at $n > \frac{2-\varepsilon}{\varepsilon}$. Hence, the approximation factor of the greedy algorithm is tight, whether or not c-local search, $c = 1, 2$, are applied. $\square$

Note that when $c \geq 3$, the above argument does not work anymore. On the other hand, a local search with $c \geq 3$ could incur high cost. Hence, after a reduced genome $\mathcal{G}'$ is obtained through the MIS approximation, we would use integer linear programming (ILP) instead.

4.3   ILP Formulation on the Reduced Instance

The genes in the reduced genome $\mathcal{G}'$ occur at most twice after using the MIS approximation in a colored interval graph on the original generic genome $\mathcal{G}$. The reason is that each gene could appear in two intervals, e.g., $(i-1, i)$ and $(i, i+1)$. Moreover, there is no guarantee that a gene in each gene family appears in $\mathcal{G}'$. Hence from $\mathcal{G}'$ we could only hope to compute a pseudo-exemplar genome (i.e., each gene appears at most once, some might never appear). In this subsection, we use integer linear programming (ILP) on the reduced genome to get a pseudo-exemplar genome based on the property that every pair of genes appear in $\mathcal{G}'$ in an adjacent form already, e.g., $(i, i + 1)$ or $(i + 1, i)$. We keep the order of genes in the intervals resulting from the greedy method. For example, $\mathcal{G}' =$

| 67 | 45 | 32 | 56 | 12 | 34 |

. Due to the space constraint, we leave out other details using ILP in this conference version. The full details will be presented in the final journal version.

5   Simulation Results

Our algorithm's simulation can be divided into two main components; namely, data generation and greedy exemplar selection written in Java, and the exemplar alignment written in Matlab using CPLEX for ILP. (The GREDU software was written in C++ which is usually much faster, and the ILP package GUROBI was used.) We ran our approximation algorithm as well as GREDU on a PC with 2.5 GHz Intel Core processor and 4 GB of memory.

Our simulated data is generated using a method similar to that in [11]. The dataset generator builds an exemplar genome, H, of size n, comprising of an integer sequence $[1 \ldots n]$ representing n unique genes. This exemplar genome is then mutated m times to produce some genome, $\mathcal{G}$, where m is a user provided integer that roughly corresponds to the number of generations between H to $\mathcal{G}$. A mutation cycle on a genome is performed by traversing each gene g_k in the genome, and mutating g_k with some probability. Our generator can perform up to eight different mutations:

- Unit Reversal: Given a gene, g_k, switch the locations of g_k and g_{k+l} with a probability of p_1.
- Unit Insertion: An arbitrary gene is inserted at location k with a probability of p_2.
- Unit Deletion: The gene, g_k, is removed with a probability of p_3.
- Unit Duplication: The gene, g_k, is copied and then inserted at location $k + 1$ in the genome with a probability of p_4.
- Segment Reversal: Given some length, l, and a gene, g_k, the ordering of the genome between genes g_k and g_{k+l} is reversed with a probability of p_5.
- Tandem Duplication: Given some length, l, and a gene, g_k, the genes between g_k and g_{k+l} are copied and placed at location $k + l + 1$ with a probability of p_6.
- Segment Deletion: Given some length, l, and a gene, g_k, the genes between g_k and g_{k+l} are removed with a probability of p_7.
- Segment Duplication: Given some length, l, and a gene, g_k, the sequence of the genome between g_k and g_{k+l} is copied and then inserted at a random location in the genome with a probability of p_8.

Every mutation cycle is performed on the mutated genome from the last mutation cycle, until m cycles have completed. Both the exemplar genome H and the genome $\mathcal{G}$ are then written to disk. These exemplar-genome pairs are then used as test datasets for both our Greedy-ILP algorithm, as well as for the GREDU software that we compare performance against.

In our simulations we use three different settings for the probabilities, P_1, P_2, and P_3.

- $P_1 = \{p_1 = 0.05, p_2 = 0.10, p_3 = 0.05, p_4 = 0.05, p_5 = 0.03, p_6 = 0.06, p_7 = 0.03, p_8 = 0.10, p_9 = 0.07, l = 5\}$
- $P_2 = \{p_1 = 0.20, p_2 = 0.15, p_3 = 0.15, p_4 = 0.10, p_5 = 0.05, p_6 = 0.08, p_7 = 0.04, p_8 = 0.12, p_9 = 0.10, l = 1\}$
- $P_3 = \{p_1 = 0.20, p_2 = 0.18, p_3 = 0.10, p_4 = 0.10, p_5 = 0.05, p_6 = 0.09, p_7 = 0.05, p_8 = 0.10, p_9 = 0.00, l = 10\}$

While these cases are not exhaustive in regards to the coverage of genome generation free variables, they do provide the three unique cases of genomes that we are interested in for our comparison of our algorithm and GREDU. The first case, P_1 is designed to not mutate aggressively, and is designed to change slowly. This means that most of the gene families will still be in the same ordering as the exemplar, and many alignment pairs should be found. The second case, P_2, is designed for rapid mutations. Higher mutations rates coupled with a small l

Table 1. Comparison results (signed genome) between the number of adjacencies n_1 from our algorithms and the number of adjacencies n_2 from GREDU. The gap - indicate that we get a core dumped warning from GREDU and can not see the results after 10 times tries.

N	M	P_1		P_2		P_3	
		n_1	n_2	n_1	n_2	n_1	n_2
500	1	247	242	216	224	204	198
	3	87	41	34	34	19	20
	5	17	12	23	8	8	2
1000	1	512	481	391	406	301	297
	3	106	96	66	61	54	47
	5	24	23	33	29	23	18
3000	1	1552	1442	1231	1244	703	756
	3	360	330	232	222	171	164
	5	80	78	97	73	64	50
5000	1	2441	2307	2044	2067	1532	-
	3	658	568	439	397	204	196
	5	163	159	272	197	110	91
7000	1	3471	3433	2495	2525	2010	-
	3	823	740	630	512	515	409
	5	223	208	249	139	151	128
9000	1	4563	4296	3507	3588	3355	-
	3	1103	969	933	828	583	-
	5	257	249	276	154	177	152
12000	1	6183	5756	5217	5322	3423	-
	3	1383	1287	1247	1112	279	237
	5	365	332	379	239	257	205

value mean that the genome that is generated will likely have a vastly different relative ordering of gene families compared to the exemplar, and few pairs should be found. The last case, P_3, is somewhat of a mixing between P_1 and P_2. The mutation rates are still set relatively high, but the value of l is also increased to 10. This means that while the genome is changing rapidly, large sections of it will also be moved and copied without changing the local ordering. We do not expect many adjacencies to be found, but we do expect slightly better results that from P_2.

Although the platforms (even the adjacency definitions) between our implementation and GREDU are quite different, we compare them using the same simulated data in Table 1. All the numbers are averaged over 10 tries. Even though that GREDU is much faster as it is written in C++, our implementation, which is based on Java and Matlab, does not need more than 20 min for any case we tested (which should be fine with this application). However, for many P_3 datasets, GREDU cannot run to completion. Our implementation is much stable and the number of adjacencies computed do not differ too much between the two (even though the definitions of adjacencies differ a bit). We also obtained similar results for unsigned genomes, but the details will have to be left out in this conference version due to the space constraint. Note that GREDU only handles signed genomes.

6 Conclusion

We approach the One-sided EAN problem by considering a general version of it, One-sided EAN+. We approach the One-sided EAN+ problem by first relaxing it as the maximum independent set in a colored interval graph, which open a new research pipeline to deal with the exemplar genomic distance problems for linear genomes. Our *greedy + ILP* algorithm can handle large scale genomic (singed and unsigned) data with deep evolution size based on the simulation results.

For the future work along this line, we can formulate the problem as MIS in a colored 2-interval graph for two generic linear genomes with duplications. We can also define similarly MIS in a colored circular-arc graph for two generic circular genomes. Note that the greedy algorithm cannot produce a 2-approximation in these settings.

Acknowledgments. This research is partially supported by NSF of China (No. 61628027), and by NSF (CNS-1644348).

References

1. Angibaud, S., Fertin, G., Rusu, I., Vialette, S.: A pseudo-boolean framework for computing rearrangement distances between genomes with duplicates. J. Comput. Biol. **14**(4), 379–393 (2007)
2. Blin, G., Fertin, G., Sikora, F., Vialette, S.: The EXEMPLARBREAKPOINTDISTANCE for non-trivial genomes cannot be approximated. In: Das, S., Uehara, R. (eds.) WALCOM 2009. LNCS, vol. 5431, pp. 357–368. Springer, Heidelberg (2009). https://doi.org/10.1007/978-3-642-00202-1_31

3. Bryant, D.: The complexity of calculating exemplar distances. In: Sankoff, D., Nadeau, J. (eds.) Comparative Genomics: Empirical and Analytical Approaches to Gene Order Dynamics, Map Alignment, and the Evolution of Gene Families, pp. 207–212. Kluwer Academic Publishers, Dordrecht (2000)

4. Chen, Z., Fu, B., Zhu, B.: The approximability of the exemplar breakpoint distance problem. In: Cheng, S.-W., Poon, C.K. (eds.) AAIM 2006. LNCS, vol. 4041, pp. 291–302. Springer, Heidelberg (2006). https://doi.org/10.1007/11775096_27

5. Chen, Z., et al.: Non-breaking similarity of genomes with gene repetitions. In: Ma, B., Zhang, K. (eds.) CPM 2007. LNCS, vol. 4580, pp. 119–130. Springer, Heidelberg (2007). https://doi.org/10.1007/978-3-540-73437-6_14

6. Chen, Z., Fu, B., Goebel, R., Lin, G., Tong, W., Xu, J., Yang, B., Zhao, Z., Zhu, B.: On the approximability of the exemplar adjacency number problem of genomes with gene repetitions. Theor. Comput. Sci. **550**, 59–65 (2014)

7. Gavril, F.: Algorithms for minimum coloring, maximum clique, minimum covering by cliques, and maximum independent set of a chordal graph. SIAM J. Comput. **1**, 180–187 (1972)

8. Jiang, M.: The zero exemplar distance problem. J. Comput. Biol. **18**(9), 1077–1086 (2011)

9. Nguyen, C.T., Tay, Y.C., Zhang, L.: Divide-and-conquer approach for the exemplar breakpoint distance. Bioinformatics **21**(10), 2171–2176 (2005)

10. Sankoff, D.: Genome rearrangement with gene families. Bioinformatics **15**(11), 909–917 (1999)

11. Shao, M., Moret, B.: A fast and exact algorithm for the exemplar breakpoint distance. J. Comput. Biol. **23**(5), 337–346 (2016)

Prediction of Type III Secreted Effectors Based on Word Embeddings for Protein Sequences

Xiaofeng Fu, Yiqun Xiao, and Yang Yang[✉]

Department of Computer Science and Engineering,
Key Laboratory of Shanghai Education Commission for Intelligent Interaction
and Cognitive Engineering, Shanghai Jiao Tong University,
Shanghai 200240, China
yangyang@cs.sjtu.edu.cn

Abstract. The type III secreted effectors (T3SEs) are virulence proteins that play an important role in the pathogenesis of Gram-negative bacteria. They are injected into the host cells by the pathogens, interfere with the immune system of the host cells, and help the growth and reproduction of the pathogens. It is a very challenging task to identify T3SEs because of the high diversity of their sequences and the lack of defined secretion signals. Moreover, their working mechanisms have not been fully understood yet. In order to speed up the recognition of T3SEs and the studies of type III secretion systems, computational tools for the prediction of T3SEs are in great demand. In this study, we regard the protein sequences as a special language. Inspired by the word2vec model in natural language processing, we convert the sequences into word embedding vectors in a similar manner with a specific segmentation strategy for protein sequences. And then we construct the T3SE predictor based on the new sequence feature representation. We conduct experiments on both mono-species data and multi-species data. The experimental results show that the new feature representation model has a competitive performance and can work together with the traditional features to enhance the identification of T3SEs.

Keywords: Type III secreted effectors · Word2vector
Feature representation

1 Introduction

The type III secreted effectors (T3SE) are virulence proteins produced by Gram-negative pathogenic bacteria to interfere with host immune signaling networks [1,2]. They are injected into the host cells through type III secretion systems (T3SS) [1], which are indispensable for the pathogenesis of a large variety of plant and animal pathogens, such as *Pseudomonas*, *Erwinia*, *Xanthomonas*, *Ralstonia*, *Salmonella*, *Yersinia*, *Shigella* and *Escherichia*, etc. [3,4].

© Springer International Publishing AG, part of Springer Nature 2018
F. Zhang et al. (Eds.): ISBRA 2018, LNBI 10847, pp. 287–298, 2018.
https://doi.org/10.1007/978-3-319-94968-0_28

The T3SEs have important functions for the virulence of pathogens, and the research achievements of T3SEs will contribute to the understanding of the work mechanism of T3SS. However, although researchers have explored for over a decade, the precise principle underlying the secretion process has not been fully uncovered, and a large proportion of T3SEs remain uncovered [5]. Moreover, T3SEs and T3SS are also found in non-pathogenic bacteria, which makes T3SEs powerful weapons for researchers to explore the immunity and functions of the host cells. Over the past decade, benefitting from the advances of high-throughput sequencing technologies, a lot of pathogenic bacterial genomes have been sequenced, and the known T3SE sequences have also accumulated rapidly. Typical T3SE databases include T3SEdb [6], T3DB [7], *etc.* The labor-intensive experimental verification methods have largely restricted the development of T3SE research, while computational methods have been demonstrated to be useful for accelerating the verification of T3SEs [8]. Some computational tools have been developed for the prediction of T3SEs [9]. The research achievements of these prediction systems could be helpful for T3SS researchers to verify T3SEs from tremendous genomic sequences efficiently.

The recognition of T3SEs is essentially a classification problem based on protein sequences. Due to the lack of defined signal/motif for known effectors, most of the existing predictors for T3SEs utilize general classification methods for protein sequences, e.g., [10–12]. Moreover, since the amino acid sequences of T3SEs have great diversity, effective feature extraction methods are highly in need to enhance the prediction performance. Many machine learning methods have been proposed for the identification of T3SEs [5,13–15]. For instance, Yang *et al.* [5,13] proposed the SSE-ACC method (amino acid composition in terms of their different secondary structures and solvent accessibility states) and topic models for T3SE recognition. Löwer and Schneider [14] used sliding-window technique to extract features. Wang *et al.* [15] proposed a position-specific feature extraction. The position-specific occurrence time of each amino acid is recorded, and then the profile is analyzed to compose features. Recently, Goldberg *et al.* proposed the pEffect method, which creates an alignment profile using PSI-BLAST method and maps the profile to feature vectors for classification in SVMs [16].

Besides the feature extraction based on statistical characteristics, sequence encoding schemes have also been adopted in biological sequence analysis. For instance, one-hot encoding is a simple and common scheme, which has been widely used in biological sequence classification [17]. Here, we regard the protein sequences as a special biological language. Instead of using traditional discrete features, like the bag-of-words model, most of the current natural language processing (NLP) tasks have adopted the continuous word representation, i.e., the word embedding, obtained by word2vec [18] or similar techniques, based on an unsupervised learning using deep models. Given the embedding vectors, it is convenient to measure word semantic correlations by calculations in the vector space. Although the concept of biological language has been proposed over a decade ago [19], the studies on continuous representation for biological sequences are very few [20]. A major reason is that defining words in biological sequences is

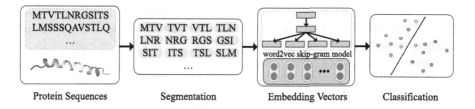

Fig. 1. Flowchart of the prediction system.

difficult. The natural languages have predefined dictionaries with defined words. Words correspond to subsequences with certain functions in the protein/DNA sequences. However, there is neither predefined word list nor spaces for separating words/phases in protein or DNA sequences. Therefore, in this study, we investigate different segmentations for protein sequences and then exploit the word2vector technique to transform the segments into embedding vectors that can be used in machine learning methods.

We represent all the T3SEs and non-T3SEs with the proposed method and build a predictor by using the word embeddings. The experiments were conducted on two datasets, including both a mono-species set and a cross-species set. We compared the new method with traditional feature extraction methods for amino acid sequences. The experimental results show that the new word embedding features have a competitive performance against traditional methods. Moreover, they can work with traditional features together, to improve the prediction accuracy.

2   Methods

In this study, we focus on the continuously distributed representation (word embedding) of protein sequences, which is based on the assumption that biological sequences can be viewed as texts written in a special language [19], where the words are k-mer subsequences, syntax and semantics may correspond to molecular structure and biological function, respectively. Analogous to Word2Vec, we can convert k-mers into word embeddings and apply them to the inference of molecular structure, dynamics, and function. Therefore, our method contains four steps: (1) segmenting protein sequences; (2) training the word embedding vectors; (3) obtaining the feature representation of protein sequences; (4) classification. Figure 1 shows the pipeline of our method. In the following subsections, we describe the four steps respectively in details.

2.1   Segmentation

Similar to Chinese sentences, there is no space for separating words in protein sequences. We assume that residue segments serve as words in this special biological language. Converting the consecutive amino acids into a concatenation of words is a key step that has a significant impact on the performance of classification/prediction. In this study, we segment the sequences using a sliding window with fixed size k, i.e., generating amino acid segments of length k in an overlapping manner (Fig. 1 shows an example when $k = 3$).

Compared to the method which segments the sequences non-overlappingly, the sliding window method could contain more sequence information. Moreover, because of the ambiguity of word boundaries, the start sites for segmentation is usually unknown, and the non-overlappingly method may result in wrong words, while the sliding window method can void this issue.

2.2 Training of the Word Embedding Vectors

In the word2vec model for natural language processing, the word embedding vectors are trained via a large corpus based on the context information in an unsupervised manner, where Wikipedia is usually used as the corpus. Thus firstly we need to define a suitable corpus, i.e., a protein database. In a previous study, Asgari and Mofrad [20] used the Swiss-Prot database as the corpus, which contains over 550,000 amino acid sequences in the current release. However, according to our experiments, the Swiss-Prot is not large enough for a sufficient training of the word vectors. In this study, we adopt UniRef50 (http://www.uniprot.org/uniref/), which is built by clustering UniRef90 seed sequences that have at least 50% sequence identity to, and 80% overlap with, the longest sequence in the cluster, including 30,000,000 sequences. The reasons for choosing this database as our corpus is that: (i) this database is large enough to satisfy the requirement of training the word vectors; (ii) this database reduces sequence redundancy to some extent, thus avoids the issues brought by high sequence identity.

The word2vec model takes full advantage of the relationship between the segment and its context and can train a continuously distributed vector for each segment. The purpose of this algorithm is to maximize the log-likelihood (Eq. 1) generated by the segment and its context,

$$L = \sum_{\omega \in C} log \quad p(\omega | Context(\omega)), \tag{1}$$

where ω is the query word (whose vector is to be estimated), $Context(\omega)$ is its context, and C is the corpus. The central problem is to find an appropriate algorithm to estimate the probability of the query word predicted by the surrounding words. According to the frequency of the segments, Mikolov *et al.* constructed a Huffman tree for the corpus [18,21], where each leaf node represents a word. The input vector X_ω in the Eq. (2) represents the sum of the vectors of the words within the context of ω, i.e., words preceding and following the query word. There is a unique path from the root node of the Huffman tree to ω, and the selection of a non-leaf node in the path can be regarded as a binary classification problem, where θ_j^ω is the parameter vector of the classification problem and d_j^ω is the result of the problem. The product of the probabilities of all $l - 1$

classification problems is used to represent the probability of ω predicted by the surrounding words (l denotes the length of the path from the root node to the leaf node),

$$p(\omega|Context(\omega)) = \prod_{j=2}^{l^{\omega}} p(d_j^{\omega}|X_{\omega}, \theta_{j-1}^{\omega}). \tag{2}$$

2.3 Protein Sequence Representation

At this stage, we already have the continuous vectors for all words/k-mers. In order to obtain the representation for a whole sequence, the k-mers' vectors need to be integrated. Here we assume that the words contribute equally to the classification of the protein sequence. A combination strategy is to concatenate all the word vectors into a $(L-1) \times d$-dimensional vector (Eq. 3), where L is the length of the sequence (there are $L-1$ words obtained by the sliding window scheme) and d is the dimensionality of the word embedding vectors. In this case, the dimensionality of the sequence is too high, and the input space will have variable sized input because the lengths of protein sequences are different, which may increase the difficulty of classification using machine learning methods. An alternative strategy is to generate d-dimensional feature representations for sequences, either by summing over all word vectors (Eq. (4)) or using the mean vectors of all the words (Eq. (5)). This strategy leads to a much lower dimensional feature space with a fixed length. In our experiments, we assess the performance of using both the sum vectors and mean vectors for representing protein sequences.

$$\chi = [V_{\omega_1}, V_{\omega_2}, \ldots, V_{\omega_{L-2}}, V_{\omega_{L-1}}] \tag{3}$$

$$\chi = \sum_i V_{\omega_i}, i \in \{1, 2, \ldots, L-1\}, \tag{4}$$

$$\chi = \frac{\sum_i V_{\omega_i}}{L-1}, i \in \{1, 2, \ldots, L-1\}, \tag{5}$$

where ω_i denotes a word, V_{ω_i} denotes the vector of ω_i, and χ is the feature vector for the whole sequence.

2.4 Classification

After we obtain the representation vectors for the whole protein sequences, the vectors are fed into classifiers. We adopt the support vector machines (SVMs) as the classifiers, which are widely used in bioinformatics because of their excellent and stable performance in the classification tasks. The RBF function is used to optimize the classifier. Its optimal parameters C and γ are determined by a grid search.

3   Results and Discussions

3.1   Data Sources

In order to assess the performance of our model on the recognition of type III secreted effectors (T3SE), we construct two datasets and conduct two groups of experiments on them, respectively. In the first group of experiments, we perform training and prediction within the same species; while in the second group, we perform a cross-species prediction. Therefore, the two datasets contain mono-species data and multi-species data, respectively.

The first dataset include T3SEs only from *Pseudomonas syringae*, which has the largest number of putative and confirmed type III secreted effectors. We have collected a total of 283 confirmed *Pseudomonas* effectors from three strains, *P. syringae* pv. tomato strain DC3000, *P. syringae* pv. syringae strain B728a and *P. syringae* pv. *phaseolicola* strain 1448A. A certain percentage of T3SEs in this dataset are obtained using homology search, which makes the similarity between the sequences very high. Considering that the high similarity of the dataset would result in overestimation of the performance, we cluster these effectors with sequence similarity over 60% and use only the representative sequence of each cluster in our dataset, resulting into 102 positive samples. The negative dataset is extracted from the genome of *P. syringae* pv. tomato strain DC3000 because it has been intensively investigated for the research of T3SS. A total of 136 non-effectors are retrieved as negative samples.

Pseudomonas syringae is a model organism in plant pathology. In order to examine the generalization performance of this method, we prepare another dataset of effectors from multiple species, including both plant and animal pathogens. Currently, there have been several T3SE databases (e.g., T3SEdb [6], Effective [22] and Bean 2.0 [23]). Among them, Bean 2.0 is the most recent and largest one. Therefore, we collect 713 T3SEs from BEAN 2.0 (http://systbio. cau.edu.cn/bean-/index.php), remove sequence redundancy by using CD-hit [24] with the sequence identity cutoff of 40%, and get 241 T3SEs. Meanwhile, 284 negative samples are randomly selected from the non-T3SE proteins released in Bean 1.0. The source pathogens of the collected T3SEs are associated with a total of 14 host species, thus it is a comprehensive dataset. The detailed data statistics are shown in Table 1. Note that in the genomes, T3SEs are much fewer than non-T3SEs. That is, the identification of T3SE is essentially an imbalanced classification problem, as mentioned in many protein-related studies [25–27]. Considering that most of the machine learning algorithms cannot handle imbalanced data well, we adopt a nearly 1:1 ratio, where the negative samples are slightly more than the positive samples.

3.2   Experimental Settings and Evaluation Criteria

There are two key parameters in our method. One is the length of words, i.e., the value of k for k-mers, and the other is the dimensionality of word vectors. When the value of k increases, the number of k-mers (20^k) increases exponentially, and

Table 1. Data statistics

Dataset	Positive samples #	Negative samples #	Total #
Dataset I	102	136	238
Dataset II	241	284	525

the k-mers have much sparse distribution in the corpus (some k-mers are even absent in the corpus). Besides, it would take much more time for training word embedding vectors in the corpus when k is large. On the contrary, if k is small, the too-short words may contain little biological meaning, like the auxiliary words in natural languages. Therefore, we should choose a proper value for k. Some previous literature has demonstrated the good performance of using 4-mers in protein classification problems, as FOUR is the typical longest distance for local interactions between amino acids [28]. Here we investigate the values around 4, i.e., $3 \leq k \leq 5$, and the result is discussed in Sect. 3.3. We set the number of dimensions for the continuous feature vectors to 100, same as the study in [20].

Our implementation of the support vector machines adopts the RBF kernel function. On both the two datasets, we conduct a 10-fold cross-validation, and the optimal values of the parameters C and γ are searched via a grid search using a nested-cross-validation. In order to provide reliable predictions for future wet-bench analysis, we use two metrics to evaluate the performance of the proposed method, including total accuracy (TA) and F_1-score (F_1). The accuracy TA is used to measure the overall prediction quality, i.e., the ratio of the test samples the system classifies correctly. The F_1-score is a common metric that takes into account both the precision and recall of the classification model.

3.3 Investigation on the Feature Representation

For the settings of word vectors, we investigate the impact of the length of words and also the integration strategy for generating sequence representation from word vectors. Figure 2 shows the results of two methods (using the sum of vectors and mean vector of words, respectively) under three values of k. As can be seen that the best performance is obtained by the mean vectors when k equals 4, which is consistent with the analysis in Sect. 3.2. And, 3-mers and 4-mers have relatively close performance, while 5-mers have much worse performance, indicating that long words do not necessarily improve the accuracy.

3.4 Result Comparisons with Other Feature Extraction Methods

We compared our model with multiple traditional feature extraction methods for amino acid sequences. The compared methods include auto covariance (AC), k-mer method, PC-PseAAC, PC-PseACC-General, SC-PseAAC and SC-PseAAC-General. They are described as follows.

- Auto covariance (AC) incorporates the correlation of the properties between two amino acids [29,30].

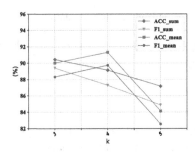

Fig. 2. Performance of different parameter settings for feature representation. 'sum' and 'mean' denote the methods using the sum and mean of word vectors, respectively.

- The k-mer method calculates the occurrence frequencies of k-mers to reflect the short-range or local sequence compositions [31].
- PC-PseAAC and SC-PseAAC methods generate the feature vectors by combining the amino acid composition and global sequence-order effects via parallel correlation and series correlation respectively [32,33].
- PC-PseAAC-General and SC-PseAAC-General are updated versions of PC-PseAAC and SC-PseAAC, respectively. They incorporate domain knowledge, such as functional domain, gene ontology and sequence evolution [32–34].

In order to assess the performance of the new method, we conduct experiments on both Datasets I and II, i.e., the *Pseudomonas syringae* dataset and the multi-species dataset. The experimental results are shown in Table 2.

Table 2. Result comparison of feature extraction methods on Datasets I and II*

Method	DataSet I		DataSet II	
	TA (%)	F_1 (%)	TA (%)	F_1 (%)
AC	71.44	64.33	71.37	55.47
1-mer	82.84	80.72	83.31	74.32
2-mer	89.58	87.70	83.59	74.22
3-mer	88.60	86.56	81.35	70.62
PC-PseAAC	89.03	87.12	**87.99**	**81.18**
PC-PseACC-General	89.47	87.40	86.64	79.69
SC-PseAAC	89.79	88.00	87.47	81.16
SC-PseAAC-General	82.94	81.01	78.71	68.86
Word vector (wordlen = 3)	90.01	88.32	83.04	73.87
Word vector (wordlen = 4)	**91.31**	**89.74**	87.30	80.60
Word vector (wordlen = 5)	84.13	82.54	84.37	75.65

*Experiments of the traditional feature extraction methods were conducted using the Pse-in-One webserver [35].

As can be seen in Table 2, the proposed method using word embedding vectors has a competitive performance compared with the traditional feature extraction methods. It has the highest accuracy on Dataset I and is very close to the best method on Dataset II. Interestingly, this method, based merely on sequence information, has obvious advantages over the methods based on the combination of sequence and domain knowledge (PC-PseAAC-General and SC-PseAAC-General), demonstrating the powerful representation ability of the continuous vectors for words. And, another reason is that the functional annotations (functional domain, gene ontology) for effectors are scarce.

The performance of our method varies as the length of words changes. Specifically, the methods using word length of 4 achieve the best results. For the k-mer method, the 2-mers have the best performance among the three values of k, suggesting that a large k does not necessarily lead to a high accuracy.

Apparently, the T3SEs within the same species have more common characteristics, while the cross-species prediction is more difficult. Generally, the accuracies of the cross-species experiments are much lower than the results of within-species experiments. As for our method, the total accuracy drops about 4% and F_1 decreases 9%, but the over 80% accuracy can still provide a valuable reference for biological researchers in the identification of T3SEs.

3.5   Results of the Combination of Word Embeddings and Traditional Features

Considering that the word embedding feature vectors are generated by using an entirely different mechanism against the previous feature extraction methods, we explore the potential performance improvement via the combination of these two kinds of features. We concatenate two different vectors together to form a new vector. The results are shown in Table 3.

Table 3. Result comparison on the combination of feature extraction methods

Method	DataSet I		DataSet II	
	Acc (%)	F_1 (%)	Acc (%)	F_1 (%)
Word vector + AC	86.42	83.92	83.85	74.06
Word vector + 1-mer	90.65	89.68	84.96	76.13
Word vector + 2-mer	90.03	87.63	87.71	80.52
Word vector + 3-mer	89.58	87.96	86.35	78.94
Word vector + PC-PseAAC	91.09	90.09	88.16	81.37
Word vector + PC-PseACC-General	**93.27**	**92.04**	87.15	80.04
Word vector + SC-PseAAC	91.53	90.25	**88.38**	**81.69**
Word vector + SC-PseAAC-General	90.11	88.28	84.54	75.71

As can be seen, the combination strategy results into the significant enhancement of performance. On Dataset I, the combination of our method and PC-

PseACC-General achieves the highest accuracy, i.e., total accuracy of 93.27% and F_1 of 92.04%, which are 2% higher than word vector method without combination, and 4% higher then PC-PseACC-General. As for Dataset II, the combination of our method and SC-PseAAC achieves the highest accuracy, which has a relatively small improvement compared with the single methods.

3.6 Discussion

The word embeddings have been widely used in natural language processing, and it has outstanding performance in the representation of word features. The proposed method for generating protein sequence embeddings will be a powerful tool in the studies of protein sequence analysis. Since the biological system is complex and can not be described in a single data type, the proposed model can only analyze the information in the sequence and neglect the spatial structure and other information of T3SEs. As a future work, more biological features can be incorporated into the predictor to enhance the discriminant ability for T3SEs. Besides, although the current experiments demonstrate the good performance of the word vectors, we haven't fully exploited this method, as the segmentation method is relatively simple, and may not be able to distinguish between useful words and useless words which brings noise to the training and prediction system.

4 Conclusion

In this paper, we propose a machine learning method to predict the type III secreted effectors. First, the protein sequences are segmented into words (k-mers) by a sliding window. Then the word embedding vectors are trained with a large protein corpus. We obtain the vector information of the sequence by combining the vectors of all the segments in the sequence. By using the state-of-art classifier, support vector machines, we construct a system to distinguish T3SEs and non-T3SEs, which outperforms the existing prediction methods based on traditional feature representation methods. Thus far, a large portion of T3SEs still remains unknown. Bioinformatics tools are of great importance for exploring the characteristics of effectors and discovering them automatically. We believe that this computational method can be applied to the prediction of T3SEs in various bacteria species, and can also assist in other sequence analysis tasks.

Acknowledgement. This work has been supported by the Shanghai Municipal Natural Science Foundation (No. 16ZR1448700).

References

1. Galán, J.E., Wolf-Wat, H.: Protein delivery into eukaryotic cells by type III secretion machines. Nature **444**(7119), 567 (2006)
2. Galán, J.E.: Common themes in the design and function of bacterial effectors. Cell Host Microbe **5**(6), 571–579 (2009)

3. He, S.Y., Nomura, K., Whittam, T.S.: Type III protein secretion mechanism in mammalian and plant pathogens. Biochimica et Biophysica Acta (BBA)-Mol. Cell Res. **1694**(1–3), 181–206 (2004)
4. Cornelis, G.R.: The type III secretion injectisome. Nat. Rev. Microbiol. **4**(11), 811 (2006)
5. Yang, Y., Zhao, J., Morgan, R.L., Ma, W., Jiang, T.: Computational prediction of type III secreted proteins from gram-negative bacteria. BMC Bioinform. **11**(1), S47 (2010)
6. Tay, D., Govindarajan, K.R., Khan, A.M., Ong, T., Samad, H.M., Soh, W., Tong, M., Zhang, F., Tan, T.W.: T3SEdb: data warehousing of virulence effectors secreted by the bacterial type III secretion system. BMC Bioinform. **11**(S-7), S4 (2010)
7. Wang, Y., Huang, H., Sun, M., Zhang, Q., Guo, D.: T3DB: an integrated database for bacterial type III secretion system. BMC Bioinform. **13**(1), 66 (2012)
8. Guttman, D.S., McHardy, A.C., Schulze-Lefert, P.: Microbial genome-enabled insights into plant-microorganism interactions. Nat. Rev. Genet. **15**(12), 797 (2014)
9. McDermott, J.E., Corrigan, A., Peterson, E., Oehmen, C., Niemann, G., Cambronne, E.D., Sharp, D., Adkins, J.N., Samudrala, R., Heffron, F.: Computational prediction of type III and IV secreted effectors in gram-negative bacteria. Infect. Immun. **79**(1), 23–32 (2011)
10. Huang, D.-S., Zhao, X.-M., Huang, G.-B., Cheung, Y.-M.: Classifying protein sequences using hydropathy blocks. Pattern Recogn. **39**(12), 2293–2300 (2006)
11. Zhao, X.-M., Du, J.-X., Wang, H.-Q., Zhu, Y., Li, Y.: A new technique for selecting features from protein sequences. Int. J. Pattern Recognit Artif. Intell. **20**(02), 271–283 (2006)
12. Zhao, X.-M., Cheung, Y.-M., Huang, D.-S.: A novel approach to extracting features from motif content and protein composition for protein sequence classification. Neural Netw. **18**(8), 1019–1028 (2005)
13. Yang, Y., Qi, S.: A new feature selection method for computational prediction of type III secreted effectors. Int. J. Data Min. Bioinform. **10**(4), 440–454 (2014)
14. Löwer, M., Schneider, G.: Prediction of type III secretion signals in genomes of gram-negative bacteria. PLoS ONE **4**(6), e5917 (2009)
15. Wang, Y., Zhang, Q., Sun, M., Guo, D.: High-accuracy prediction of bacterial type III secreted effectors based on position-specific amino acid composition profiles. Bioinformatics **27**(6), 777–784 (2011)
16. Goldberg, T., Rost, B., Bromberg, Y.: Computational prediction shines light on type III secretion origins. Sci. Rep. **6**, 34516 (2016)
17. Baldi, P., Brunak, S.: Bioinformatics: The Machine Learning Approach. MIT Press, Cambridge (2001)
18. Mikolov, T., Chen, K., Corrado, G., Dean, J.: Efficient estimation of word representations in vector space. arXiv preprint arXiv:1301.3781 (2013)
19. Klein-Seetharaman, J., Reddy, R.: Biological language modeling: convergence of computational linguistics and biological chemistry. Converg. Technol. Improv. Hum. Perform. 378 (2002)
20. Asgari, E., Mofrad, M.R.K.: Continuous distributed representation of biological sequences for deep proteomics and genomics. PLoS ONE **10**(11), e0141287 (2015)
21. Mikolov, T., Sutskever, I., Chen, K., Corrado, G., Dean, J.: Distributed representations of words and phrases and their compositionality. CoRR, abs/1310.4546 (2013)
22. Jehl, M.-A., Arnold, R., Rattei, T.: Effective—a database of predicted secreted bacterial proteins. Nucleic Acids Res. **39**(Suppl_1), D591–D595 (2010)

23. Dong, X., Lu, X., Zhang, Z.: Bean 2.0: an integrated web resource for the identification and functional analysis of type III secreted effectors. Database **2015**, bav064 (2015)

24. Li, W., Godzik, A.: CD-HIT: a fast program for clustering and comparing large sets of protein or nucleotide sequences. Bioinformatics **22**(13), 1658–1659 (2006)

25. Song, D., Yang, Y., Yu, B., Zheng, B., Deng, Z., Lu, B.-L., Chen, X., Jiang, T.: Computational prediction of novel non-coding RNAs in Arabidopsis thaliana. BMC Bioinform. **10**(1), S36 (2009)

26. Zhao, X.-M., Wang, Y., Chen, L., Aihara, K.: Gene function prediction using labeled and unlabeled data. BMC Bioinform. **9**(1), 57 (2008)

27. Zhao, X.-M., Li, X., Chen, L., Aihara, K.: Protein classification with imbalanced data. Proteins: Struct. Funct. Bioinform. **70**(4), 1125–1132 (2008)

28. Cheng, B.Y.M., Carbonell, J.G., Klein-Seetharaman, J.: Protein classification based on text document classification techniques. Proteins: Struct. Funct. Bioinform. **58**(4), 955–970 (2005)

29. Dong, Q., Zhou, S., Guan, J.: A new taxonomy-based protein fold recognition approach based on autocross-covariance transformation. Bioinform. **25**(20), 2655–2662 (2009)

30. Guo, Y., Yu, L., Wen, Z., Li, M.: Using support vector machine combined with auto covariance to predict protein-protein interactions from protein sequences. Nucleic Acids Res. **36**(9), 3025–3030 (2008)

31. Liu, B., Wang, X., Lin, L., Dong, Q., Wang, X.: A discriminative method for protein remote homology detection and fold recognition combining top-n-grams and latent semantic analysis. BMC Bioinform. **9**(1), 510 (2008)

32. Chou, K.-C.: Prediction of protein cellular attributes using pseudo-amino acid composition. Proteins: Struct. Funct. Bioinform. **43**(3), 246–255 (2001)

33. Chou, K.-C.: Using amphiphilic pseudo amino acid composition to predict enzyme subfamily classes. Bioinformatics **21**(1), 10–19 (2004)

34. Du, P., Gu, S., Jiao, Y.: PseAAC-general: fast building various modes of general form of chous pseudo-amino acid composition for large-scale protein datasets. Int. J. Mol. Sci. **15**(3), 3495–3506 (2014)

35. Liu, B., Liu, F., Wang, X., Chen, J., Fang, L., Chou, K.-C.: Pse-in-One: a web server for generating various modes of pseudo components of DNA, RNA, and protein sequences. Nucleic Acids Res. **43**(W1), W65–W71 (2015)

Extending the Evolvability Model to the Prokaryotic World: Simulations and Results on Real Data

Sagi Snir[1(✉)] and Ben Yohay[2]

[1] Department of Evolutionary and Environmental Biology,
University of Haifa, Haifa, Israel
`ssagi@research.haifa.ac.il`
[2] Department of Computer Science, University of Haifa, Haifa, Israel
`benyohay91@gmail.com`

Abstract. In 2006, Valiant introduced a variation to his celebrated PAC model to Biology, by which he wished to explain how such complex life mechanisms evolved in that short time by two simple mechanisms - random variation and natural selection. Soon after, several works extended and specialized his work to more specific processes. To the best of our knowledge, there is no such extension to the prokaryotic world, in which gene sharing is the prevailing mode of evolution.

Here we extend the evolvability framework to accommodate horizontal gene transfer (HGT), the transfer of genetic material between unrelated organisms. While in a separate work we focused on the theoretical aspects of this extension and its learnability power, here the focus is on more practical and biological facets of this new model. Specifically, we focus on the evolutionary process of developing a trait and model it as the *conjunction* function. We demonstrate the speedup in learning time for a variant of conjunction to which learning algorithms are known. We also confront the new model with the recombination model on real data of *E. coli* strains under the task of developing pathogenicity and obtain results adhering to current existing knowledge.

Apart from the sheer extension to the understudied prokaryotic world, our work offers comparisons of three different models of evolution under the same conditions, which we believe is unique and of a separate interest.

The code of the simulations is freely available at: https://github.com/byohay/LearningModels.git.

Keywords: Prokaryotic evolution · Evolvability
Horizontal gene transfer · PAC learning · E. coli

1 Introduction

One of the most fundamental tasks in biology is deciphering the history of life on Earth. In the past it was generally thought that the history of life is best

© Springer International Publishing AG, part of Springer Nature 2018
F. Zhang et al. (Eds.): ISBRA 2018, LNBI 10847, pp. 299–313, 2018.
https://doi.org/10.1007/978-3-319-94968-0_29

described using a tree structure. The leaves of this tree represent extant species (taxa) and tree branches - evolutionary relationships. This vertical inheritance framework of evolution, frequently dubbed as *Darwinian evolution* to commemorate Darwin's influential part in our understanding of evolution. It is of no wonder that Darwin's only drawing in his *On the Origin of Species* [4] is a sketch of a tree structure. An important implication of this theory, is that all life forms (at least the ones known to Darwin) evolved through a process of mutational events occurring in an ancestor-descendant basis since the beginning of life on Earth.

Despite its intriguing appeal to fields in exact sciences like information theory or computational learning, it was only in 2006 that this theory was put by Valiant [23,24] under such a rigorous framework attempting to explore the learning power of this mechanism. A genome is viewed as a function, reacting to a set of signals from the outer world, where this function improves towards a hypothetical, hidden, *ideal function* through a mechanism generating variations and natural selection. Then, the formal question asked by Valiant is what are the complexity of functions that can be efficiently learned. To address this question, he introduced a computational model of evolution, that he denoted *evolvability,* that captures the central ideas of the mechanism. The goal of computational learning theory is to separate concept classes that can be efficiently learned under a certain mechanism, from those that cannot. Examples of concept classes of interest include *Decision Trees, Parities,* etc. The question of evolvability can be asked in the language of computational learning theory: For what classes of ideal functions, can one expect to find an evolutionary mechanism that gets arbitrarily close to the ideal, within feasible computational resources?

Apart from the computational appeal of this new field, it is especially interesting to fit biological processes into Valiant's model. Kanade [12] considered the mechanism of recombination occurring through a sexual reproduction, and suggested a model extending the basic evolvability to capture recombination. Angelino and Kanade [2] asked what is the suitable representation of transcription and under what conditions it is evolvable.

Except for the recombination model of Kanade, all previous works, and in particular "evolvability", assumed a vertical mode of evolution where the only variation introduced to the parent's inherited genome is by random mutations. However the advent of High-throughput sequencing revealed strong signals that stand in absolute conflict to the classical Darwinian theory. It appeared that a big part of life on Earth does not adhere to the principals of vertical evolution. Specifically, prokaryotic evolution is characterized by extensive gene mobility between species that is crucial for their functionality [9,13]. The principal mechanism accounting for gene mobility is horizontal gene transfer (HGT) [6,11,14,17,18] in which a gene (or a group of genes) of a donor species being acquired by a recipient organism. Recent studies have shown that HGT stands out as the dominant factor in bacterial evolution, responsible for phenomena such as adaptation to niches, development of antibiotic resistance, and pathogenicity. Despite the importance of each of these phenomena, our knowledge on HGT is very limited, partly due to the lack of use of analytic tools and models in the field [7].

In this work we extend the Evolvability theory to account for this less studied, yet very important, mode of evolution. In particular, we define the notion of *population-based models* where a *population* is evolving simultaneously and information is shared between members of this population. We define such a computational learning model, which is an extension to the basic evolvability model. The new model accommodates phenomena such as HGT where genetic information in the form of DNA is passed between individuals. While in a recent separate work [21], we focused on the theoretical aspects of the newly defined model and showed that it provides an asymptotic acceleration in learning, here the emphasis is on more biological practical points. We first give the formal necessary definitions as required by the new extension. Subsequently, we use the *conjunction* function to model the process of acquiring a character or a property by an organism. Focusing on this important function class, we show experimental results regarding the new and existing models. Finally, we analyze real data in the form of developing virulence in *E. coli*. We demonstrate that this function can be depicted by a *monotone conjunction* and compare between the models on this data. Our results, in the form of actual concrete times, give convincing evidence to the superiority of the HGT process over more conservative mechanisms, explaining the vastness of the gradually discovered world of mobile elements - *the mobilome* [20].

2   Preliminaries

We now give necessary definitions required for the extension. For the sake of compatibility with previous works, we tried to use original definitions from existing works in places where this was possible. In such cases we give explicit reference to the appropriate work inside the definition.

2.1   The Evolvability Model

A description of the evolvability framework is necessary for our extension and is given hereby. For the full definition of the evolvability model the reader is referred to [24].

We start with a notational comment. Throughout the paper we abuse notation $P - r$ to mean $P \backslash \{r\}$. We assume a finite set of *conditions* organisms have to respond to. A condition may represent a certain disease, availability of food or water, etc. All the possible combinations of conditions are given by the set $X = \{x_1, ..., x_n\}$. For simplicity, we represent a single condition as a boolean attribute with a low value of -1 and a high value of $+1$. We define D to be a *probability distribution* over X that describes the relative frequency with which the various combinations of values for x_i are generated.

The functions discussed in this work are boolean functions with domain X and possible outputs of -1 and $+1$. A concept class C over X is a set of boolean functions. Suppose that $f \in C$ is the ideal function, i.e. the desirable response for that organism. In the context of evolution, the ideal function can be thought of as the organism that is most adapt to all environmental conditions. The goal is to learn the ideal function and produce a hypothesis within computationally-bounded resources that depend on a polynomial in n and on an accuracy parameter ϵ.

The hypothesis shall be viewed as a *representation* of a function since it should be represented concretely in the organism. A *representation class* R is a set of representations, such that every $r \in R$ is a boolean function, $r : X \to \{-1, +1\}$. In this paper we also allow randomized boolean functions and use $r(x)$ to also denote $\mathbb{E}[r(x)]$. Thus, r can be viewed as a real-valued function with range $[-1, 1]$.

The goal of the evolvability learning algorithm is to evolve some representation into a hypothesis that gets arbitrary close to the ideal function, using evolutionary processes. The learning occurs throughout discrete generations.

2.2 Evolvability with HGT

We now define a model that captures the central mechanisms in the process of HGT.

Prokaryotic genomes are characterized by extensive gene mobility. Each individual may receive genes, in a rather short period of time from many different individuals. We capture this process by allowing a representation to mutate in dependence of other representations (in Biology the other representations typically represent organisms that are in close proximity to that representation). We encompass this idea by defining the following neighborhood function:

Definition 1 (HGT Neighborhood Function). *For polynomial $p(\cdot, \cdot)$, a p-bounded HGT neighborhood function is a randomized Turing machine that takes as input a representation $r \in R$, a set of representations, $P' \subset R$ and a constant ϵ and outputs a multi-set of representations $Neigh(r, P', \epsilon) \subseteq R$. The running time of the Turing machine is bounded by $p(n, \epsilon^{-1})$ and $|Neigh(r, P', \epsilon)| \leq p(n, \epsilon^{-1})$. If $Neigh(r, P', \epsilon)$ is empty, it is interpreted as the representation r cannot continue to the next generation.*

Remark 1. We have defined $Neigh(r, P', \epsilon)$ as a multi-set since it can be populated by an algorithm that chooses to insert the same representation multiple times to the set $Neigh(r, P', \epsilon)$.

Remark 2. Note that a mutation in the classical sense of Valiant's model is the sole way to add new variation in the process of evolution. The HGT neighborhood function is allowed the power of an efficient Turing machine, and is assumed to model both mutation and HGT.

Thus, r is mutated according to the set P'. This models the process in which a genome (typically of a prokaryote) is receiving genes from other nearby genomes.

This extension is based on Kanade's recombination model [12] to the basic evolvability model [24], however where in the recombination model the mutation depends on at most two representations, the HGT neighborhood function allows the mutation to depend on a polynomial number of representations.

We now define the performance. The performance is used to give a quantitative measurement of the how well a representation approximates the ideal function.

Definition 2 (Performance [24]). *The performance of a representation r with respect to target function f and distribution D is defined as:*

$$Perf_{f,D}(r) = \mathbb{E}_D[f(x) \cdot r(x)] \in [-1,1]. \tag{1}$$

The evolutionary algorithm will have access to an oracle that given a representation returns its estimated performance. We define the estimated performance according to Valiant's basic evolvability definition, with a slight modification: Instead of observing the sample size s, we observe the estimation error (noise) τ. A conversion from s to τ can be done using Hoeffding-Chernoff bound.

Definition 3 (Estimated Performance). *The estimated performance function takes as input a representation r and outputs τ-$Perf_{f,D}(r)$ which satisfies $|\tau$-$Perf_{f,D}(r) - Perf_{f,D}(r)| \leq \tau$. We require that τ^{-1} is bounded by a polynomial in n, ϵ^{-1}.*

We now define the tolerance function. We will use the tolerance function to separate the mutations that performed good from the other mutations in relation to the current representation's performance.

Definition 4 (Tolerance function [24]). *A tolerance function t takes as input a representation $r \in R$ and an accuracy parameter ϵ and outputs $t(r, \epsilon) \in [0,1]$ that is bounded above and below by two polynomially-related polynomials. That is, there exist polynomials $tl(\cdot, \cdot), tu(\cdot, \cdot)$ such that for every $r \in R$, n and ϵ, $1/tu(n, \epsilon^{-1}) \leq t(r, \epsilon) \leq 1/tl(n, \epsilon^{-1})$ and that there exists a constant a such that $tu(n, \epsilon^{-1}) \leq (tl(n, \epsilon^{-1}))^a$. Furthermore, t can be computed in polynomial time in n, ϵ^{-1}.*

We define a *probability function* $\rho(r, r', \epsilon)$ that returns the probability that r mutates into r'. It is required that the sum of the probabilities $\rho(r, r', \epsilon)$ over all r' is 1.

A selection rule Sel selects a (possibly random) representation of the neighborhood function based on its estimated performance. In this work we use the selection rule used by Valiant, which we denote by SelNB.

Definition 5. *For an error parameter ϵ, a tolerance t, noise τ, and probability function ρ, the selection rule SelNB is an algorithm that for any representation r, any population P', outputs a random variable r' determined as follows:*

1. *Let $Bene_{t,\tau}(r, P', \epsilon) = \{r' \in Neigh(r, P', \epsilon) | \tau$-$Perf_{f,D}(r') \geq \tau$-$Perf_{f,D}(r) + t(r, \epsilon)\}$ and $Neut_{t,\tau}(r, P', \epsilon) = \{r' \in Neigh(r, P', \epsilon) | |\tau$-$Perf_{f,D}(r') - \tau$-$Perf_{f,D}(r)| \leq t(r, \epsilon)\}$.*

2. If $Bene_{t,\tau}(r, P', \epsilon) \neq \varnothing$, output one from it according to the relative probability $\rho(r, r', \epsilon) / \sum_{r'' \in Bene_{t,\tau}(r,P',\epsilon)} \rho(r, r'', \epsilon)$.
3. Otherwise, if $Neut_{t,\tau}(r, P', \epsilon) \neq \varnothing$, output one from it according to the relative probability $\rho(r, r', \epsilon) / \sum_{r'' \in Neut_{t,\tau}(r,P',\epsilon)} \rho(r, r'', \epsilon)$.
4. Otherwise, output $\perp$.

Thus, `SelNB` chooses a beneficial mutation if one exists, and otherwise chooses a neutral representation.

The notion of a *population* was suggested by Kanade [12] to allow the process of recombination between individuals. Indeed in an environment where information is shared between organisms, it is more natural to look at a set of representations across generations. Thus, we assume the existence of a finite population with polynomial size.

An HGT mutator (or evolutionary step) takes a population P_i to population P_{i+1} at the next generation. This involves taking variants of representations in P_i using the neighborhood function, and inserting them to P_{i+1} using the selection rule. The transition is completed when the size of the next population equals the size of the current population. Formally,

Definition 6 (Modified from [12]). *An HGT mutator takes as input a starting population $P_i \subseteq R$, and using an HGT neighborhood function that defines $Neigh(r, P', \epsilon)$ for every $r \in R, P' \subset R$, and a selection rule `Sel` outputs a population P_{i+1} as follows:*

1. *Let $P_{i+1} = \varnothing$.*
2. *While $|P_{i+1}| < |P_i|$*
   2.1. *Select randomly $r \in P_i$.*
   2.2. *Consider the mutations $Neigh(r, P_i - r, \epsilon)$.*
   2.3. *Activate the selection rule function $Sel(r, P_i - r, \epsilon)$ which returns a representation r'.*
   2.4. *If $r' \neq \perp$, put r' in P_{i+1}.*

Remark 3. Note that the same function $r \in P_i$ can be chosen multiple times during the evolutionary step. This may seem unnatural because the same genome cannot mutate several times and still remain in the original population. We say that this is due to the process of prokaryotic replication. Prokaryotes divide very rapidly; Their population may double itself in a single day. We assume that evolutionary processes take much longer than replication and thus, we assume that replication may have occurred during an evolutionary step.

Definition 7 (Modified from [12]). *For a polynomial $p(\cdot, \cdot)$, a p-bounded evolutionary algorithm consists of a representation class R, an HGT neighborhood function operator Neigh, a tolerance function t, a probability function ρ and has access to a performance oracle τ-$Perf_{f,D}$. An evolutionary algorithm starting with population P_0 is a sequence of evolutionary steps (activations of the mutator), that successively produce populations $P_0, P_1, P_2 \ldots$ It is required that Neigh is p-bounded, $|P_0| \leq p(n, \epsilon^{-1})$ and $\tau^{-1} \leq p(n, \epsilon^{-1})$.*

Remark 4. Note that basic evolvability is a special case where the population consists only of a single representation, i.e. $P = \{r\}$. In this case, neighborhood will be of the form $Neigh(r, \varnothing, \epsilon)$ which is essentially equivalent to $Neigh(r, \epsilon)$ in Valiant's model.

Finally, we define the notion of evolvability with HGT in g generations.

Definition 8 (Modified from [12]). *We say that a concept class C is evolvable with HGT with respect to distribution D over X in g generations, if for some polynomial $p(n, \epsilon^{-1})$ there exists a p-bounded evolutionary algorithm, that for every $\epsilon > 0$, from any starting population P_0 and for every target function $f \in C$, with probability at least $1 - \epsilon$ for some $k < g$ reaches a population P_k containing a member $r \in P_k$ such that $Perf_{f,D}(r) \geq 1 - \epsilon$.*

In our recent work [21], we have showed the equivalence between evolvability and evolvability with HGT, by proving the following straightforward theorem:

Theorem 1. *A concept class C is evolvable with HGT extension if and only if it is evolvable.*

The equivalence shows that the models can learn the same range of concept classes efficiently (i.e. within polynomially-bounded resources). Nevertheless, different models can learn the same problem in different speed, i.e. different number of generations. The main result of our previous work [21] manifests that the HGT extension allows an acceleration in terms of the number of generations. This is done using a general reduction from the *parallel CSQ model* that we define below.

2.3   Parallel CSQ Models

We define a model for parallel correlational statistical-query learning with a τ-CSQ oracle. The model was first introduced in Kanade's paper. A parallel CSQ algorithm has a polynomial number of processors p. We assume that there is a common clock which defines the parallel time steps. During every time step, each processor may ask a query from the oracle, perform polynomially bounded computation, and send a message that any other processor can read. The oracle answers all the queries in parallel.

Definition 9 (Parallel CSQ Learning [12]). *A concept class C over an instance space X is (τ, T)-parallel CSQ learnable using p processors under distribution D, if there exists a parallel CSQ algorithm that uses p processors and for every $\epsilon > 0$ and target function $f \in C$, after at most T parallel steps and with access to a $\tau - CSQ$ oracle, outputs a hypothesis h such that $Perf_{f,D}(h) \geq 1 - \epsilon$. Each query φ must be polynomially (in n, ϵ^{-1}) evaluatable and τ^{-1} must be bounded by a polynomial in n, ϵ^{-1}. Each parallel step must be completed by each processor in polynomial time.*

Now we can state the theorem showing a speedup in terms of the number of generations.

Theorem 2. [21] *Suppose concept class C is (τ, T)-parallel CSQ learnable using p processors. Then C is evolvable with HGT starting with an initialized population P_0 within polynomially bounded resources in $O(T)$ generations, using the selection rule* SelNB.

For the sake of completeness, an outline of the proof is provided in the full version of the paper.

3 Results

In this section we show several implications of the theoretical results described in previous sections. We model a biological trait that depends on multiple parameters as a *conjunction* function and acquisition of that trait as learning of the function. Due to the importance of this class, we derive several analytical results on it that appear in the full version of the paper, and strengthen these results with experiments. We run simulations in which the conjunctions concept class is learned, and show that these results affirm our analytical results. We end this section by applying these models to real biological data regarding the development of pathogenicity in microbes.

3.1 The Conjunction Function and Concept Class

Let the sample space consist of n boolean variables (literals), $X = \{-1, 1\}^n$. A *conjunction* function [16] (class) f is defined by a subset $S \subseteq X$. Given a sample $x \in X$, f outputs 1 if each literal in f is consistent with the literals of x. Otherwise, f outputs -1. The *concept class* of conjunctions is the set of all conjunctions, denoted by C.

The concept class of *monotone conjunctions* is the set of conjunction classes such that the classes do not contain a negated literal.

Conjunction is biologically relevant as many biological processes or characters can be seen as a result of the simultaneous existence of a set of genes and absence of another set (see for example microbial pathogenicity in our real data part Subsect. 3.3 below). The same also holds for an expression of a certain protein, that is conditioned on the existence and absence of some other proteins [1].

The values $1/-1$ at index i of a sample indicate that the ith gene is expressed/not expressed. A conjunction function is a combination of genes (genome), where the presence/absence of the ith literal indicates the presence/absence of the ith gene. Thus, learning the ideal function (genome) can be viewed as acquiring/losing certain genes in the representations. The learning stops when a sufficiently close genome was found (i.e. the performance of a representation has an error rate of ϵ).

3.2 Simulation Study

We conducted simulation study to illustrate the difference between the various models under realistic size problems. The processes (models) examined are: the mutation process (i.e. evolvability), recombination, and HGT. As each such process uses another technique to adapt to the environment, we ask, how fast, in terms of generations, a given function is learned, under each such evolutionary process.

Roughly speaking, we can divide the models into two very distinct groups: individum-based model, where only a single representation is examined, and population-based models, where a population of representations are generated each generation and relations between them are enabled. Even though recombination and HGT are both population-based, recombination allows merging only between two individuals, while HGT allows information sharing between the whole population. We will see that this variance makes a large difference in the results.

In reality, HGT is mandated by *HGT rate* that determines probability of HGT events. Therefore, the first experiment measures the effect of HGT rate on the speed of evolution. We model the HGT as a Poisson process [10] operating on a genome through time [8, 19]. This allows easy conversion from rate to event probability. We executed the HGT algorithm with HGT rate varies from $P_{HGT} = 0$ to $P_{HGT} = 1$. We then compared the results with the results from the other models.

In the second experiment, we examined the interplay between the two processes occurring in the population simultaneously. Underneath the *population processes* - recombination and HGT - an underlying mutation process operates individually on every element. We therefore set to test that interplay under the two population models. Specifically, we varied the underlying *mutation factor* while maintaining all other parameters constant, including the population parameters.

The third experiment deals with the role of the size of the population in learning. We start with some population and increase it up to a size of 5 times that starting population.

Learning and Models Description

We have conducted the experiments with three models: the basic evolvability model described by Valiant, evolvability with recombination as described by Kanade, and the model we described in Subsect. 2.1, evolvability with HGT. We start by describing the common setup and goal of the models. Our overall goal is to draw a distinction between the three learning models. As the learning processes are computationally very heavy, we selected fairly small parameter values, however the general trend is still reflected.

The parameters in the experiments are thus: The number of boolean variables was set to 40 ($n = 40$). Under this value an exhaustive search for the ideal function will not be possible. We chose the approximation parameter ϵ to be 2^{-31} guaranteeing that the performance of a random starting representation is low with high probability[1]. Thus, if no starting representation has high

---

[1] We need to calculate the expected performance of two random monotone conjunction functions. This is done by using a combinatorial computation. For n variables, the expected performance rate is almost $1 - 2^{-\frac{3n}{4}}$. Thus if ϵ is smaller than $2^{-\frac{3n}{4}}$, the chance of a high performance at generation 0 is relatively low. For $n = 40$, we need to choose a number smaller than 2^{-30}, therefore we chose ϵ to be 2^{-31}.

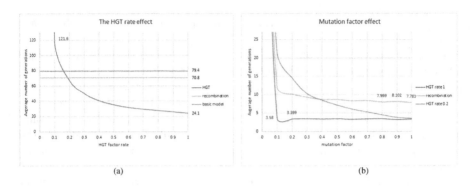

Fig. 1. (a) Experimental results of HGT effect on the HGT model. Note that in the case of recombination and the basic evolvability model, the HGT rate doesn't affect these models at all and therefore the value is constant. (b) The results of the experiment of the affect of an increasing mutation factor on recombination and HGT model. When the mutation factor is 0, the algorithm can't always learn the ideal function so the number of generations is ∞. This is due to the fact that recombination and HGT don't introduce new variation to the population.

performance, learning is performed. D was chosen to be the uniform distribution. Valiant [24] proved that monotone conjunctions is evolvable and described an algorithm for evolving this concept class. The tolerance function and the noise of the performance oracle are derived from that algorithm.

A run of the experiment starts by choosing a random ideal function f and a representation r (or a population of such), and trying to learn f throughout generations. The next generation is obtained by applying the *mutator* (or *recombinator*) to the current generation.

The learning stops when at least one representation r in the current population satisfies $Perf_{f,D}(r) \geq 1 - \epsilon$. For completeness, a more detailed description of the models is given in the full version of the paper.

Experimental Results

We now describe the experiments done. In any experiment, the number of runs was set to 10 for any value of the independent parameter (x) and the average result (generations) is plotted.

Our first experiment focused on the effect of HGT on the speed of learning. Obviously, this parameter is effective only to the HGT model. In the population-based models, mutation factor was set to 0.1 so it does not interfere with the overlying processes recombination and HGT (in the basic evolvability model mutation is the only process so we set it to 1). The results of the experiment are shown in Fig. 1. With an HGT rate of 0.2 or higher, the HGT model outraces the other models. Actually, under HGT rate 1, the model learns in almost third of the generations that took the recombination model. Alternatively, under HGT rate 0, learning is confined only to natural mutation with a factor of 0.1, which explains why it is very slow, even slower than the basic evolvability model. We can therefore infer that HGT rate plays a major role in the learning process of the HGT model.

In the next experiment we examined the role of the underlying mutation factor under the three models. For HGT, we considered two rates: 0.2 and 1 In the experiment, mutation factor varied from 0 to 1. Ideal function length and ϵ were set to 10 and 2^{-8} respectively. The results appear in Fig. 1. We comment that under no mutations (i.e. factor zero), there were runs where the models couldn't learn the ideal function, due to the fact that the models without natural mutation do not introduce new variation to the population. Under recombination, from 0.1 to 0.6, we can see a gradual decrease in the generations. However, around a mutation factor of 0.6, it has almost no affect on it. In the case of HGT with a rate of 1, we see that the mutation factor has almost no affect on the model as the overlying model introduce enough variation. However, under HGT rate of 0.2, the mutation factor has a large impact. A mutation factor of 1 makes the model seven times faster than under 0.1. Finally, as the two HGT curves meet, we hypothesize that under certain mutation rate, the HGT model is faster than recombination for any HGT rate.

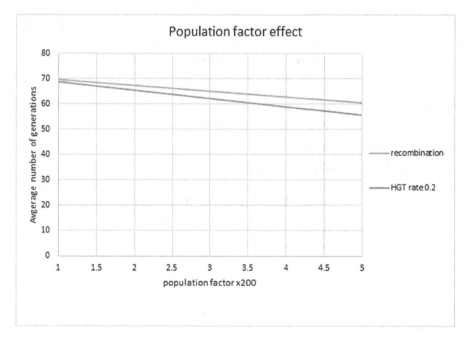

Fig. 2. Experimental results of the effect of the size of the population on the number of generations.

Our final experiment examined the effect of population size on speed of learning under the two population models - recombination and HGT. We therefore varied population size from 100 up to 1000. The results (see Fig. 2) show interestingly a constant decrease in learning time (generations) under both population models, however HGT decreases faster than recombination.

3.3   Real Data Analysis

In this section we apply the same models from the previous section on real data. We chose to focus on the pathogenicity of the E. coli bacteria. The *virulence* of an organism is the degree of pathology caused by that organism. In order to use the evolvability framework, we need a quantitative trait of pathogenicity, which is why we focus on virulence in this section. A *virulence gene* is a gene whose existence in the genome of an organism affects its virulence. A genome is considered pathogenic if it has an appropriate virulence gene combination [3].

We do not however claim that these models represent the processes exactly as they occur in nature. The use of real data in the rigorous framework of evolvability and the comparison between the models grants a realistic aspect to this framework and, in particular, to evolvability with HGT, and hence its importance.

We now show an example of how to deduce the boolean variables from the genes: The pathogenic strain *Enterotoxigenic Escherichia coli* was identified in [5] by carrying either the gene combination *fedA, estII* or *faeG, estI, estII, eltA*. First, we enumerate the genes $x_1 = fedA$, $x_2 = estII$, $x_3 = faeG$, $x_4 = estI$, $x_5 = eltA$. Then, assuming there are no more virulence genes, we can model the virulence of the genome by the following conjunction: $x_1 \wedge x_2 \wedge x_3 \wedge x_4 \wedge x_5$. Thus, the more virulence genes an organism has, the more virulent it becomes. Note that because we consider only the existence of genes, we can limit the concept class to be the class of monotone conjunctions.

Simulation of Real Data Results

We compare between the power of two biological processes: the process of HGT and recombination. 58 virulence genes were observed in [3]. The number of boolean variables was set to 100, where 58 of them were randomly chosen to represent the virulence genes and their values were set to 1. The other variables were chosen randomly. A value of 1 in one of the other variables represents a virulence gene that has yet been discovered.

The size of the population is chosen to be 75, corresponding to the 75 E. coli isolates that were taken in Chapman et al. paper [3]. The approximation parameter ϵ was chosen to be 2^{-70}. We believe that this approximation of the ideal function is reasonable for a representation to be considered highly virulent.

An estimation of the HGT rate for Escherichia coli showed that 17.6% of the genes have been transferred since its divergence from Salmonella lineage 100 million years ago [15]. We consider a single generation in our models as a million year, and set the HGT and recombination rate to be 0.176% (A recombination rate of x signifies that if recombination does not occur, one of the two representations is chosen randomly).

Over a total of 20 runs, the recombination model learned in 256 generations average, and the HGT model learned in 35 generations average.

In conclusion, if evolution would have occurred according to our parameters, a highly virulent organism would have emerged in 35 million years using HGT process. If instead the recombination process would have taken place, the pathogenic organism would have emerged in 256 million years.

4   Conclusions

In his seminal work from 2006 [24], Leslie Valiant extended his celebrated PAC model [22] to the biological world and denoted it *evolvability*. Evolvability quantifies the process of evolution in terms of computational learning power. Subsequently, Kanade [12] extended evolvability to higher level organisms' reproduction by combining the mechanism of recombination. Nevertheless, to the best of our knowledge, no such extension to evolvability was suggested for prokaryotes. Prokaryotic evolution largely proceeds by exchanging DNA between unrelated organisms, a mechanism denoted *horizontal gene transfer* (HGT). It is mainly due to HGT that microbes adapt to new ecological conditions, develop resistance to antibiotics, and so forth. In this work, we define a conceptual model encompassing this phenomena. The model is conceptual/mathematical and therefore isn't limited to a specific biological phenomena. Valiant's evolvability is a special case in this new model. The evolutionary advantage of the new model is by allowing (genetic) information sharing between individuals in an evolving population, similarly to the recombination model.

While in a recent work we have focused on the theoretical aspects of the newly suggested model, here the emphasis is on its biological application. Specifically, we focus at apparently a biologically relevant function - *conjunction*, modeling existence of an entity contingent on the presence of several other entities. We show that for conjunction, the new model achieves asymptotic acceleration in learning time over evolvability and recombination. We corroborate these findings in simulation where we use a randomized learning algorithm for conjunction under evolvability. We conclude with learning a real data function of virulence in the *E. coli* bacterium.

The primary contribution of this work is the extension of the Evolvability framework to account to prokaryotes. We believe its importance stems from the lag of application of rigorous computational learning tools to model evolution in this life domain. The comparison between several modes of real life evolutionary mechanisms under a common ground, from a rigorous computational learning perspective, provides another explanation to the versatility of these increasingly discovered world.

There are several future directions to take from this work that we consider of interest. Conjunction may be relevant for the case discussed here, but other biological mechanisms may require other concept classes. This may give insight into these mechanisms. We think that evolvability is a powerful yet flexible tool and can be used extensively to analyze more real biological data, using the scheme described in this work. Finally, modeling more evolutionary phenomena with this framework is interesting both from the aspect of computer science and from the aspect of biology.

References

1. Alon, U.: An Introduction To Systems Biology: Design Principles of Biological Circuits. CRC Press, Boca Raton (2006)
2. Angelino, E., Kanade, V.: Attribute-efficient evolvability of linear functions. In: Innovations in Theoretical Computer Science, ITCS 2014, Princeton, NJ, USA, 12–14 January 2014, pp. 287–300 (2014)
3. Chapman, T.A., Wu, X.-Y., Barchia, I., Bettelheim, K.A., Driesen, S., Trott, D., Wilson, M., Chin, J.J.-C.: Comparison of virulence gene profiles of Escherichia coli strains isolated from healthy and diarrheic swine. Appl. Environ. Microbiol. **72**(7), 4782–4795 (2006)
4. Darwin, C.: On the Origin of Species. John Murray, London (1859)
5. Do, T., Stephens, C., Townsend, K., Wu, X., Chapman, T., Chin, J., McCormick, B., Bara, M., Trott, D.: Rapid identification of virulence genes in enterotoxigenic Escherichia coli isolates associated with diarrhoea in Queensland piggeries. Aust. Vet. J. **83**(5), 293–299 (2005)
6. Doolittle, W.F.: Phylogenetic classification and the universal tree. Science **284**(5423), 2124–2128 (1999)
7. Doolittle, W.F., Bapteste, E.: Pattern pluralism and the tree of life hypothesis. Proc. Nat. Acad. Sci. **104**(7), 2043–2049 (2007)
8. Galtier, N.: A model of horizontal gene transfer and the bacterial phylogeny problem. Syst. Biol. **56**(4), 633–642 (2007)
9. Gogarten, J.P., Townsend, J.P.: Horizontal gene transfer, genome innovation and evolution. Nature Reviews. Microbiology **3**(9), 679–87 (2005)
10. Grimmett, G., Stirzaker, D.: Probability and Random Processes. Oxford University Press, Oxford (2001)
11. Jin, G., Nakhleh, L., Snir, S., Tuller, T.: Inferring phylogenetic networks by the maximum parsimony criterion: a case study. Mol. Biol. Evol. **24**(1), 324–337 (2007)
12. Kanade, V.: Evolution with recombination. In: IEEE 52nd Annual Symposium on Foundations of Computer Science, FOCS 2011, Palm Springs, CA, USA, 22–25 October 2011, pp. 837–846 (2011)
13. Koonin, E.V., Galperin, M.Y.: Sequence - Evolution - Function: Computational Approaches in Comparative Genomics. Kluwer Academic Publishers, Norwell (2002)
14. Koonin, E.V., Makarova, K., Aravind, L.: Horizontal gene transfer in prokaryotes - quantification and classification. Annu. Rev. Microbiol. **55**, 709–742 (2001)
15. Lawrence, J.G., Ochman, H.: Molecular archaeology of the Escherichia coli genome. Proc. Nat. Acad. Sci. **95**(16), 9413–9417 (1998)
16. Mendelson, E.: Introduction to Mathematical Logic, 4th edn. Chapman and Hall, London (1997)
17. Nakamura, Y., Itoh, T., Matsuda, H., Gojobori, T.: Corrigendum: biased biological functions of horizontally transferred genes in prokaryotic genomes. Nat. Genet. **36**(10), 1126 (2004)
18. Ochman, H., Lawrence, J.G., Groisman, E.A.: Lateral gene transfer and the nature of bacterial innovation. Nature **405**(6784), 299–304 (2000)
19. Roch, S., Snir, S.: Recovering the tree-like trend of evolution despite extensive lateral genetic transfer: a probabilistic analysis. CoRR, abs/1206.3520 (2012)
20. Siefert, J.L.: Defining the mobilome. In: Gogarten, M., Gogarten, J., Olendzenski, L. (eds.) Horizontal Gene Transfer: Genomes in Flux. Methods in Molecular Biology. Humana Press, New York (2009). https://doi.org/10.1007/978-1-60327-853-9_2

21. Snir, S., Yohay, B.: Prokaryotic evolutionary mechanisms accelerate learning (submitted)
22. Valiant, L.G.: A theory of the learnable. In: Proceedings of the 16th Annual ACM Symposium on Theory of Computing, Washington, DC, USA, 30 April–2 May 1984, pp. 436–445 (1984)
23. Valiant, L.G.: Evolvability. Electronic Colloquium on Computational Complexity (ECCC), **13**(120), (2006)
24. Valiant, L.G.: Evolvability. J. ACM **56**(1), 3 (2009)

Predicting Opioid Epidemic by Using Twitter Data

Yubao Wu[1][✉], Pavel Skums[1,2], Alex Zelikovsky[1], David Campo Rendon[2], and Xueting Liao[1]

[1] Department of Computer Science, Georgia State University, Atlanta, GA, USA
{ywu28,pskums}@gsu.edu, alexz@cs.gsu.edu, xliao3@student.gsu.edu
[2] Centers for Disease Control and Prevention, Atlanta, GA, USA
fyv6@cdc.gov

Abstract. Opioid crisis was declared as a public health emergency in 2017 by the President of USA. According to the Centers for Disease Control and Prevention, more than 91 Americans die every day from an opioid overdose. Nearly $4B is provided to address the opioid epidemic in the 2018 spending bill and help fulfill the President's Opioid Initiative.

How to monitor and predict the opioid epidemic accurately and in real time? The traditional methods mainly use the hospital data and usually have a lag of several years. Even though they are accurate, the long lag period prevents us from monitoring and predicting the epidemic in real time. We observe that people discuss things related to the epidemic a lot in social media platforms. These user behavior data collected from social media platforms can potentially help us monitor and predict the epidemic in real time.

In this paper, we study how to use Twitter to monitor the epidemic. We collect the historic tweets containing the set of keywords related to the epidemic. We count the frequency of the tweets posted at each month and each state. We compare the frequency values with the real-world death rates at each month and each state. We identify high correlation between tweet frequency values and real-world death rates. The statistical significance demonstrates that the Twitter data can be used for predicting the death rate and epidemic in future.

1 Introduction

According to the Centers for Disease Control and Prevention, more than 91 Americans die every day from an opioid overdose. Opioid crisis is killing more people than car crashes and it is the deadliest drug crisis in American history. President Trump has declared the opioid crisis a public health emergency in October 2017. The 2018 spending bill provides nearly $4B to address the opioid epidemic and help fulfill the President's Opioid Initiative.

How to predict and monitor the epidemic accurately and in real time? This is a fundamental question that needs to be addressed urgently. The traditional methods mainly use the real-world death data collected from hospitals and usually have a lag of several years. We observe that people discuss things related to

© Springer International Publishing AG, part of Springer Nature 2018
F. Zhang et al. (Eds.): ISBRA 2018, LNBI 10847, pp. 314–318, 2018.
https://doi.org/10.1007/978-3-319-94968-0_30

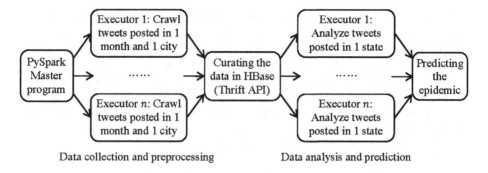

Fig. 1. Software architecture.

the epidemic and drugs a lot in social media platforms. "No family should EVER have to go through this. My cousin dies from a drug overdose" and "Apparently I am very into heroin and I overdose every night" are two real tweets about drug overdose. If we can infer the drug usage behaviors of users from the posted texts by using artificial intelligence, we potentially can monitor and predict the epidemic in real time.

In this project, we study how to use Twitter to monitor the opioid epidemic. We design and implement a novel distributed software system. The system collects the historic tweets and curates them in a distributed database. It then performs further data analysis to monitor and predict the epidemic. We have collected millions of historic tweets and compare them with the real-world death rates. We find that the frequency values of tweets significantly correlate with the real-world death rates. This demonstrates that social media data can be used to monitor and predict the epidemic in real time.

Modern web scraping and data streaming techniques are used to collect the data from Twitter. Big data programming tools such as Spark and HBase are used to preprocess and curate the data. We also design distributed computing algorithms to perform the analysis on the large amount of tweets. The overall contributions are summarized as follows.

- We empirically prove that Twitter data can generally be used for monitoring and predicting the opioid epidemic.
- We design and implement a software system which can collect the historic tweets, perform analysis, and monitor and predict the epidemic.

The proposed software system is designed to monitor and predict the opioid epidemic. But it can also be used for other types of epidemics such as HIV, HCV, flu, and alcoholism.

2   Software Architecture

In order to process large amount of tweet data, we implement the software by using Apache Spark. Therefore, the software system can be deployed in a computer cluster. Figure 1 shows the software architecture of the system. From Fig. 1, we can see that there are two stages. In the first stage, the system collects and pre-processes tweets from Twitter platform. The collected data is stored in HBase, a distributed database system. In the second stage, the system analyzes the tweets and predict the epidemic. In both stages, multiple computer nodes run in parallel thus the system can process large amount of data efficiently.

All programs are written in Python. HBase Thrift APIs are used to access the database. Selenium is used to automate the Google Chrome web browser and crawl tweets from Twitter advanced search websites.

3   Query Conditions

We aim at crawling the historic tweets posted in USA related to opioid epidemic. To achieve this goal, we specify three concrete conditions.

- 30 keywords: opioid, alprazolam, amphetamine, antidepressant, benzodiazepine, buprenorphine, cocaine, diazepam, fentanyl, heroin, hydrocodone, meth, methadone, morphine, naloxone, narcan, opana, opiate, overdose, oxycodone, oxymorphone, percocet, suboxone, subutex, pill, rehab, sober, withdrawal, shooting up, track marks
- 144 cities in USA with large populations and minimum two cities in each state; the diameter is set to 45 miles
- 145 months: from March 21, 2006, to March 26, 2018.

The set of 30 keywords are chosen by domain experts and are shown empirically to be associated with opioid epidemic. The selected 144 cities are chosen based on the population size. Minimum two cities are chosen in each state. The dates are from the date of first tweet to a recent date. There are 145 months in total. The algorithm thus submits one query for each city and each month. Since infinite scrolling is used in the Twitter advanced search website, the algorithm scrolls down the website in order to collect all tweets.

4   Results and Discussion

The system crawls $1,896,961$ tweets in total. We design the distributed algorithm and count the number of tweets posted in each month and in each state. Figure 2 shows the frequency values. The x-axis represents the month from October 2009 to March 2018. There are no tweets satisfying the query conditions from March 2006 to September 2009 in any cities. Therefore, the x-axis starts from October 2009. The y-axis represents the number of tweets published in one month and in one state. Each line in the figure represents the number of tweets posted in one state. There are 50 states in total.

From Fig. 2, we can see that the number of tweets are generally increasing in the past 10 years. We also observe a peak from Dec 2012 to August 2014. In 2015, press reports that there was an outbreak of HIV caused by the use of

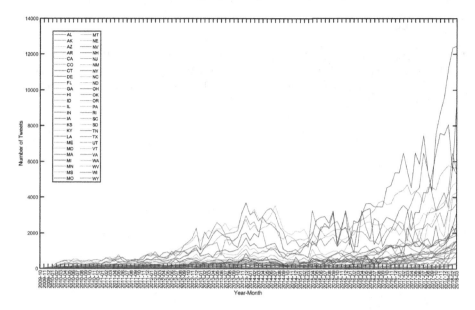

Fig. 2. Frequency values of tweets.

Opana as an injectable recreational drug [1]. This event correlates with the peak in that period.

We also downloaded the national overdose death rate data. We aggregate the data into year and state level. Similarly, we also aggregate the tweet frequency values into year and state level. Then we compare the two sets of data by calculating the Pearson correlation. The overall correlation coefficient value is 0.83. This demonstrates that the overdose death rates correlate with the frequency values of tweets. Therefore, we can use the frequency of tweets to monitor and predict the death rates and the epidemic in real time.

5   Conclusion

Opioid overdose crisis is an urgent societal issue to be solved. In this paper, we explore the possibility of monitoring and predicting the epidemic by using Twitter, a social media platform. We design the distributed software system and crawl millions of tweets containing the keywords about the epidemic. We observe meaningful patterns from the crawled tweets. We compare the tweet data with real-world overdose deaths and find that they highly correlate with each other. This demonstrates that the tweets can be used to predict the epidemic in real time.

Acknowledgements. AZ was partially supported by NSF Grant CCF-1619110.

References

1. Peters, P.J., Pontones, P., Hoover, K.W., Patel, M.R., Galang, R.R., Shields, J., Blosser, S.J., Spiller, M.W., Combs, B., Switzer, W.M., et al.: HIV infection linked to injection use of oxymorphone in Indiana, 2014–2015. New Engl. J. Med. **375**(3), 229–239 (2016)

Analysis and Visualization Tools

Cluster Matching Distance for Rooted Phylogenetic Trees

Jucheol Moon[1]($\boxtimes$) and Oliver Eulenstein[2]

[1] Department of Computer Science and Computer Engineering,
California State University Long Beach, Long Beach, CA 90840, USA
jucheol.moon@csulb.edu
[2] Department of Computer Science, Iowa State University,
Ames, IA 50011, USA
oeulenst@iastate.edu

Abstract. Phylogenetic trees are fundamental to biology and are bene-fitting several other research areas. Various methods have been developed for inferring such trees, and comparing them is an important problem in computational phylogenetics. Addressing this problem requires tree measures, but all of them suffer from problems that can severely limit their applicability in practice. This also holds true for one of the old-est and most widely used tree measures, the Robinson-Foulds distance. While this measure is satisfying the properties of a metric and is effi-ciently computable, it has a negatively skewed distribution, a poor range of discrimination and diameter, and may not be robust when comparing erroneous trees. The cluster distance is a measure for comparing rooted trees that can be interpreted as a weighted version of the Robinson-Foulds distance. We show that when compared with the Robinson-Foulds distance, the cluster distance is much more robust towards small errors in the compared trees, and has a significantly improved distribution and range.

Keywords: Evolutionary trees · Bipartite perfect matching
Robinson-Foulds distance · Cluster matching distance

1 Introduction

Phylogenetic trees depict the phylogenetic relationships of entities (e.g., molec-ular sequences, genomes, or species), and are a fundamental tool for organizing our knowledge of biological diversity. Through these relationships, we are able to understand how entities have evolved over time the way they are today, and analyzing them benefits a vast variety of fundamental research areas such as biology, ecology, epidemiology, and conservation biology [12,14,18,26].

Analyzing phylogenetic trees requires the comparative evaluation of their differences, similarities, and distances, which has become a fundamental task in computational phylogenetics [11,30]. To compare phylogenetic trees a large

© Springer International Publishing AG, part of Springer Nature 2018
F. Zhang et al. (Eds.): ISBRA 2018, LNBI 10847, pp. 321–332, 2018.
https://doi.org/10.1007/978-3-319-94968-0_31

variety of measures has been considered and analyzed [7,19,20]. However, all of these measures have shortcomings that can severely limit their applicability in practice, which range from intrinsic exponential time complexities [1,5,9] to negatively skewed distributions [31], and several measures suffer from topology biases [34] or do not satisfy the properties of a metric [24]. For example, computing distance measures based on edit distances under the classic tree edit operations nearest neighbor interchange (NNI), subtree pruning and regrafting (SPR), and tree bisection and reconnection (TBR), are NP-Hard [1,9,15,16]. Gene tree parsimony costs that rely on evolutionary models to compare trees do not satisfy the properties of a metric [24] and suffer from topology biases [33,34]. While the in practice widely-used Robinson-Foulds (RF) distance can be computed in linear time [10], this distance is not robust when a small error is present in the compared trees. A distance is *robust* when a small error in the compared trees that is modeled by successive applications of the classic tree operations will not cause abrupt distance changes [23]. In addition, the RF distance has a negatively skewed distribution [8,31] where most distances are close to the maximum possible distance (i.e., the diameter of the RF distance). A weighted version of the RF distance is the *bipartition matching distance* [23], which is robust and has a significantly better discrimination than the RF distance due to a less skewed distribution. Unfortunately, in contrast to the RF distance, the bipartition matching distance is limited to the comparison of unrooted trees, while at the same time many tree comparisons involve rooted trees [19]. The *cluster matching distance* [4] can be seen as a rooted version of the bipartition matching distance [23]. Like the bipartition matching distance, the cluster matching distance relies on matching and can be computed in polynomial time. However, the behavior of the cluster matching distance when small error is present has not been analyzed.

Here, we show the robustness of the cluster matching distance by proving asymptotical bounds on the change of the cluster matching distance caused by the tree edit operations NNI, SPR, and TBR. Furthermore, in our experimental studies, we demonstrate that the cluster matching distance is less sensitive to the classic tree edit operations than the rooted RF distance. Thus, unlike the rooted RF distance, the cluster matching distance is robust when small error is present in the compared trees. Finally, we demonstrate that the cluster matching distance is bell-shaped under two classic tree sampling models, and offers a significantly better discrimination when compared to the rooted RF distance.

1.1   Related Work

The need to compare phylogenetic trees has given rise to the proliferation of various measures for the pairwise comparison of such trees [19]. Here we describe distance-based measures for the pairwise comparison of trees over the same label set that are closely related to the presented work and discuss their advantages and shortcomings.

While all of the presented measures induce a metric on the tree space, which is not true for measures that rely on biological models [24], they largely differ in their asymptotic computation times and distributions. In addition, these measures vary in terms of their diameters, and gradients regarding the classic tree edit operations NNI, SPR, and TBR. The *diameter* of a measure for the tree space of all n taxa trees is the maximum distance between any pair of trees in this space [19]. In practice, such diameters are often used to normalize their corresponding measures in order to compare them when analyzing distances between trees [17]. The *gradient* of a tree edit operation with respect to a given distance metric is the maximum distance between all pairs of trees that can be transformed into each other by one edit operation. Errors in trees can be expressed in terms of the tree edit operations [35], and thus the gradient of an edit operation for a measure can be used to describe the robustness regarding the error of this measure [23]. In the following, we overview the measures of interest for this work, which are (i) tree edit based measures, (ii) the RF measure, and (iii) the bipartition matching distance.

Tree Edit Based Measures. Maybe the most natural tree measures are based on the classic tree edit operations that are informally described for an unrooted and full-binary tree T over n taxa as follows:

- *Nearest neighbor interchange (NNI)*: This operation selects an internal edge in T (i.e., an edge that is not incident to a leaf), and exchanges a subtree on one side of the selected edge with a subtree on the other side of the edge.
- *Subtree prune and regraft (SPR)*: This operation prunes a subtree from T by cutting an edge and redrafts the subtree to a new vertex obtained by subdividing an edge of the edited tree.
- *Tree bisection and reconnection (TBR)*: This operation divides tree T into two subtrees by removing an edge, and then reconnects these subtrees by creating a new edge between the midpoints of edges in them.

The *NNI, SPR, and TBR measures* are defined to count the minimal number of corresponding edit operations required to change a given pair of trees into each other. The NNI distance has been introduced independently by Das Gupta [9] and Li et al. [21], and computing this distance is NP-hard [22]. Later on, the SPR distance and TBR distance were introduced for unrooted and rooted trees, and their NP-harness was shown eventually [1,5]. All of these measures induce metrics or distances on the space of trees [19]. The diameter of the NNI distance is $\Theta(n \log n)$ [9,21], and the diameters for the SPR and TBR distance are $\Theta(n)$ [1].

The RF Measure. The RF measure (also referred to as symmetric difference measure [29], or partition measure [31]) is a popular and widely used measure in the literature [6,10,25,29]. Originally proposed by Bourque [6] in 1978, this measure was generalized by Robinson and Foulds [27,29] later on, and is inducing a distance on the space of trees, which can be computed in linear time in the size of the trees that are compared [10]. However, the RF distance has a negatively skewed distribution, and in practice, this distance is only useful when the compared trees are "very similar" [31]. Further, the RF distance is not robust towards small changes, as reattaching a single leaf elsewhere in one of the compared trees can maximize the distance.

Here, we are focusing on the *(rooted) RF distance* for a pair of rooted and full binary trees over the same taxon set that is the normalized count of the symmetric difference of the cluster representations of these trees [29]. The *cluster representation* of a tree is the set containing a cluster for each of the vertices of the tree, and the *cluster* of a vertex is the set of taxa of the subtree rooted at this vertex. The cluster presentation of a tree (represented as a graph) are equivalent representations of each other [30]. Observe that the diameter of the rooted RF distance for the space of trees over n taxa is $n - 2$ [20].

The Bipartition Matching Distance. Lin et al. [23] have proposed the bipartition matching distance. Similar to the clusters in a rooted tree, every internal edge e in an unrooted tree T defines a nontrivial bipartition σ_e on the leaves, and the tree T is uniquely represented by the set of bipartitions $\Sigma(T) = \{\sigma_e | e \in E(T)\}$, where $E(T)$ is the set of internal edges in T. A bipartition can be represented by the binary vector, i.e., if $\sigma = (1, \ldots, k | k + 1, \ldots n)$, then the corresponding binary representation is $[1, \ldots, 1, 0, \ldots, 0]$ where the number of 1's is equal to k. Given two trees, T_1 and T_2 on the same set of leaves, a complete weighted bipartite graph $G(X, Y, E)$ with $X = \Sigma(T_1)$ and $Y = \Sigma(T_2)$ is denoted by $B(T_1, T_2)$. The weight of each edge $e = \{u, v\}$ in $B(T_1, T_2)$ is set to $W(u, v) = \min\{HD(V_u, V_v), HD(V_u, \overline{V_v})\}$ where V_u and V_v are the two binary vector representations of the bipartition u and v, $\overline{V}$ is the complement vector representation of V, and HD is the Hamming distance. The bipartition matching distance $BM(T_1, T_2)$ between trees T_1 and T_2 is the weight of the minimum-weight perfect matching in $B(T_1, T_2)$ with the weighting scheme W. Considering the space of unrooted and full binary trees over n taxa, the bipartition matching distance for a pair of trees can be computed in $O(n^{2.5} \log n)$ time [23]. Further, experimental studies suggest that this distance has a better discrimination when compared to the unrooted RF distance [23].

1.2   Contribution

Here we prove that the gradients of the edit operations NNI, SPR, and TBR with respect to the bipartition matching distance are bound by $\Theta(n)$, $\Theta(n^2)$, and $\Theta(n^2)$ respectively.

In an experimental study, we demonstrate the distribution of the cluster matching distance between randomly generated binary trees using Yule-Harding model and birth-death process model. For both models, the cluster matching distance is more broadly distributed in the form of a bell-shape and has a wider range than the RF distance. We also compare how the tree distance metrics are correlated with the number of classical tree edit operations. When compared to the RF distance, the cluster matching distance is gradually saturated towards its maximum value.

2   Preliminaries and Basic Definitions

A *(phylogenetic) tree* T is a connected acyclic graph that has exactly one distinguished vertex of degree two, called the *root* of T, and where all of the remaining vertices are either of degree three or one. The vertices of degree larger than one are the *internal* vertices of T, and the remaining vertices are the *leaves* of T. For a tree T we denote its vertex set, edge set, leaves, internal vertices, and root, by $V(T), E(T), \mathcal{L}(T), V_{int}(T)$, and $r(T)$, respectively.

Let T be a tree. Given $X \subseteq \mathcal{L}(T)$ we define the *leaf complement* of X as $\overline{X} := \mathcal{L}(T) \setminus X$. The *subtree* of T induced by X, denoted by $T(X)$, is the minimal connected subgraph of T that contains X. The *restricted subtree* of T induced by X, denoted by $T|X$, is the tree obtained from $T(X)$ by suppressing all vertices of degree two with the exception of the root.

In the following we introduce needed terminology relating to the semi-order represented by T. We define $\leq_T$ to be the partial order on $V(T)$, where $x \leq_T y$ if y is a vertex on the path between $r(T)$ and x. If $x \leq_T y$, we call x a *descendant* of y, and y an *ancestor* of x. We also define $x <_T y$, if $x \leq_T y$ and $x \neq y$. If $\{x, y\} \in E(T)$ and $x \leq_T y$, then we call y the *parent* of x and x a *child* of y. The *cluster* of x is defined by $\mathcal{C}_T(x) := \mathcal{L}(T(x))$, and the set of all clusters of T is defined by $\mathcal{H}(T) = \bigcup_{y \in V(T)} \mathcal{C}_T(y)$. $X \in \mathcal{H}(T)$ is called a *trivial* cluster if $X = \mathcal{L}(T)$ or $|X| = 1$, it is called *non-trivial* otherwise.

Let T_1 and T_2 be trees, then the *(rooted) Robinson-Foulds (rRF) distance* [29] is defined as $RF(T_1, T_2) := \frac{1}{2}((\|\mathcal{H}(T_1) \setminus \mathcal{H}(T_2)\|) + (|\mathcal{H}(T_2) \setminus \mathcal{H}(T_1)|))$.

The cluster matching distance [4] for T_1 and T_2 is based on the complete weighted bipartite graph $B(T_1, T_2) := G(X, Y, E)$ with $X = V_{int}(T_1) \setminus \{r(T_1)\}$ and $Y = V_{int}(T_2) \setminus \{r(T_2)\}$ where the weight of each edge $\{u, v\}$ is $W(u, v) := |\mathcal{C}_{T_1}(u) \ominus \mathcal{C}_{T_2}(v)| = |(\mathcal{C}_{T_1}(u) \setminus \mathcal{C}_{T_2}(v)) \cup (\mathcal{C}_{T_2}(v) \setminus \mathcal{C}_{T_1}(u))|$. Now, the *cluster matching distance* $CM(T_1, T_2)$ between T_1 and T_2 is defined to be the weight of the minimum weight perfect matching in $B(T_1, T_2)$.

Let $\mathfrak{T}(n)$ be the space of all trees on n leaves. Then the *diameter* of $\mathfrak{T}(n)$ with respect to a distance metric D on $\mathfrak{T}(n)$ is defined as $\Delta(D, n) := \max\{D(T_1, T_2) \mid T_1, T_2 \in \mathfrak{T}(n)\}$. The diameter of the CM distance is $\Delta(CM, n) = \Theta(n^2)$ [4].

3   Gradients to the Tree Edit Operations

Let T be a tree and $\phi(T)$ be the set of trees derived by applying the edit operation ϕ to T, then $\phi(T)$ is called the *(local) neighborhood* of T under ϕ [30]. We provide the definitions for the classic tree edit operations in their rooted settings.

– rNNI [28]: Let $T_2 \in rNNI(T_1)$. An internal vertex u of a rooted binary tree T_1 has two incident edges that connects its children l and r. A rooted binary tree T_2 is obtained from T_1 by deleting $e = \{u, l\}$ (or $e' = \{u, r\}$), adding the edge between l (or r) and the vertex subdivides the edge that is incident with $Pa_{T_1}(u)$ and u's sibling, and then suppressing any degree-two vertices.

- rSPR [1]: Let $T_2 \in rSPR(T_1)$. $e = \{u, v\}$ and $u \leq_{T_1} v$. A rooted binary tree T_2 is obtained from T_1 by deleting e, adding the edge between u and the vertex that subdivides the edge of $T_1 \setminus e$, and then suppressing any degree-two vertices.
- rTBR [1]: Let $T_2 \in rTBR(T_1)$. Analogous to rSPR, a rooted binary tree T_2 is obtained from T_1 by deleting e, adding an edge between vertices such that each of the vertices subdivides the edge of one and the other component of $T_1 \setminus e$, and then suppressing any degree-two vertices.

Definition 1. *The gradient of a tree edit operation ϕ with respect to a distance D on $\mathfrak{T}(n)$ is $\mathcal{G}(T(n), D, \phi) := \max\{D(T_1, T_2) \mid T_1, T_2 \in \mathfrak{T}(n) \wedge T_2 \in \phi(T_1)\}$.*

Proposition 1. $\mathcal{G}(n, RF, rNNI) = 1$, $\mathcal{G}(n, CM, rNNI) = \Theta(n)$.

Proof. Consider two trees T_1 and T_2 as shown in Fig. 1. Suppose that T_1 in Fig. 1 is a caterpillar tree, $\mathcal{C}_{T_1}(u_1) = \{1, \ldots, k-2, k-1\}$, $\mathcal{C}_{T_1}(u_2) = \{1, \ldots, k-2, k-1, k\}$, $\mathcal{C}_{T_2}(v_1) = \{k-1, k\}$, and $\mathcal{C}_{T_2}(v_2) = \{1, \ldots, k-2, k-1, k\}$ where $3 \leq k \leq n$. The rNNI operation replaces the cluster $\mathcal{C}_{T_1}(u_1) = \{1, \ldots, k-2, k-1\}$ in $\mathcal{H}(T_1)$ with the cluster $\mathcal{C}_{T_2}(v_1) = \{k-1, k\}$ in $\mathcal{H}(T_2)$. Hence, $RF(T_1, T_2) = \mathcal{G}(T(n), RF, NNI) = 1$. For the CM distance, in $B(T_1, T_2)$, we have $W(u_2, v_2) = 0$, since $\mathcal{C}_{T_1}(u_2) = \mathcal{C}_{T_2}(v_2) = \{1, \ldots, k-2, k-1, k\}$. The edge weight $W(u_1, v_1)$ between u_1 and v_1 in $B(T_1, T_2)$ is $k-1$, since $|\mathcal{C}_{T_1}(u_1) \ominus \mathcal{C}_{T_2}(v_1)| = |\{1, \ldots, k-2, k-1\} \ominus \{k-1, k\}| = |\{1, \ldots, k-2\} \cup \{k\}| = k-1$. Therefore, $\mathcal{G}(T(n), CM, rNNI) = \Theta(n)$, since $3 \leq k \leq n$.

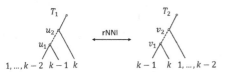 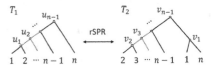

Fig. 1. An rNNI operation where $T_1 \in rNNI(T_2)$ and $T_2 \in rNNI(T_1)$.

Fig. 2. An rSPR operation where $RF(T_1, T_2) = n - 2$.

Proposition 2. $\mathcal{G}(n, RF, rSPR) = n - 2$, $\mathcal{G}(n, CM, rSPR) = \Theta(n^2)$

Proof. Consider two trees T_1 and T_2 as shown in Fig. 2. The bound for the RF distance is derived by prune one leaf (1) at one end of T_1 and regraft it to the other end (n) of the tree. Hence, $\mathcal{G}(T(n), RF, rSPR) = n - 2$. For the CM distance, consider two trees T_1 and T_2 as shown in Fig. 3. By the rSPR operation from T_1 to T_2, the edge $\{u_k, u_{k+1}\}$ is deleted, and the subtree $T_1(u_k)$ is grafted between u_k and u_{k+1} where $1 < l < k < n$. Note that $\mathcal{C}_{T_2}(v_m) = \{l+1, \ldots, m\}$ for $v_{l+1} <_{T_2} v_m \leq_{T_2} v_k$. Suppose that $\frac{n}{4} \leq l$ and $\frac{3}{4}n \leq k$, then in $B(T_1, T_2)$,

$$W(u, v_m) = \begin{cases} m & u \leq_{T_1} u_l \\ m - 1 & u = u_{l+1} \\ l + \delta & otherwise \ (\delta > 0) \end{cases}$$

$W(u, v_m) > \frac{n}{4}$ for $v_{l+1} <_{T_2} v_m \leq_{T_2} v_k$ because $m > l + 1$ and there are at least $\frac{n}{2}$ such a vertex v_m. Hence, any matching in $B(T_1, T_2)$ has a weight of at least $\frac{n}{4} \times \frac{n}{2} = \Omega(n^2)$. The upper bound is trivial by $\Delta(CM, n) = \Theta(n^2)$.

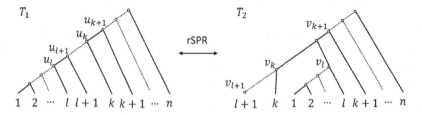

Fig. 3. An example of an rSPR operation where $CM(T_1, T_2) = \Theta(n^2)$, and $T_1 \in rSPR(T_2)$ and $T_2 \in rSPR(T_1)$.

Proposition 3. $\mathcal{G}(n, RS, rTBR) = n - 2$, $\mathcal{G}(n, RSM, rTBR) = \Theta(n^2)$

Proof. Since rSPR is a special case of rTBR, the results follow from Proposition 2.

4 Experiments

We demonstrate the characteristics of the rRF distance in comparison with the CM distance using simulated datasets. First, we compare the distances of pairs of trees under these measures when randomly sampled under two classic models. Then, we compare how the rRF distance and the CM distance are correlated with the number of consecutive tree edit operations that are either rNNI, rSPR, or rTBR. For the experiments, we define a *profile* to be a tuple of trees over the same leaf set. All of the experiments were performed on a workstation with an Intel® Xeon® CPU E7-8837 @2.66 GHz with 128 GB RAM.

4.1 Distribution of the Tree Distance Metrics

We compared the distributions of the rRF distance and the CM distance for randomly sampled trees under two classic models, the Yule-Harding model [13] and the birth-death process model [2].

Dataset. We generated the profiles $P_k := \{p_1, \ldots, p_l\}$ and $Q_k := \{q_1, \ldots, q_l\}$ of random trees over k leaves for each $k \in \{100, 1000\}$, and $l := 10^5$ separately under the Yule-Harding model [13] and the birth-death process model [2]. The trees under each of these models were sampled as follows.

Yule-Harding Model. The following procedure is sampling trees, for each k under the Yule-Harding model as shown in [3].

i. an initial list of k single-vertex trees (representing the leaves of the final tree) is generated.

ii. two randomly chosen trees are merged into a new tree by making the roots of these trees the children of a new root.

iii. this process is repeated until the list contains only one tree.

Birth-Death Process Model. The trees were sampled under the birth-death model using the software DendroPy version 3.10 [32] with the parameters 0.1 and 0 for the birth rate and death rate, respectively.

Experimental Setting. For the profiles P_k and Q_k ($k \in \{100, 1000\}$) generated under each of the two models, we computed the rRF distance and the CM distance for each pair p_i and q_i, where $1 \le i \le l$.

Results and Discussion. We discuss the results for each of the two models.

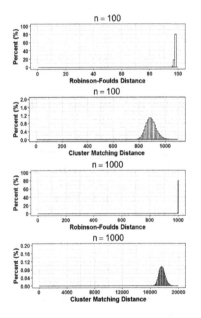

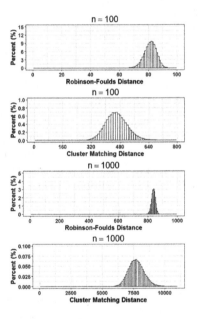

Fig. 4. Distribution of the rRF and the CM distances between a pair of randomly generated binary trees (Yule-Harding model) on 100 and 1000 leaves.

Fig. 5. Distribution of the rRF and the CM distances between a pair of randomly generated binary trees (birth-death process model) on 100 and 1000 leaves.

Yule-Harding Model. The distributions of the rRF distance and the CM distance between the pairs of randomly generated trees are depicted in Fig. 4, and the corresponding descriptive statistics are shown in Table 1. The rRF distances show left-skewed distribution for the sampled pairs of trees with 100 and 1000 leaves, and thus the range of these distributions is very narrow. In addition, also the minimum value and mean value for both of these distributions are very close to the theoretical maximum values (i.e., diameters). Furthermore, the standard deviation and the range of the rRF distances are similar for trees that have between 100 and 1000 leaves, suggesting that they are not proportional to the number of leaves.

In contrast, the CM distances are more broadly and bell-shape like distributed, which have also much wider ranges than the corresponding rRF distributions for both 100 and 1000 leaves.

Table 1. Descriptive statistics of the RF distance and the CM distance between a pair of randomly generated binary trees on 100 and 1000 leaves.

Distance	Leaves	Yule-Harding					Birth-death				
		Mean	SD	Min	Max	Range	Mean	SD	Min	Max	Range
RF	100	97.77	0.48	93	98	5	81.37	4.16	61	96	35
CM	100	891.6	38.28	760	1123	363	460.52	59.41	249	851	602
RF	1000	997.78	0.47	994	998	4	837.8	12.89	788	883	95
CM	1000	17659.27	423.3	16253	20031	3778	7705.55	623.8	5257	11327	6070

Birth-Death Process Model. Table 1 summarizes the descriptive statistics and Fig. 5 shows the distributions of the RF distance and the CM distance between a pair of randomly generated trees. Unlike the Yule-Harding model, the distributions of the RF distance and the CM distance are both in the form of a bell-shape. However, the distribution of the CM distance shows a wider range than the one of the RF distance, for both 100 and 1000 leaves.

4.2   Distance Metrics Under the Tree Edit Operations

We demonstrate how the rRF distances and the CM distances correlate with the number of consecutive rNNI, rSPR, and rTBR edit operations. From the result of Sect. 4.1, the rRF distance is expected to be saturated faster than the CM distance by repeating the number of tree edit operations.

Dataset. We generated a profile P consisting of 1000 random trees on 500 leaves, where each tree in P was sampled under the Yule-Harding model using the procedure described in Sect. 4.1. For each of the rooted tree edit operations rNNI, rSPR, and rTBR we generated the profiles $Q(i) := \{q(i)_1, \ldots, q(i)_{1000}\}$ for (i) every $i \in \{1, \ldots, 2000\}$ for the rNNI operation, and (ii) every $i \in \{1, \ldots, 500\}$ for the rSPR and rTBR operations. The profiles were generated as follows.

i. Given a tree edit operation, the initial profile $Q(1)$ is set to profile P. If this operation is rNNI the range r is set to 2000, and is set to 500 otherwise.

ii. For each $i \in \{2,\dots,r\}$ the profile $Q(i+1)$ is generated from profile $Q(i)$. The tree $q(i+1)_j$ is created by applying the input tree edit operation to tree $q(i+1)_j$ (for each $j \in \{1,\dots,1000\}$), where the selection of edges in $q(i)_j$ that is needed to specify the operation is chosen randomly. E.g., for the rSPR operation, two edges (possibly including a root edge) of tree $q(i)_j$ are randomly chosen, where the first edge determines the pruning location of the subtree, and the second edge the regrafting location of the subtree.

Experimental Setting. We computed the distances between the tree pairs $q(1)_j$ and $q(i)_j$ averaged over all $j \in \{1,\dots,1000\}$, under the rRF and the CM distance measures, for every $i \in \{1,\dots,r\}$ ($r = 2000$ for rNNI, and $r = 500$ for rSPR and rTBR) using the profiles that were generated for each of the edit operations. Similarly, the maximum of the distances between the tree pairs $q(1)_j$ and $q(i)_j$ over all $j \in \{1,\dots,1000\}$ was computed to finally compute the ration of the averages to their corresponding maximum distances ratio.

(a) rNNI operations   (b) rSPR operations   (c) rTBR operations

Fig. 6. The average RF distance and CM distance of 1000 trees on 500 leaves as a function of the number of consecutive tree edit operations.

4.3   Results and Discussion

Figure 6 shows the average RF distance and CM distance between the initial tree and rNNI operation applied trees. The gradient of the rRF distance curve is very steep between 0–1200 operations, and the inclination of the curve is gradual after 1600 operations. However, after 1600 operations the CM distance still has an increasing trend. Figure 6 shows the average RF distance and CM distance between the initial tree and rSPR operation applied trees. While the gradient of the sRF distance curve is gradual after 400 operations, but the gradient of the CM distance is in an increasing trend. Figure 6 shows the average rRF distance and CM distance between the initial tree and rTBR operation applied trees. Unlike the rSPR operation, the gradients of the RF and the CM distance curves are both steep between 0–200 operations, and they are gradual after 350 operations.

5   Conclusion

There may not be an optimal tree comparison measure, and one or more measures may be used by the practitioner depending on the application. However, such choices can be guided by the strengths and weaknesses of such measures. The CM measure induces a metric on the space of rooted trees and that this metric is low polynomial time computable. In contrast, many other distance measures used in comparative phylogenetics, are intrinsically difficult to compute, such as the edit distances under the classic tree edit operations.

We showed that the CM distance offers a variety of desirable features, in particular when compared to the popular rooted RF distance, which is poorly distributed, shows insufficient discrimination, and is too sensitive to tree edit operations. In particular, we demonstrated that the CM metric is distributed much more broadly and is less biased when compared to the rRF distance. In addition, our experiments suggest that the CM metric is less sensitive to a tree edit operation than the rRF distance. Thus, the CM distance appears to be a viable alternative to the classic RF metric for rooted trees.

Acknowledgments. This material is based upon work supported by the National Science Foundation under Grant No. 1617626.

References

1. Allen, B.L., Steel, M.: Subtree transfer operations and their induced metrics on evolutionary trees. Ann. Comb. **5**(1), 1–15 (2001)
2. Arvestad, L., et al.: Gene tree reconstruction and orthology analysis based on an integrated model for duplications and sequence evolution. In: RECOMB, pp. 326–335. ACM (2004)
3. Betkier, A., Szczesny, P., Górecki, P.: Fast algorithms for inferring gene-species associations. In: Harrison, R., Li, Y., Măndoiu, I. (eds.) ISBRA 2015. LNCS, vol. 9096, pp. 36–47. Springer, Cham (2015). https://doi.org/10.1007/978-3-319-19048-8_4
4. Bogdanowicz, D., Giaro, K.: On a matching distance between rooted phylogenetic trees. Int. J. Appl. Math. Comput. Sci. **23**(3), 669–684 (2013)
5. Bordewich, M., Semple, C.: On the computational complexity of the rooted subtree prune and regraft distance. Ann. Comb. **8**(4), 409–423 (2005)
6. Bourque, M.: Arbres de Steiner et réseaux dont varie l'emplacement de certains sommets. Ph.D. thesis, University of Montréal Montréal, Canada (1978)
7. Bryant, D.: Hunting for trees, building trees and comparing trees: theory and method in phylogenetic analysis. Ph.D. thesis, University of Canterbury, New Zealand (1997)
8. Bryant, D., Steel, M.: Computing the distribution of a tree metric. IEEE/ACM Trans. Comput. Biol. Bioinf. **6**(3), 420–426 (2009)
9. Das Gupta, B., et al.: On distances between phylogenetic trees. In: SODA 1997, pp. 427–436 (1997)
10. Day, W.H.E.: Optimal algorithms for comparing trees with labeled leaves. J. Classif. **2**(1), 7–28 (1985)
11. Felenstein, J.: Inferring Phylogenies. Sinauer, Sunderland (2003)

12. Forster, P., Renfrew, C.: Phylogenetic Methods and the Prehistory of Languages. McDonald Institute of Archeological, Cambridge (2006)
13. Harding, E.F.: The probabilities of rooted tree-shapes generated by random bifurcation. Adv. Appl. Probab. **3**(1), 44–77 (1971)
14. Harris, S.R., et al.: Whole-genome sequencing for analysis of an outbreak of meticillin-resistant staphylococcus aureus: a descriptive study. Lancet. Infect. Dis. **13**(2), 130–136 (2013)
15. Hein, J., et al.: On the complexity of comparing evolutionary trees. Discret. Appl. Math. **71**(1–3), 153–169 (1996)
16. Hickey, G., et al.: SPR distance computation for unrooted trees. Evol. Bioinform. Online **4**, 17–27 (2008)
17. Huber, K.T., Spillner, A., Suchecki, R., Moulton, V.: Metrics on multilabeled trees: interrelationships and diameter bounds. IEEE/ACM Trans. Comput. Biol. Bioinf. **8**(4), 1029–1040 (2011)
18. Hufbauer, R.A., et al.: Population structure, ploidy levels and allelopathy of Centaurea maculosa (spotted knapweed) and C. diffusa (diffuse knapweed) in North America and Eurasia. In: ISBCW, pp. 121–126. USDA Forest Service (2003)
19. Katherine, S.J.: Review paper: the shape of phylogenetic treespace. Syst. Biol. **66**(1), e83–e94 (2017)
20. Kuhner, M.K., Yamato, J.: Practical performance of tree comparison metrics. Syst. Biol. **64**(2), 205–214 (2015)
21. Li, M., Tromp, J., Zhang, L.: On the nearest neighbour interchange distance between evolutionary trees. J. Theor. Biol. **182**(4), 463–467 (1996)
22. Li, M., Zhang, L.: Twist-rotation transformations of binary trees and arithmetic expressions. J. Algorithms **32**(2), 155–166 (1999)
23. Lin, Y., Rajan, V., Moret, B.M.E.: A metric for phylogenetic trees based on matching. IEEE/ACM Trans. Comput. Biol. Bioinf. **9**(4), 1014–1022 (2012)
24. Ma, B., Li, M., Zhang, L.: From gene trees to species trees. SIAM J. Comput. **30**(3), 729–752 (2000)
25. Makarenkov, V., Leclerc, B.: Comparison of additive trees using circular orders. J. Comput. Biol. **7**(5), 731–744 (2000)
26. Nik-Zainal, S., et al.: The life history of 21 breast cancers. Cell **149**(5), 994–1007 (2012)
27. Robinson, D.F., Foulds, L.R.: Comparison of weighted labelled trees. In: Horadam, A.F., Wallis, W.D. (eds.) Combinatorial Mathematics VI. LNM, vol. 748, pp. 119–126. Springer, Heidelberg (1979). https://doi.org/10.1007/BFb0102690
28. Robinson, D.F.: Comparison of labeled trees with valency three. J. Comb. Theory Ser. B **11**(2), 105–119 (1971)
29. Robinson, D.F., Foulds, L.R.: Comparison of phylogenetic trees. Math. Biosci. **53**(1–2), 131–147 (1981)
30. Semple, C., Steel, M.A.: Phylogenetics. Oxford (2003)
31. Steel, M.A., Penny, D.: Distributions of tree comparison metrics. Syst. Biol. **42**(2), 126–141 (1993)
32. Sukumaran, J., Holder, M.T.: DendroPy: a python library for phylogenetic computing. Bioinformatics **26**(12), 1569–1571 (2010)
33. Than, C.V., Rosenberg, N.A.: Mathematical properties of the deep coalescence cost. IEEE/ACM Trans. Comput. Biol. Bioinf. **10**(1), 61–72 (2013)
34. Wilkinson, M., et al.: The shape of supertrees to come: tree shape related properties of fourteen supertree methods. Syst. Biol. **54**(3), 419–431 (2005)
35. Wu, Y.-C., et al.: TreeFix: statistically informed gene tree error correction using species trees. Syst. Biol. **62**(1), 110–120 (2013)

RNA-Seq Data Analysis

Truncated Robust Principal Component Analysis and Noise Reduction for Single Cell RNA-seq Data

Krzysztof Gogolewski[1(✉)], Maciej Sykulski[2,3], Neo Christopher Chung[1], and Anna Gambin[1]

[1] Institute of Informatics, Faculty of Mathematics, Informatics and Mechanics, University of Warsaw, Warsaw, Poland
{k.gogolewski,aniag}@mimuw.edu.pl, nchchung@gmail.com
[2] Department of Medical Genetics, Warsaw Medical University, Warsaw, Poland
macieksk@gmail.com
[3] genXone Ltd., Poznań, Poland

Abstract. The development of single cell RNA sequencing (scRNA-seq) has enabled innovative approaches to investigating mRNA abundances. In our study, we are interested in extracting the systematic patterns of scRNA-seq data in an unsupervised manner, thus we have developed two extensions of robust principal component analysis (RPCA). First, we present a truncated version of RPCA (tRPCA), that is much faster and memory efficient. Second, we introduce a noise reduction in tRPCA with L_2 regularization (tRPCAL2). Unlike RPCA that only considers a low-rank L and sparse S matrices, the proposed method can also extract a noise E matrix inherent in modern genomic data. We demonstrate its usefulness by applying our methods on the peripheral blood mononuclear cell (PBMC) scRNA-seq data. Particularly, the clustering of a low-rank L matrix showcases better classification of unlabeled single cells. Overall, the proposed variants are well-suited for high-dimensional and noisy data that are routinely generated in genomics.

Keywords: Principal component analysis · Robust PCA
Truncated singular value decomposition · Matrix decomposition
Unsupervised learning · Single cell RNA-seq

1 Introduction

Single cell RNA sequencing (scRNA-seq) present new opportunities to elucidate systematic patterns of variation underlying biological processes and complex phenotypes. Conventionally, bulk RNA-seq data provide mean gene expression values from a large number of cells in that sample. However, a mixture of multiple cells that often have different functions or origins may hide relevant information, carry high variance related to their cellular composition, and might not be

---

K. Gogolewski and M. Sykulski—These authors equally contributed to this work.

F. Zhang et al. (Eds.): ISBRA 2018, LNBI 10847, pp. 335–346, 2018.
https://doi.org/10.1007/978-3-319-94968-0_32

reproducible in separate studies [1–3]. With scRNA-seq, we can overcome these challenges by measuring gene expression at a single cell resolution [4,5]. However, scRNA-seq data present new challenges for unsupervised learning methods because of unlabeled samples, higher dimensionality, dropouts, and sparsity.

Unsupervised learning techniques have become increasingly popular and useful for exploring and analyzing scRNA-seq data. In particular, principal component analysis (PCA) is most frequently used to reduce dimensions that enable a number of downstream statistical and machine learning [6]. Furthermore, closely related to factor analysis and latent variable models, principal components (PCs) help us to identify hidden and unmeasured structure that arise from biological and technical sources of variation [7–9]. Some of biological applications include tracking Definitive Endoderm Cells (DfE) to explain their linage from Embryonic Stem Cells [10], classifying sensory neuron types [11], and identifying potentially damanged cells [12]. To account for an underlying sparse component (e.g., sparsely corrupted data or sparse latent structure), [13] proposed robust PCA (RPCA) that can decompose the observed data into low-rank and sparse components.

We build on the strength of RPCA [13] to introduce an computationally efficient truncated version and a noise reduction using L_2 regularization. In high dimensional genomic data, the systematic variation is likely contained in a small number of PCs, whereas lower-ranked PCs only contain noise. Therefore, our truncated RPCA (tRPCA) uses the top k singular vectors to estimate low-rank and sparse components. Noise reduction of scRNA-seq data was possible by introducing an error component, in addition to low-rank and sparse components that were originally introduced in [13]. Advancements of matrix decomposition have a long history, including non-negative matrix factorization [14], sparse PCA [15], penalized matrix decomposition [16], and more. Inspired by these methods, our innovation enables separation of low-rank and sparse components, while imposing a L_2 penalty on a noise term inherent in large scale genomic data.

The paper is organized as follows. In Sect. 2 *Methods*, we present two proposed methods based on RPCA, namely its truncated version and noise reduction with L_2 regularization. We provide the algorithms and their characteristics. Section 3 contains the description and processing procedures for the scRNA-seq datasets used as the case study. In Sect. 4 *Results and Discussion*, we present the main results of our analysis, as well as provide some interpretations of low-rank, sparse, and noise components. Finally, in Sect. 5. *Conclusions and Further Research*, we summarize our work and discuss the future steps concerning the proposed methods.

2   Robust PCA and Extensions

Robust PCA. In our work we start from the decomposition algorithm proposed by Candes et al. [13] called robust PCA (RPCA). The aim of the RPCA is to decompose the input matrix M, into low-rank matrix L and sparse matrix S

components. Simultaneously, the algorithm should minimize the following optimization problem:

$$\min_{L,S} ||L||_* + \lambda_1 ||S||_1, \text{ where } M = L + S$$

Here we denote $||A||_*$ as the nuclear norm of matrix A and $||A||_1$ as the first norm of a vectorized A matrix which are given by the following formulas:

$$||A||_* = \sum \sigma_i = \text{tr}\left(\sqrt{AA^T}\right), \text{ and } ||A||_1 = \sum_{i,j} |a_{ij}|$$

In their work, authors discuss the assumptions that matrix M should follow for the decomposition to exist. Moreover they prove that the parameter λ_1 can be set to $1/\sqrt{\min(m,n)}$, where m, n are dimensions of the input matrix M, which guarantees proper decomposition into low-rank and sparse components as $m, n \to \infty$ under weak probabilistic assumptions, however, the spectrum of feasible values of λ_1 parameters is broad.

In order to solve the above problem, as proposed in [17], we use an implementation of a special case of the Alternating Directions method, which belongs to a more general class of augmented Lagrangian (AGL) multiplier algorithms. The approach is based on minimizing the following AGL operator with respect to L and S matrix alternately:

$$l(L, S, Y) = ||L||_* + \lambda_1 ||S||_1 + \langle Y, M - L - S \rangle + \frac{\mu}{2} ||M - L - S||_F^2$$

where Y is the Lagrange multiplier matrix, the inner product of matrices $\langle \cdot, \cdot \rangle$ is defined as the trace of their product, i.e. $\langle A, B \rangle = \text{tr}(AB^T)$, $||A||_F$ is the Forbenius norm of the form $||A||_F = \sqrt{\sum_{i,j} a_{i,j}^2}$ and μ is the penalty coefficient.

The outline of the solution is presented in the Algorithm 1, in which we use two shrinkage operators $\mathcal{S}_\tau(x) = \text{sgn}(x) \cdot \max(|x| - \tau, 0)$ and $\mathcal{D}_\tau = U\mathcal{S}_\tau(\Sigma)V^*$, where $U\Sigma V^*$ is a SVD of X and τ is the shrinkage threshold value. In case of initialization of the μ parameter and convergence condition, we set $\mu = \frac{m \cdot n}{4 \cdot ||M||_1}$, as suggested in [17] and terminate the algorithm when $||M - L - S||_F \leq \delta ||M||_F$ and $\delta = 10^{-7}$. The base implementation of the algorithm that we have extended in this work is publicly available as a R package [18] (https://cran.r-project.org/web/packages/rpca/).

Truncated Version of RPCA. First, we consider a truncated version of the algorithm, which calculates the L matrix in the $L + S$ decomposition in such a way that it is of a given rank k or the lowest possible rank grater than k_0, for which the problem has a solution that meet all its criteria. In order to achieve that behavior we use the truncated version of SVD (implementation from the irlba R package [19]) instead of a full SVD and iteratively modify the μ parameter with

$$\mu_{i+1}^{-1} = \max(c \cdot \mu_i^{-1}, \sigma_{k+1})$$

Algorithm 1. RPCA by Alternating Directions

1: **procedure** RPCA(λ_1)
2:      $S_0, Y_0 \leftarrow 0; \mu > 0$
3:      **while** not converged **do**
4:          compute $L_{i+1} = \mathcal{D}_{\mu^{-1}}(M - S_i + \mu^{-1}Y_i)$
5:          compute $S_{i+1} = \mathcal{S}_{\lambda_1\mu^{-1}}(M - L_{i+1} + \mu^{-1}Y_i)$
6:          compute $Y_{i+1} = Y_i + \mu \cdot (M - L_{i+1} - S_{i+1})$

Algorithm 2. truncated-RPCA

1: **procedure** TRPCA(λ_1, k_0, c)
2:      $S_0, Y_0 = 0; \mu_0 > 0; k = k_0$
3:      **while** not converged **do**
4:          compute $L_{i+1} = \mathcal{D}_{\mu_i^{-1}}(M - S_i + \mu_i^{-1}Y_i)$
5:          compute $S_{i+1} = \mathcal{S}_{\lambda_1\mu_i^{-1}}(M - L_{i+1} + \mu_i^{-1}Y_i)$
6:          compute $Y_{i+1} = Y_i + \mu_i \cdot (M - L_{i+1} - S_{i+1})$
7:          compute $\mu_{i+1}^{-1} = \max(c \cdot \mu_i^{-1}, \sigma_{k+1})$
8:          **if** $\mu_i^{-1} == \sigma_{k+1}$ **then** increase k

---

where σ_k is the k-th singular value from the truncated SVD and $c < 1$ is the AGL constraints penalty growth rate.

As μ_i^{-1} decreases the penalty coefficient for $M = L + S$ constraints increases, which also speeds up the algorithm's convergence, however, in theory, AGL algorithm converges to the constraint problem even when $\mu_i \nrightarrow \infty$. Simultaneously, when μ_{i+1}^{-1} is set to the value of σ_{k+1} we increase k, i.e. the number of computed SVD vectors, which is at same time the approximate expected rank of L matrix.

The above modification (see Algorithm 2) reduces significantly the computation time of the algorithm compared to the original RPCA preserving its accuracy. However, in the case of real data (e.g. biomedical) the decomposition into low-rank and sparse matrices is not always feasible or easily obtainable. The data matrix usually has significant singular values (below the first k low-rank important ones) that may come from biological activity, technical reasons, or other unknown sources. These prevent the recovery of low-rank component as when subtracted from M they do not constitute a sparse matrix. We interpret these perturbations in the L matrix as a noise or low-importance information. Since it does not have a sparse nature we extend the decomposition into $L + S + E$, where the matrix E contains a dense noise controlled for using the L_2 norm on vectorized matrix M (i.e. Frobenius norm).

Noise Reduction. In order to relax the assumptions on the input matrix we introduce the E matrix to the decomposition. Now, the decomposition problem can be reformulated as follows:

$$M = L + S + E$$
$$\min_{L,S,E} ||L||_* + \lambda_1||S||_1 + \lambda_2||E||_F$$

The E matrix is meant to contain the information of low importance or noise, which is carried by the lowest singular values in the SVD of L matrix. To solve this problem we extend the Alternating Directions approach and we minimize the following AGL operator with respect to the E matrix:

$$l(L, S, E, Y) = \quad ||L||_* + \lambda_1||S||_1 + \lambda_2||E||_F$$
$$+ \langle Y, M - L - S - E \rangle + \frac{\mu}{2}||M - L - S - E||_F^2$$

Solving $\frac{\partial l}{\partial E} = 0$ results in:

$$E\left(\frac{\lambda_2}{||E||_F} + \mu\right) = Y + \mu(M - L - S)$$

Let $C = Y + \mu(M - L - S)$, then $\exists_{d \in \mathbb{R}} E = d \cdot C$. Assuming that $C \neq 0$ we determine the value of d. Since $d < 0$ results in a contradiction, we assume that $d \geq 0$ we have:

$$d = \frac{||C||_F - \lambda_2}{\mu||C||_F} = \frac{1}{\mu}\left(1 - \frac{\lambda_2}{||C||_F}\right) \geq 0$$

which holds for $||C||_F \geq \lambda_2$. We define the operator:

$$\mathcal{E}_\tau(X) = \max\left(0, 1 - \frac{\tau}{||X||_F}\right) \cdot X$$

which describes how to determine the matrix E which minimizes l.

Finally, we extend the algorithm of truncated-PCA by applying the defined operator $\mathcal{E}_\tau$. In our approach we apply the operator twice, both, after minimization with respect to the L, and S matrix. It is worth to emphasize, that in the case of large $\lambda_2 > ||C||_F$ we end up with the previously introduced truncated-RPCA procedure. Moreover, in every iteration we adjust k parameter to be of minimal value such that $\mathcal{D}_{\mu^{-1}}$ operator can be properly applied. Algorithm 3 presents the pseudo-code of the whole decomposition procedure.

3   Single Cell Transcriptomic Data

In this study we use the publicly available scRNA-seq datasets provided by the 10x Genomics company (https://www.10xgenomics.com/solutions/single-cell/). Specifically, our results, that are presented in the next section, were obtained using the scRNA-seq datasets experiments performed on peripheral blood mononuclear cells (PBMCs) from a healthy donor. PBMCs are primary cells with relatively small amounts of RNA (1pg RNA/cell). The final dataset contains 2.7k individual single cells, sequenced on Illumina NextSeq 500 with approx. 69k reads per cell. Amplification was performed on 98bp read1 (transcript), 8bp I5 sample barcode, 14bp I7 GemCode barcode and 10bp read2 (UMI).

Algorithm 3. truncated-RPCA with L2 regularization

1: **procedure** TRPCAL2(λ_1, λ_2, k_0, c)
2:     $S_0, Y_0, E_0 = 0$; $\mu_0 > 0$; $k = k_0$
3:     **while** not converged **do**
4:         compute $L_{i+1} = \mathcal{D}_{\mu^{-1}}(M - S_i - E_i + \mu^{-1}Y_i)$
5:         compute $E_{i+1}^* = \mathcal{E}_{\lambda_2\mu^{-1}}(M - S_i - L_{i+1} + \mu^{-1}Y_i)$
6:         compute $S_{i+1} = \mathcal{S}_{\lambda_1\mu^{-1}}(M - E_{i+1}^* - L_{i+1} + \mu^{-1}Y_i)$
7:         compute $E_{i+1} = \mathcal{E}_{\lambda_2\mu^{-1}}(M - S_{i+1} - L_{i+1} + \mu^{-1}Y_i)$
8:         compute $Y_{i+1} = Y_i + \mu \cdot (M - E_{i+1} - L_{i+1} - S_{i+1})$
9:         compute $\mu_{i+1}^{-1} = \max(c \cdot \mu_i^{-1}, \sigma_{k+1})$
10:        **if** $\mu_i^{-1} == \sigma_{k+1}$ **then** increase k
11:        **else** $k = 1 + \text{argmax}_j (\sigma_j > \mu_{i+1}^{-1})$

Along with the 2.7k PBMCs dataset, we have used the scRNA-seq data retrieved from homogeneous samples of specific cell types that constitute the PBMC sample. Each type-specific dataset has over 90% of purity for each subtype by Fluorescence Activated Cell Sorting (FACS) [20]. The transcriptomes were used in [21] and described the following types and subtypes: CD14$^+$ Monocytes, CD56$^+$ Natural Killer cells, CD19$^+$ B cells, CD34$^+$ cells and subfamilies of T cells: CD8$^+$ Cytotoxic T cells, CD8$^+$/CD45RA$^+$ Naive Cytotoxic T cells, CD4$^+$/CD45RO$^+$ Memory T cells, CD4$^+$ Helper T cells.

Each of the above datasets is given in the form of a counts matrix A i.e. i-th row represents a gene and j-th column represents an individual cell. The value of a_{ij} is the number of counts of the i-th gene for the j-th cell. Since our method is meant to filter out the sparse signal in S and the dense noise in E we do not apply the typical quality control step. All cells are used in the analysis and we expect all perturbations that break the linear behavior (e.g. biological or technical outliers or fluctuations) to remain in $S + E$ component of the decomposition. Additionally, for each dataset we filter out genes that had zero number of counts for all cells in a given set. Finally, the number of counts for each cell was normalized by its total number of counts and log-scaled. Further on we denote the processed 2.7k PBMCs data matrix by M.

Test Set Construction. In order to test our method we first set the labeling of cells from the PBMCs dataset. For each available type-specific dataset we calculate its average transcriptome. However, since the correlation between averaged subtype-specific transcriptomes within T-cell family is relatively high, for the purpose of this work, we label the cells with one of the five possible types: (i) Monocytes, (ii) Natural Killers, (iii) B cells, (iv) T cells, (v) Unknown. T cells family transcriptome is designated as an average among all T cells subtypes transcriptomes.

The criterion for labeling consists of two conditions. First, a cell is assumed to be of an unknown type if it does not correlate with any of the given profiles at least at the level of 0.5. Next, the cell is assumed to be of a specific type if the separation between its correlation and correlations with other types is statistically significant (p-value < 0.05) otherwise it is assumed to be unknown.

Even though there are no transcriptomes available for other cell types, e.g. Megakaryocytes we are aware that they may also exist in our dataset and thus expect to find them using our decomposition method.

4   Results and Discussion

By definition our final extension of RPCA explains the input matrix data (M) in terms of compressed, low-rank information (L), sparse signal (S) and noise (E). In order to validate our method on real data and evaluate its suitability for biomedical data analysis, we investigate the scRNA-seq 2.7k PBMCs data. We report, that the tRPCAL2 algorithm converged after 49 iterations, exec. time: 97 s (compared to 20 s PCA from R prcomp). As expected, due to the high background variance tRPCA and RPCA did not converge before 1000 iterations.

Clustering via Low-Rank Matrix. First we validate the quality of the dimension reduction by clustering cells basing on their low-rank representation kept in the L matrix. Using the unsupervised, hierarchical clustering algorithm we determined 5 clusters, which we visualized using t-SNE approach [22] (see Fig. 1). In contrast with expected cell types (derived from correlation with

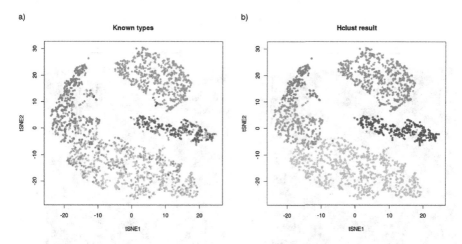

Fig. 1. Clustering of 2.7k PBMCs. In both panels, cells are visualized using t-SNE dimension reduction algorithm (perplexity $= 35$) run on the 10-dimensional representation of the original data (M) derived from L matrix. (a) Colors correspond to the cell types inferred from the correlation of each cell original transcriptome (columns of M) with type-specific PBMCs transcriptomes. We have determined: 630 Monocytes (orange), 251 B-cells (pink), 437 Natural Killer cells (blue) and 700 T-cells (yellow). Remaining 682 (gray) we assume to be of an unknown or tentative type. (b) Colors correspond to 5 clusters determined by hierarchical clustering method. Colors of the clusters are set such that the correspondence between predicted and original clusters is noticeable. Additionally, increased activity of CD8 in the gray cluster suggests that it can be mostly composed of cytotoxic T-cells. (Color figure online)

type-characteristic transcriptomes) we observe that the obtained clustering well determines all 4 main families of cells from PBMCs. Additionally, one more cluster separating NK and T cell family clusters was discovered. The cluster is described by increased activity of CD8A and CD8B (Bonferroni adjusted p-value $< 10^{-3}$) and regular activity of CD4, CD45 and CD25 genes in contrast to other cells, which suggests a cluster of mostly $CD8^+$ T cytotoxic cells and explains its similarity to NK cells [21,23].

Next, we compared our way of dimension reduction with the method analogous to the one used in [21]. With SVD we calculate top 10 singular values (in pursuance of the L matrix rank) of the PBMC data matrix (M) using R irlba package. Then, we approximate the original data through the reduced 10-dimensional space. We perform the hierarchical clustering of all cells on the most characteristic marker genes per cell type (selected from the literature) from the described and the L matrices. The aim is to verify how well the dimensionality reduction preserved the most reliable, biological information related to type-specific marker genes. It appeared that not only the L matrix guarantees more accurate clustering, but also it contains more pronounced differences of the signal between clusters of both cells and genes, c.f. Fig. 2.

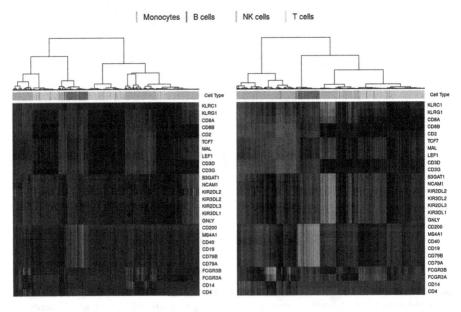

Fig. 2. Marker gene based clustering comparison. The figure compares clustering of cells of known type with literature-based marker genes characterizing the analyzed types of PBMC cells. The left panel is related to the signal represented in terms of the truncated PCA (10 highest singular values used). The right panel corresponds to the signal stored in the L matrix of the $L + S + E$ decomposition. Top bars encode the original correlation-inferred cell types. Colors in the heatmap describe the activity level of a gene from lowest (red) through average (black) up to highest (green). (Color figure online)

Monocyte Subtypes and Co-expression Detection. The literature suggests existence of at least three subtypes of monocytes in PBMCs [24]. Their characterization can be based on the presence of CD14 (coded by CD14 gene) and CD16 (coded by FCGR3A, FCGR3B genes) clusters of differentiation: (i) the classical monocyte with high activity of CD14 (CD14^{++} FCGR3A^{-}); (ii) the intermediate monocyte with high activity of CD14 and low activity of FCGR3A (CD14^{++}FCGR3A^{+}); (iii) the non-classical monocyte with low activity of CD14 and co-expressed FCGR3A (CD14^{+}FCGR3A^{++}).

Interestingly, such classification of subtypes can be found using the low rank signal from the L matrix (see Fig. 3). The activity of CD14 is almost uniquely distributed among cluster of monocyte cells and, simultaneously, the activity of FCGR3A changes with the gradient defining the cell subtype progression among all monocytes. Moreover, the Fig. 3c shows how the original expression values are distributed among decomposition matrices. The sparse peaks of activity are stored in S and the linear part in L. Finally, E contains remaining noise of mean 0 and the standard deviation of order 10^{-4} for both CD14 and FCGR3A.

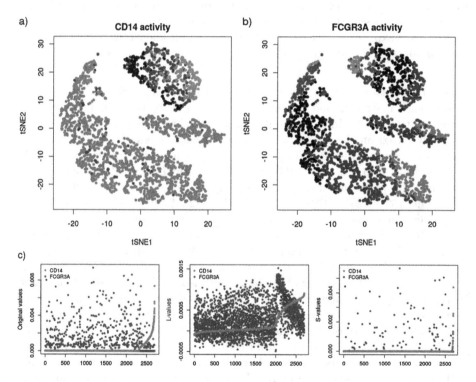

Fig. 3. CD14 and FCGR3A activity levels. Panels present the activity of monocytes marker genes. (a) and (b) Figures present the activity of CD14 and FCGR3A genes among all cells, respectively. The level of activity (low to high) is spanned on the red to green color scale. (c) Consecutive panels present: (i) the log-transformed data from M; (ii) low-rank signal in L matrix; (iii) sparse signal in S matrix. In each panel cells (x-axis) are sorted by the activity level of CD14 (y-axis). (Color figure online)

The other interesting property of the L matrix is the fact that it recovers co-expression patterns between genes. Namely, the activity of B cells can be detected e.g. by the presence of CD79 heterodimer composed of CD79A and CD79B proteins [25]. Their co-expression measured in terms of correlation was at the level of 0.227, while after the decomposition their low-rank signal had correlation of level 0.995. Similarly, the correlation between FCGR3A and GNLY characterizing Natural Killer cells increased from 0.400 to 0.949. Naturally, these observations result from filtering out the sparse and noise signals. Nonetheless, it is worth to emphasize that this type of information is retrieved by the method, because it can help in suggesting new co-expression patterns.

Sparse Signal Interpretation. The presence of megakaryocytes in our PBMC dataset, that was reported in the population of PBMCs sample from [21], was not evident using the low-rank L matrix. Even though, a small cluster of cells of unknown type was separated by t-SNE method (see Fig. 1) it was not straightforward conclusion from clustering results. However, with the hierarchical clustering performed on the subset of unknown type cells and genes that had at least one non-zero entry form the sparse S matrix we recover well-separated cluster of 9 cells. Further analysis confirmed that the cluster is characterized by high over-expression of PF4 gene, which is a well known marker for mature megakaryocytes [26], in comparison to other unknown cell types.

Noise Reduction Level. Finally, we want to discuss briefly the importance of the noise matrix E and setting of both λ_1 and λ_2 parameters. The final decomposition quality in terms of information distribution among three matrices is mainly based on the choice of these crucial parameters. For the purpose of this study, we performed a grid-based search through the parameter space which resulted in $\lambda_1 = 0.016$, $\lambda_2 = 10.0$ and consequently $L + S + E$ decomposition with the following norms of the (vectorized) matrices: $||\cdot||_* : 5.753, 60.289, 57.881$; $||\cdot||_1 : 3398.162, 60.289, 2670.012$; $||\cdot||_2 : 4.265, 2.826, 1.440$ (L, S, E respectively).

To determine the order of magnitude for both parameters we have made use of the theory described in [13] as well as estimations based on the properties of the MM^T matrix trace operator. Because tRPCAL2 algorithm mixes L_1 and L_2 norms, and because of their nature, final decomposition depends not only on relative or absolute values of lambdas, but also on distributions of elements in the decomposed matrices. We investigated the influence of λ_1 and λ_2 on decomposition properties such as: the rank of the resulting L matrix, relative and absolute sparsity of the S matrix and the variance of the noise level in E matrix. We report few observations on PBMCs data: (i) relative sparsity of S (i.e. proportion between sparsity of S and original matrices) decreases exponentially with respect to $\frac{\lambda_1}{\lambda_2}$; the rank of matrix L increases; (ii) sub-linearly as a function of λ_1 and fixed λ_2; (iii) polynomially as a function of λ_2 and fixed λ_1.

Finally, in theory it seems intuitive to expect elements of the E matrix to be normally distributed with 0 mean. However, the computational experiments showed, that this distribution is a mixture of the zero-centered Gaussian and another low-variance Gaussian concentrated around non-negative,

$lambda_2$-dependent value. We assume that this is a linear dependency, nonetheless its precise description is still unexplored.

More formal investigation of theoretical properties of the tRPCAL2 decomposition with respect to lambdas and L, S, E matrices could be of high interest in terms of future research.

5 Conclusions and Further Research

Concluding, in this paper we introduce an extension of the robust PCA matrix decomposition method. We propose a $L + S + E$ decomposition of the matrix into: low-rank L, sparse S and noise E components. Thanks to the reduction of L_2 moderated noise we restore the inner structure of the matrix, which approximates the original data with high accuracy, as well as recognize the sparse perturbation signal of the data. We present the case study based on the scRNA-seq data from 2.7k PBMCs. The method provides relatively fast and accurate dimension reduction and clustering of the high-dimensional data detecting different subtypes within a given cell type, co-expression patterns and novel subtypes.

One possible direction for the further research is to derive precise formulas for λ_1 and λ_2 parameters that guarantee optimal solutions of the decomposition problem. So far, simulation-based selection of the parameters is time consuming. Ideally, a lambda parameter selection method would result with the most *natural* $L + S + E$ decomposition, taking into account user's expectations in terms of, for example, Bayesian priors to relative magnitudes, and to other components' statistics. This and the applicability of our method to other types of data we see as a promising direction of research. Preliminary results of video and image analysis, not described in this paper, suggest that the method can be successfully harnessed in the field of video-surveillance and image analysis. The most recent implementation of the tRPCAL2 algorithm as an R package is available under https://github.com/macieksk/rpca.

Acknowledgments. This work was supported by the Polish National Science Centre grant no. 2016/21/N/ST6/01507 and no. 2016/23/D/ST6/03613. The authors thank B. Miasojedow, Ph.D. for comments and suggestions.

References

1. Novelli, G., Ciccacci, C., Borgiani, P., Amati, M.P., Abadie, E.: Genetic tests and genomic biomarkers: regulation, qualification and validation. Clin. Cases Miner. Bone Metab. **5**(2), 149–154 (2008)
2. Wills, Q.F., et al.: Single-cell gene expression analysis reveals genetic associations masked in whole-tissue experiments. Nat. Biotechnol. **31**(8), 748–752 (2013)
3. Gogolewski, K., Wronowska, W., Lech, A., Lesyng, B., Gambin, A.: Inferring molecular processes heterogeneity from transcriptional data. Biomed Res. Int. **2017**, 14 p. (2017). https://doi.org/10.1155/2017/6961786. Article no. 6961786
4. Wang, Y., Navin, N.E.: Advances and applications of single-cell sequencing technologies. Mol. Cell **58**(4), 598–609 (2015)

5. Ramskold, D., et al.: Full-length mRNA-Seq from single-cell levels of RNA and individual circulating tumor cells. Nat. Biotechnol. **30**(8), 777–782 (2012)
6. Jolliffe, I.T.: Principal Component Analysis. Springer, Heidelberg (2002). https://doi.org/10.1007/b98835
7. Bartholomew, D.J., Knott, M., Moustaki, I.: Latent Variable Models and Factor Analysis: A Unified Approach. Wiley Series in Probability and Statistics (2011)
8. Chung, N.C., Storey, J.D.: Statistical significance of variables driving systematic variation in high-dimensional data. Bioinformatics **31**(4), 545–554 (2015)
9. Leek, J.T.: Asymptotic conditional singular value decomposition for high-dimensional genomic data. Biometrics **67**, 344–352 (2010)
10. Chu, L.F., et al.: Single-cell RNA-seq reveals novel regulators of human embryonic stem cell differentiation to definitive endoderm. Genome Biol. **17**(1), 173 (2016)
11. Usoskin, D., et al.: Unbiased classification of sensory neuron types by large-scale single-cell RNA sequencing. Nat. Neurosci. **18**(1), 145–153 (2015)
12. Ilicic, T., et al.: Classification of low quality cells from single-cell RNA-seq data. Genome Biol. **17**, 29 (2016)
13. Candès, E.J., Li, X., Ma, Y., Wright, J.: Robust principal component analysis? J. ACM **58**(3), 11:1–11:37 (2011)
14. Lee, D.D., Seung, H.S.: Learning the parts of objects by non-negative matrix factorization. Nature **401**(6755), 788–791 (1999)
15. Zou, H., Hastie, T., Tibshirani, R.: Sparse principal component analysis. JCGS **15**(2), 262–286 (2006)
16. Witten, D.M., Tibshirani, R., Hastie, T.: A penalized matrix decomposition, with applications to sparse principal components and canonical correlation analysis. Biostatistics **10**(3), 515–534 (2009)
17. Yuan, X., Yang, J.: Sparse and Low-Rank Matrix Decomposition Via Alternating Direction Methods. optimization-online.org (2009)
18. Sykulski, M.: RPCA: RobustPCA: Decompose a Matrix into Low-Rank and Sparse Components (2015). R package version 0.2.3
19. Baglama, J., Reichel, L., Lewis, B.W.: irlba: Fast Truncated Singular Value Decomposition and Principal Components Analysis for Large Dense and Sparse Matrices (2018). R package version 2.3.2
20. Basu, S., Campbell, H.M., Dittel, B.N., Ray, A.: Purification of specific cell population by fluorescence activated cell sorting (FACS). J. Vis. Exp. **10**(41) (2010)
21. Zheng, G.X., et al.: Massively parallel digital transcriptional profiling of single cells. Nat. Commun. **8**, 14049 (2017)
22. van der Maaten, L.: Accelerating t-SNE using tree-based algorithms. J. Mach. Learn. Res. **15**, 3221–3245 (2014)
23. Ohkawa, T., et al.: Systematic characterization of human CD8+ T cells with natural killer cell markers in comparison with natural killer cells and normal CD8+ T cells. Immunology **103**(3), 281–290 (2001)
24. Ziegler-Heitbrock, L., et al.: Nomenclature of monocytes and dendritic cells in blood. Blood **116**(16), 74–80 (2010)
25. Chu, P.G., Arber, D.A.: CD79: a review. Appl. Immunohistochem. Mol. Morphol. **9**(2), 97–106 (2001)
26. Adachi, M., Ryo, R., Sato, T., Yamaguchi, N.: Platelet factor 4 gene expression in a human megakaryocytic leukemia cell line (CMK) and its differentiated subclone (CMK11-5). Exp. Hematol. **19**(9), 923–927 (1991)

Locality Sensitive Imputation
for Single-Cell RNA-Seq Data

Marmar Moussa$^{(\boxtimes)}$ and Ion I. Măndoiu

Computer Science and Engineering Department,
University of Connecticut, Storrs, CT, USA
{marmar.moussa,ion}@engr.uconn.edu

Abstract. One of the most notable challenges in single cell RNA-Seq
data analysis is the so called drop-out effect, where only a fraction of the
transcriptome of each cell is captured. The random nature of drop-outs,
however, makes it possible to consider imputation methods as means of
correcting for drop-outs. In this paper we study some existing scRNA-Seq
imputation methods and propose a novel iterative imputation approach
based on efficiently computing highly similar cells. We then present the
results of a comprehensive assessment of existing and proposed methods
on real scRNA-Seq datasets with varying per cell sequencing depth.

Keywords: Single cell RNA-Seq · Imputation

1 Introduction

Emerging single cell RNA sequencing (scRNA-Seq) technologies enable the analysis of transcriptional profiles at single cell resolution, bringing new insights into
tissue heterogeneity, cell differentiation, cell type identification and many other
applications. The scRNA-Seq technologies, however, suffer from several sources
of significant technical and biological noise, that need to be addressed differently
than in bulk RNA-Seq.

One of the most notable challenges is the so called drop-out effect. Whether
occurring because of inefficient mRNA capture, or naturally due to low number
of RNA transcripts and the stochastic nature of gene expression, the result is
capturing only a fraction of the transcriptome of each cell and hence data that
has a high degree of sparsity. The drop-outs typically do not affect the highly
expressed genes but may affect biologically interesting genes expressed at low
levels such as transcription factors. Combining cells as a measure to compensate
for the drop-out effects could be defeating the purpose of performing single cell
RNA-Seq. In this paper we take advantage of the random nature of drop-outs and
develop imputation methods for scRNA-Seq. In next section we briefly discuss
some existing scRNA-Seq imputation methods and propose a novel iterative
imputation approach based on efficiently computing highly similar cells. We then
present the results of a comprehensive assessment of the existing and proposed
methods on real scRNA-Seq datasets with varying sequencing depth.

© Springer International Publishing AG, part of Springer Nature 2018
F. Zhang et al. (Eds.): ISBRA 2018, LNBI 10847, pp. 347–360, 2018.
https://doi.org/10.1007/978-3-319-94968-0_33

2 Methods

2.1 Existing Single Cell RNA-Seq Imputation Methods

DrImpute [4]. The DrImpute R package implements imputation for scRNA-Seq based on clustering the data. First DrImpute computes the distance between cells using Spearman and Pearson correlations, then it performs cell clustering based on each distance matrix, followed by imputing zero values multiple times based on the resulting clusters, and finally averaging the imputation results to produce a final value for the drop-outs.

scImpute [8]. The scImpute R package makes the assumption that most genes have a bimodal expression pattern that can be described by a mixture model with two components. The first component is a Gamma distribution used to account for the drop-outs, while the second component is a Normal distribution to represent the actual gene expression levels. Thus, in [8], the expression level of gene i is considered a random variable with density function $f_{X_i}(x) = \lambda_i Gamma(x; \alpha_i; \beta_i) + (1 - \lambda_i) Normal(x; \mu_i; \sigma_i)$, where λ_i is the drop-out rate of gene i, α_i and β_i are shape and rate parameters of its Gamma distribution component, and μ_i and σ_i are the mean and standard deviation of its Normal distribution component. The parameters in the mixture model are estimated using Expectation-Maximization (EM). The authors' intuition behind this mixture model is that if a gene has high expression and low variation in the majority of cells, then a zero count is more likely to be a drop-out value than when the opposite occurs, i.e., when a gene has constantly low expression or medium expression with high variation, then a zero count reflects real biological variability. According to [8] this model does not assume an empirical relationship between drop-out rates and mean expression levels and thus allows for more flexibility in model estimation.

KNNImpute [16]. Weighted K-nearest neighbors (KNNimpute), a method originally developed for microarray data, selects genes with expression profiles similar to the gene of interest to impute missing values. For instance, consider a gene A that has a missing value in cell 1, KNN will find K other genes which have a value present in cell 1, with expression most similar to A in cells $2 - N$, where N is the total number of cells. A weighted average of values in cell 1 for the K genes closest in Euclidean distance is then used as an estimate for the missing value for gene A.

There are also some methods for clustering with implicit imputation, like BISCUIT [1,13] and CIDR [9]. These however are out of scope of this paper, as we are focusing on stand-alone imputation methods yielding imputed gene expression profiles that can be used for downstream analyses beyond unsupervised clustering, like dimensionality reduction, counting cells that express known markers, and differential gene expression analysis.

2.2   Proposed Method: Locality Sensitive Imputation (LSImpute)

We propose a novel algorithm that uses similarity between cells to infer missing values in an iterative approach. The algorithm summary is as follows:

Step 1. Given a set S of n cells (represented by their scRNA-Seq gene expression profiles), start by selecting pairs of cells with highest similarity level until at least m_{min} distinct cells ($m_{min} = 6$ in our implementation) are selected or the highest pair similarity drops below a given threshold. This process guarantees that each selected cell has highest pairwise similarity level to at least one other selected cell.[1]

Step 2. Cluster the m cells selected in Step 1 using a suitable clustering algorithm (our implementation uses spherical K-means with $k = \sqrt{m}$). The clusters formed in this step are expected to be "tight", with each selected cell having high similarity to the other cells in its cluster.

Step 3. For each of the clusters identified in step 2, replace zero values for each gene j with values imputed based on the expression levels of gene j in all the cells within the cluster.

Step 4. The selected cells now have imputed values and the clusters they form are collapsed into their respective centroids. The centroids are pooled together with unselected cells to form a new set S, and the process is repeated starting again at Step 1.

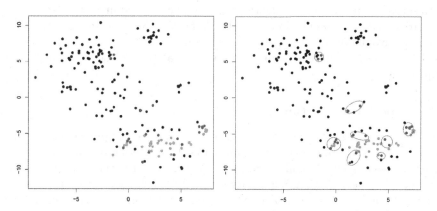

Fig. 1. Illustration of Steps 1 (left) and 2 (right) of LSImpute. Gray dots represent already processed cells and collapsed centroids from previous iterations. Pink dots represent cells in pairs with highest similarity level which are selected for clustering. (Color figure online)

---

[1] Note that, unlike KNN, which uses similarity between genes, LSImpute uses similarity between cells. Also, the number of nearest cells used for imputation is not fixed but depends on the minimum similarity threshold.

Note that, naturally, in Step 3 expression levels are imputed only for original cells and not for centroids but centroid expression levels are used in the imputation process if they are selected in Step 1. The expression levels used to replace the zero expression values can be inferred via different models. In Sect. 3 we give results for two simple approaches, namely using the mean, respectively the median of all expression values for gene j in cells belonging to the cluster (these variants are referred to as *LSImputeMean*, respectively *LSImputeMed* in Sect. 3). Using the median of both zero and non-zero values first, decides implicitly whether a zero is a drop-out event or a true biological effect, and prevents large but isolated expression values from driving imputation of nearby zeros, while collapsing into centroids in each iteration limits the propagation of potential imputation errors. Figure 1 illustrates the first two steps of the algorithm.

The worst case number of iterations taken by the algorithm is $O(n)$ as the total number of remaining cells and centroids starts at n and decreases by at least one in each iteration. In practice the number of iterations is much smaller. Our current implementation has two options for finding the pairs of cells with highest similarity level in Step 1. The first option is to use Cosine similarity and the $O(n \log n)$ algorithm of [3]. Alternatively, this could be done in $O(n)$ time using Jaccard similarity and Locality Sensitive Hashing [6]. Both similarity metrics are available in the Shiny app available at http://cnv1.engr.uconn.edu: 3838/LSImpute/, where the user can also adjust the minimum similarity threshold used in Step 1. It is recommended however to use a high similarity threshold, which will restrict the imputation to only highly similar cells as a way of being conservative with imputation to avoid the risk of over-imputation. A low similarity threshold can lead to imputing more values and can be used when the data set is of particularly low depths. All results presented in Sect. 3 use Cosine similarity and a minimum similarity threshold of 0.85 for all sets regardless of depth to avoid over-fitting. Using Jaccard similarity based on the R package LSHR [15] resulted in similar imputation levels as the Cosine similarity based implementation.

2.3   Experimental Setup

Data Sets. To assess the performance of the compared imputation methods, we used multiple evaluation metrics on data sets consisting of real scRNA-Seq reads down-sampled to simulate varying sequencing depths per cell. Specifically, we used ultra-deep scRNA-Seq data generated for 209 somatosensory neurons isolated from the mouse dorsal root ganglion (DRG) and described in [7]. An average of 31.5M 2×100 read pairs were sequenced for each cell, leading to the detection of an average of $10,950 \pm 1,218$ genes per cell. To simulate varying levels of drop-out effects we down-sampled the full dataset to 50K, 100K, 200K, 300K, 400K, 500K, 1M, 5M, 10M, respectively 20M read pairs per cell. At each sequencing depth *transcript per million (TPM)* gene expression values were estimated for each neuron using the IsoEM2 package [10]. As ground truth we used TPM values determined by running IsoEM2 on the full set of reads. For clustering accuracy evaluation, we used as ground truth the cluster assignment

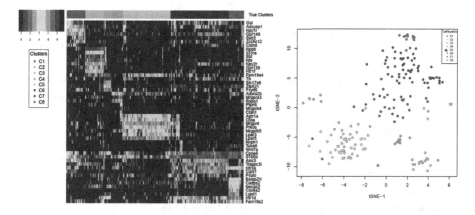

Fig. 2. Heatmap of log-transformed TPM values of marker genes identified for DRG neurons in [7] (left) and t-SNE plot showing the 8 clusters from [7] (right).

from [7], focusing on the 8 cell populations identified using scRNA-Seq data and not its refinement based on neuron sizes (see Fig. 2). The C1-C8 clusters we use in this paper correspond to the following cell populations identified by their most prominent marker genes as indicated by [7]: C1: Gal; C2: Nppb; C3: Th; C4: Mrgpra3 & Mrgprb4; C5:Mrgprd-high; C6:Mrgprd-low & S100b-high; C7: S100b-low; C8: Ntrk2 & S100b-high.

Evaluation Metrics. We used the following metrics to evaluate the imputation methods' performance at different sequencing depths:

- **Detection fraction accuracy.** A common application of single cell analyses is to estimate the percentage of cells expressing a given marker gene, for instance $CD4+$ or $CD8+$ tumor infiltrating lymphocytes [2]. A gene is considered to be detected in a cell if the (imputed or ground truth) TPM is positive. For each imputation method, the detection fraction is defined as the number of cells in which the cell is detected divided by the total number of cells. This was compared to the 'true' detection ratio, defined based on ground truth TPM values.
- **Median percent error (MPE).** As defined in [12], the *Median Percentage Error (MPE)* is the median of the set of relative errors for the gene metric examined, in this case the detection fraction. If a gene has predicted detection fraction y and a ground truth detection fraction of x, the gene's relative error is defined as $\frac{|y-x|}{x}$. For each sequencing depth we computed MPE relative to all genes as well as subsets of genes corresponding to the four quartiles defined by gene averages of non-zero ground truth TPM values over all cells (ranges of mean non-zero TPM values for the four quartiles were $[0, 2.3]$ $(2.3, 6.744]$, $(6.744, 24.517]$, and $(24.517, 18576.98]$, respectively. Full error curves plotting the percentage of genes with relative error above varying thresholds were also used for a more detailed comparison of imputation methods.

– **Gene detection accuracy.** This metric views gene detection as a binary classification problem. For each imputation method, *true positives (TP)* are the (gene, cell) pairs for which both imputed and ground truth TPM values are positive, while *true negatives (TN)* are (gene, cell) pairs for which both TPM values are zero. The accuracy is computed as the number of true predictions $(TP+TN)$ divided by the product between the number of genes and the number of cells.

– **Clustering micro-accuracy.** For each sequencing depth and imputation method we clustered imputed TPM values using several clustering algorithms and assessed the effect of imputation on clustering accuracy using the micro-accuracy measure [5,17] defined by $\sum_{i=1}^{K} C_i / \sum_{i=1}^{K} N_i$, where K is the number of classes, N_i is the size of class i, and C_i is the number of correctly labeled samples in class i relative to the ground truth from [7].

3   Results and Discussion

To assess imputation accuracy on data sets with varying amounts of drop-outs we sub-sampled the ultra-deep DRG scRNA-Seq data to simulate sequencing depths between 50K and 20M read pairs per cell. For each sequencing depth the metrics described in Sect. 2.3 were computed for three previous methods (DrImpute, scImpute and KNNImpute), the two variants of our locality sensitive imputation method described in Sect. 2.2 (LSImputeMean and LSImputeMed), and, as a reference, for the 'Raw Data' consisting of TPM values without any imputation.

Detection Fraction Accuracy. Figure 3 plots the true detection fraction (x-axis) against the detection fraction in the raw data, respectively after imputation with each of the five compared methods (y-axis) at three selected sequencing depths (100K, 1M, respectively 10M read pairs per cell; high resolution plots for all ten evaluated sequencing depths are available at https://doi.org/10.1101/291807. Each dot in the scatter plots represents one gene. Dot color shades are based on the four quartiles as defined above. For an ideal imputation method all dots would lie on the main diagonal, which represents perfect agreement between predicted and true detection fractions. Dots below the diagonal correspond to genes for which the detection fraction is under-estimated, while dots above the diagonal correspond to genes for which the detection fraction is over-estimated. Drop-outs in the raw data yield severe under-estimation of the detection fraction for most genes at sequencing depths of 100K and 1M read pairs per cell, but at 10M read pairs per cell detection fractions computed based on raw data are very close to the true fractions for nearly all genes. Existing methods over-impute detection fractions for most genes, even at low sequencing depths. At 100K read pairs per cell LSImputeMed under-estimates detection fractions, improving very little over raw values, while LSImputeMean gives most accurate detection fractions. At higher sequencing depths LSImputeMean begins over-imputing, while LSImputeMed yields most accurate detection fractions at 1M read pairs per cell and only slightly over-imputes at 10M read pairs per cell.

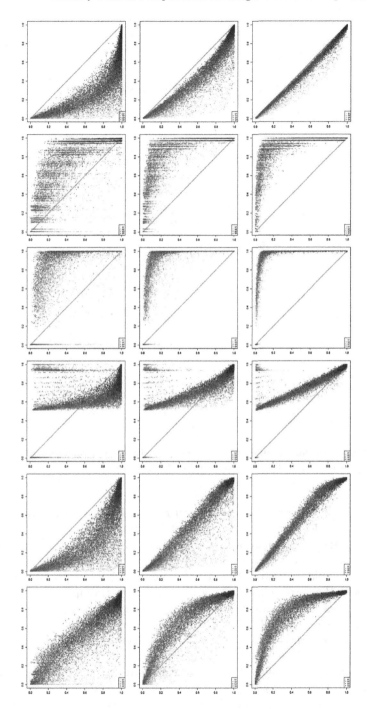

Fig. 3. True vs. imputed detection fractions (left to right: 100K, 1M, 10M read pairs per cell; top to bottom: Raw data, DrImpute, scImpute, KNNImpute, LSImputeMed, and LSImputeMean). (Color figure online)

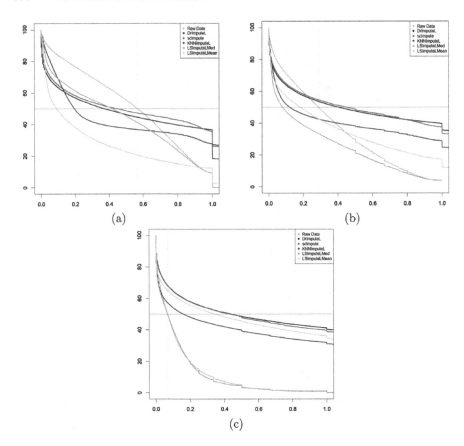

Fig. 4. Error curves for (a) 100K, (b) 1M, respectively (c) 10M read pairs per cell. The abscissa of dashed vertical lines correspond to MPE of raw data.

Detection Fraction Error Curves and MPE Comparison. While dot-plots in Fig. 3 give a useful qualitative comparison of detection fraction accuracy of different methods, for a more quantitative comparison of detection fraction accuracy Fig. 4 gives the so called *error curve* of each method. The error curve plots, for every threshold x between 0 and 1, the percentage of genes with a relative error above x. The error curves in Fig. 4 confirm that LSImputeMean has highest detection fraction accuracy of the compared methods at a sequencing depth of 100K read pairs per cell, while LSImputeMed significantly outperforms the other methods at 1M read pairs per cell and matches raw data accuracy at 10M read pairs per cell. The relative performance of the methods can be even more concisely captured by their MPE values, which are the abscissae of the points where the horizontal line with an ordinate of 0.5 crosses the corresponding error curves. The surface plots in Fig. 5 display MPE values (y-axis, on a logarithmic scale) as a function of both sequencing depth (x-axis) and mean non-zero expression quartile (z-axis). The only imputation methods that do not result in

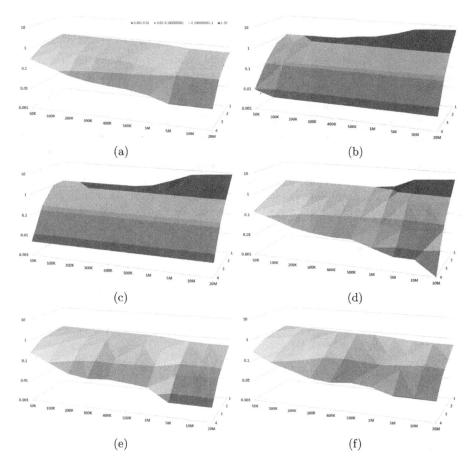

Fig. 5. Surface plots indicating median percent error values in log scale (y-axis) for each depth (x-axis) in each quantile (z-axis) for each method: (a) Raw data, (b) DrImpute, (c) scImpute, (d) KNNImpute, (e) LSImputeMed, and (f) LSImputeMean (Color figure online)

MPE values over 100%, depicted in red in the surface plot, are LSImputeMed and LSImputeMean. At all sequencing depths and for all assessed imputation methods genes in the lowest quartile (Q1) have very high MPE, suggesting that detection fractions based on imputed values should not be used for these genes.

Gene Detection Accuracy and Relation to MPE. Table 1 shows the gene detection accuracy achieved by the compared imputation methods, with the highest accuracy at each sequencing depth typeset in bold. We assessed gene detection accuracy both based on fractional ground truth and imputed TPM values, as well as after rounding both to the nearest integer, which is equivalent to using a TPM of 0.5 as the detection threshold. For the results without rounding, DrImpute has the highest gene detection accuracy at 50K and 100K read pairs per

Table 1. Gene detection accuracy

Data	Not rounded						Rounded					
	Raw	Dr.	sc.	KNN.	LSMd	LSMn	Raw	Dr.	sc.	KNN.	LSMd	LSMn
50K	0.676	**0.822**	0.700	0.799	0.687	0.693	0.752	**0.866**	0.748	0.700	0.762	0.765
100K	0.740	**0.810**	0.778	0.713	0.772	0.797	0.816	**0.876**	0.720	0.712	0.841	0.850
200K	0.800	0.778	0.754	0.726	0.836	**0.839**	0.872	0.878	0.689	0.722	**0.892**	0.884
300K	0.829	0.772	0.740	0.732	**0.864**	0.861	0.899	0.880	0.673	0.726	**0.909**	0.892
400K	0.847	0.762	0.731	0.736	**0.872**	0.868	0.915	0.882	0.663	0.730	**0.918**	0.895
500K	0.859	0.759	0.725	0.738	**0.878**	0.878	0.927	0.883	0.655	0.732	**0.928**	0.909
1M	0.891	0.737	0.703	0.747	**0.899**	0.896	**0.952**	0.882	0.634	0.738	0.947	0.937
5M	**0.918**	0.705	0.661	0.762	0.902	0.910	**0.980**	0.894	0.621	0.772	0.940	0.960
10M	**0.920**	0.768	0.692	0.648	0.896	0.887	**0.987**	0.907	0.627	0.800	0.947	0.939
20M	**0.921**	0.690	0.635	0.774	0.892	0.901	**0.994**	0.921	0.634	0.825	0.959	0.970

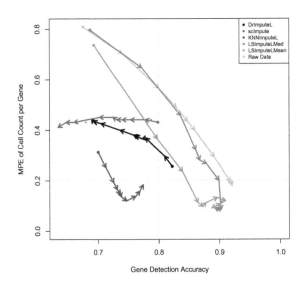

Fig. 6. Gene detection accuracy vs. MPE at varying sequencing depths.

cell. LSImputeMean has highest gene detection accuracy for 200K read pairs per cell, while LSImputeMed outperforms the other methods for 300K–1M read pairs per cell. Raw data (no imputation) gives best gene detection accuracy at 5M read pairs per cell and higher depths. For the rounded data sets, DrImpute also has the highest gene detection accuracy at 50K and 100K read pairs per cell, while LSImputeMed outperforms the other methods for 200K–500K read pairs per cell. For sequencing depth of 1M read pairs per cell and higher the raw data gives best detection accuracy followed by LSImpute methods.

At very low sequencing depth it is possible for some methods to impute values that are not detected in the ground truth. This could lead to good performance in detection fraction accuracy despite low performance in gene detection accuracy. Furthermore, although one would expect all accuracy measures to improve with increased sequencing depth, this may not necessarily be the case for methods that over-impute. To illustrate the relation between MPE and gene detection accuracy and the effect of sequencing depth increase, in Fig. 6 we plot for each method the gene detection accuracy and MPE achieved without rounding at each sequencing depth from 50K up to 20M read pairs per cell, with consecutive depths connected by arrows pointing in the direction of sequencing depth increase. Since high accuracy and low MPE are preferable, the points near the lower right corner of the plot and arrows pointing towards it indicate better results. For some methods like scImpute and DrImpute, although the starting point (50K read pairs per cell) shows considerable improvement over raw data, as sequencing depth increases one or both of the accuracy measures substantially worsen due to over-imputation. Both LSImputeMed and LSImputeMean start with improvement over raw data in both MPE as and Gene Detection Accuracy and continue in the right direction for higher depths until as mentioned before, the raw data without any imputation gives slightly better gene detection accuracy at 5M read pairs per cell and higher, which suggests that imputation at such high depths comes with the risk of over-imputation for all methods tested.

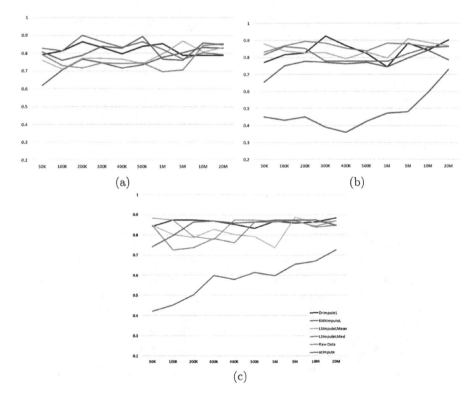

Fig. 7. Micro-accuracy on inputed data for (a) PCA-based hierarchical clustering using Spearman correlation, (b) TF-IDF_Top_C [11], and (c) PCA-based spherical k-means.

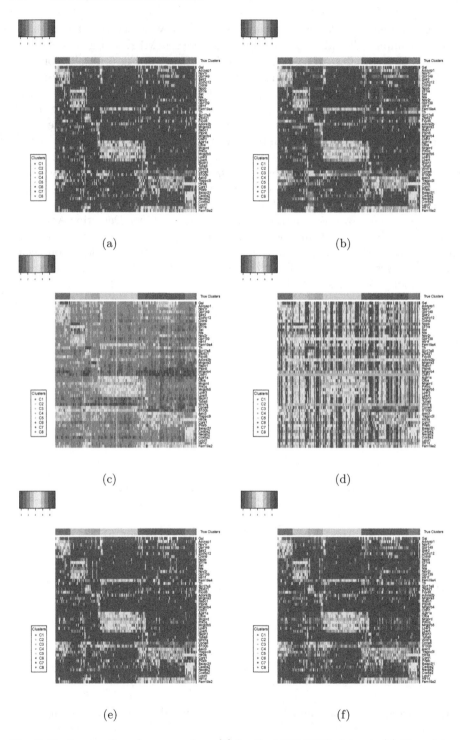

Fig. 8. Heatmaps of marker genes from [7] for the 100K DRG dataset: (a) Raw data, (b) DrImpute, (c) scImpute, (d) KNNImpute, (e) LSImputeMed, (f) LSImputeMean.

Clustering Accuracy. In order to assess the impact of imputation on clustering results, we tested each of the imputation methods in combination with following clustering methods: PCA based hierarchical clustering using Spearman correlation, the TF-IDF_Top_C clustering approach from [11], and PCA based spherical k-means clustering. The micro-accuracy results in Fig. 7 suggest that the effect of imputation varies when combined with different clustering approaches. We also tested Seurat [14] k-means clustering of genes and cells (using $k = 8$ with default parameters), however there was very little change in clustering accuracy for different depths. Although the MPE and detection accuracy of some imputation methods suggest the imputation radically alters gene expression profiles, the similarity between cells of a cluster could still hold when all cell profiles are changed in a consistent manner. This can very well lead to no or little change in clustering accuracy, when in fact cell expression profiles are far from the ground truth as the MPE and gene detection accuracy results suggest. As seen in Fig. 8 featuring the $\log(x + 1)$ expression levels of the marker genes for the DRG 100K data set, although the expression levels of most genes are changed through imputation, the clusters driven by high expression levels of several marker genes can still be the prominent signal for clustering and in most cases this signal remains visually apparent in the heatmaps. Clustering accuracy is hence not recommended as the sole performance evaluation metric when assessing imputation methods.

4   Conclusion

Although imputation can be a useful step in scRNA-Seq analysis pipelines, it can become a two-edged sword if expression values are over-imputed. In this paper we evaluated the performance of several existing imputation R packages and presented a novel approach for imputation. LSImpute, especially the variant based on median imputation, tends to impute more conservatively than existing methods resulting in improved performance based on a variety of metrics. Overall, LSImpute is more likely to reduce drop-out effects and reduce sparsity of the data without introducing false expression patterns or over-imputation. Cosine and Jaccard similarity based implementations of LSImpute are available as a Shiny app at http://cnv1.engr.uconn.edu:3838/LSImpute/.

References

1. Azizi, E., Prabhakaran, S., Carr, A., Pe'er, D.: Bayesian inference for single-cell clustering and imputing. Genomics Comput. Biol. **3**(1), 46 (2017)
2. Duan, F., Duitama, J., Al Seesi, S., Ayres, C.M., Corcelli, S.A., Pawashe, A.P., Blanchard, T., McMahon, D., Sidney, J., Sette, A., et al.: Genomic and bioinformatic profiling of mutational neoepitopes reveals new rules to predict anticancer immunogenicity. J. Exp. Med. **211**(11), 2231–2248 (2014)
3. Hornik, K., Feinerer, I., Kober, M., Buchta, C.: Spherical k-means clustering. J. Stat. Softw. **50**(10), 1–22 (2013)
4. Kwak, I.Y., Gong, W., Koyano-Nakagawa, N., Garry, D.: DrImpute: imputing dropout events in single cell RNA sequencing data. bioRxiv, p. 181479 (2017)

5. Lee, C., Măndoiu, I.I., Nelson, C.E.: Inferring ethnicity from mitochondrial DNA sequence. In: BMC Proceedings, vol. 5, p. S11. BioMed Central (2011)
6. Leskovec, J., Rajaraman, A., Ullman, J.D.: Mining of Massive Datasets. Cambridge University Press, Cambridge (2014)
7. Li, C.L., Li, K.C., Wu, D., Chen, Y., Luo, H., Zhao, J.R., Wang, S.S., Sun, M.M., Lu, Y.J., Zhong, Y.Q., et al.: Somatosensory neuron types identified by high-coverage single-cell rna-sequencing and functional heterogeneity. Cell Res. **26**(1), 83 (2016)
8. Li, W.V., Li, J.J.: scImpute: accurate and robust imputation for single cell RNA-seq data. bioRxiv, p. 141598 (2017)
9. Lin, P., Troup, M., Ho, J.W.: CIDR: ultrafast and accurate clustering through imputation for single-cell RNA-seq data. Genome Biol. **18**(1), 59 (2017)
10. Mandric, I., Temate-Tiagueu, Y., Shcheglova, T., Al Seesi, S., Zelikovsky, A., Măndoiu, I.I.: Fast bootstrapping-based estimation of confidence intervals of expression levels and differential expression from RNA-seq data. Bioinformatics **33**(20), 3302–3304 (2017)
11. Moussa, M., Mandoiu, I.: Single cell RNA-seq data clustering using TF-IDF based methods. BMC-Genomics (2018, to appear)
12. Nicolae, M., Mangul, S., Mandoiu, I.I., Zelikovsky, A.: Estimation of alternative splicing isoform frequencies from RNA-seq data. Algorithms Mol. Biol. **6**(1), 9 (2011)
13. Prabhakaran, S., Azizi, E., Carr, A., Peer, D.: Dirichlet process mixture model for correcting technical variation in single-cell gene expression data. In: International Conference on Machine Learning, pp. 1070–1079 (2016)
14. Satija, R., Farrell, J.A., Gennert, D., Schier, A.F., Regev, A.: Spatial reconstruction of single-cell gene expression data. Nat. Biotechnol. **33**(5), 495 (2015)
15. Selivanov, D.: dselivanov/LSHR. https://github.com/dselivanov/LSHR
16. Troyanskaya, O., Cantor, M., Sherlock, G., Brown, P., Hastie, T., Tibshirani, R., Botstein, D., Altman, R.B.: Missing value estimation methods for DNA microarrays. Bioinformatics **17**(6), 520–525 (2001)
17. Van Asch, V.: Macro-and micro-averaged evaluation measures. Technical report (2013)

Author Index

Printed in the United States
By Bookmasters